PROGRESS IN AQUATIC ECOSYSTEMS RESEARCH

PROGRESS IN AQUATIC ECOSYSTEMS RESEARCH

A. R. Burk
Editor

Nova Science Publishers Inc.
New York

For permission to use material from this book please contact us:
Telephone 631-231-7269; Fax 631-231-8175
Web Site: http://www.novapublishers.com

Library of Congress Cataloging-in-Publication Data

Progress in aquatic ecosystem research / A.R. Burk (editor).
p. cm.
Includes index.
ISBN 1-59454-383-6 (hardcover)
1. Aquatic ecology--Research. I. Burk, A. R.
QH541.5.W3P76 2005
577.6--dc22 2005005424

Published by Nova Science Publishers, Inc. ✦ New York

CONTENTS

PREFACE

Ecology is the study of the interrelationships between organisms and their environment, including the biotic and abiotic components. There are at least six kinds of ecology: ecosystem, physiological, behavioral, population, and community. Specific topics include: Acid Deposition, Acid Rain Revisited, Biodiversity, Biocomplexity, Carbon Sequestration in Soils, Coral Reefs, Ecosystem Services, Environmental Justice, Fire Ecology, Floods, Global Climate Change, Hypoxia, and Invasion. This new book presents new research on ecology from around the world.

From the beginning of the 20th century to the 1970s water management in Portugal was essentially regarded as a non-integrated activity. Various hydraulics projects were undertaken whose main objective was to make an inventory of works that were required to develop specific economic sectors. In that context, the energy and agricultural sectors were considered the most relevant. Schemes were developed within a water policy that set hydraulic works above environmental concerns; this is just one of the things discussed in Chapter 1.

Chapter 2 aims to give a synthesized yet comprehensive account of the main trends and approaches on monitoring ecological indicators for river research and management, and to show the importance of the links between society's values and environmental water quality. Human pressure on freshwater ecosystems has increased along with a greater awareness of environmental quality concepts like river health and ecosystem integrity. Several industrialized countries have already integrated these concepts in recent water quality guidelines, giving more emphasis on studying and regenerating biotic communities. Such is the case of the European Water Framework Directive or the Environmental Monitoring and Assessment Program of the U.S. Environmental Protection Agency

In response to a perceived decline in the quality of aquatic resources and associated changes in public policy, a great deal of recent ecological research has investigated the feasibility of restoring aquatic ecosystems. In 1992, the National Research Council (NRC) on Aquatic Ecosystem Restoration called for a more systematic and holistic approach to aquatic restoration that addresses not only structural aspects of morphology and hydrology but also attributes of ecosystem function. In this chapter the authors review the status of lake, river, and wetland restoration since the NRC recommendations. In particular, the authors focus on the restoration of trophic functionality in these three ecosystems and the degree to which it has been achieved by structural modification and by biomanipulation. Trophic restoration of lakes, while typically featuring the reduction of external and internal nutrient loads, has successfully incorporated biomanipulation of fish populations to manage phytoplankton through trophic cascade. River and wetland restoration efforts have included trophic function to a lesser degree. In these ecosystems, hydrologic regime and landscape connectivity have necessarily been given far more attention according to Chapter 3.

A described in Chapter 4, developed catchments riparian wetlands can intercept runoff and potentially act as buffer zones protecting downstream waterways from excess N enrichment. However, understanding of the processes involved in N transformation and the sustainability of the N removal function are far from complete. Due to its low natural abundance the stable isotope of nitrogen (^{15}N) can be used as a tracer to investigate N transformation processes in ecological systems. In separate laboratory experiments the authors investigated the fate of $^{15}NH_4^+$ in lake-edge wetland microcosms inhabited by *Salix cinerea* (grey willow) and *Typha orientalis* (raupo) plants and the fate of $^{15}NO_3^-$ in stream-side wetland microcosms inhabited by *Glyceria declinata* (glaucous sweetgrass) relative to unplanted microcosms. The authors found that *Salix* and *Typha* plants readily assimilated NH_4^+ (9-11% of applied $^{15}NH_4^+$), enhanced the natural diffusion of $^{15}NH_4^+$ up through organic soil into the overlying water (by 7 to 18-fold) and, via litterfall, provided a flocculent substrate for sequential mineralisation-nitrification-dentrification processes in the overlying water layer.

Long-term changes in net ecosystem metabolism (NEM) and net denitrification (ND) in the Ohta River estuary of northern Hiroshima Bay were analyzed based on the phosphorus (P) and nitrogen (N) budgets, using the method recommended by the Land-Ocean Interaction in the Coastal Zone (LOICZ) Working Group. From 1987 to 1997, data were collected as part of several monitoring programs conducted by governmental organizations. The NEM of the estuary was calculated to be 0.52 g C m^{-2} day^{-1} on average, with a positive average value in the upper layer (0.71 g C m^{-2} day^{-1}) and a negative average value in the lower layer (-0.20 g C m^{-2} day^{-1}), thus indicating production in the upper layer and active decomposition in the lower layer. The ND was calculated to be -90 and 72 mg N m^{-2} day^{-1} for the upper layer and lower layer, respectively, suggesting that nitrogen fixation was occurring in the upper layer while denitrification dominated the lower layer, perhaps in the surficial sediments as discussed in Chapter 5.

The objective of chapter 6 is to review current knowledge of the design of horizontal subsurface flow constructed wetlands planted with common reed when constructed to reach secondary treatment. Observations included in that paper are based on our own previous studies and those to be found in the literature. TSS removal in these systems is usually high and effluents have concentrations lower than 35 mg/L. To avoid clogging and ensure effluent quality the surface TSS loading rate should be lower than 20 g TSS/m^2.d. BOD_5 removal is highly dependent on the surface organic loading rate, which should be lower than 4-6 g BOD_5/m^2.d to reliably obtain effluents with a concentration of less than 25-30 mg/L. Nutrient removal is not significant (10 to 40%) when the systems are designed for secondary treatment. Microbial removal ranges from 2 to 4 log-units for faecal bacteria and 1 to 3 for faecal viruses. Further discussed in Chapter 6 are the improvements to both nutrient and microbial removal require additional steps beyond those provided by an unique horizontal subsurface flow constructed wetlands.

In Spain, very little interest has been shown in gaining knowledge of the purifying potential of aquatic macrophytes and testing their adequacy and viability as opposed to the conventional systems of urban sewage treatment. This has not been the case in the United States and various European countries, where this issue has received special attention since over ten years. According to Chapter 7 the treatment pilot systems are located in the Experimental Low Cost Wastewater Treatment Station of León (NW Spain).

In Chapter 8 the two principal forms of logic that can be used as a basis for spatial modelling are Boolean and fuzzy logic. This paper documents research investigating the use of fuzzy logic modelling to predict the distribution of dryland secondary salinity. In considering the modelling of salinity, classical set theory would demand that all areas be classified as being saline or non-saline. Fuzzy logic allows for this unrealistic division to be removed, modelling the uncertainty of the boundary between these regions and the measurements on which they are based by allowing degrees of membership to classes, that better reflect the natural variation of soil properties. To this end, an existing approach for predicting the distribution of salinity, Fuzzy Landscape Analysis GIS (FLAG), developed by Roberts *et al.* (1997) is implemented. An attempt is made to optimize the predictive power of FLAG through the inclusion of geological and remote sensing data. FLAG models salinity distribution within the framework of fuzzy logic. Results from this investigation are compared with the outputs of a predictive model of salinity based on probability theory and Boolean logic.

In: Progress in Aquatic Ecosystems Research
Editor: A. R. Burk, pp. 1-34

ISBN 1-59454-383-6

Chapter 1

A COHERENT AND COMPREHENSIVE RIVER BASIN MANAGEMENT STRATEGY APPLICATIONS

J. S. Antunes do Carmo and* J. C. Marque
IMAR – University of Coimbra, FCT, Coimbra, Portugal
R. M. V. Cortes
University of Trás-os-Montes e Alto Douro, Vila Real, Portugal

ABSTRACT

From the beginning of the 20th century to the 1970s water management in Portugal was essentially regarded as a non-integrated activity. Various hydraulics projects were undertaken whose main objective was to make an inventory of works that were required to develop specific economic sectors. In that context, the energy and agricultural sectors were considered the most relevant. Schemes were developed within a water policy that set hydraulic works above environmental concerns.

Regulations published in the early 1980s established a consistent legal target for water management, for the first time in Portugal. More recently, the Water Framework Directive came into force, creating a coherent and comprehensive framework to protect European Union (EU) waters. This constitutes the pillar of water policy for the 21st century. The newly-introduced concepts and regulations thus established have led to the setting up of an interface with users, with the preparation of reports and field information about water use proposals and hydrographic network changes.

Current regulations in Portugal, which are necessary under EU legislation, attribute Water Resources Planning per Hydrographic Basin, or Hydrographic Region, and establish the contents of the above-mentioned plans. All Portugal's River Basin Management Plans relating to the international river basins, based on the Convention on Cooperation for Portuguese-Spanish River Basins, which establishes a framework for cooperating in the protection of shared water basins, were concluded in 2001. These plans are structured around five main systems: environmental, socio-economic, infrastructural, institutional and financial-fiscal. The systematization, interpretation and

* IMAR – University of Coimbra, Department of Civil Engineering, Pólo II, 3030-290 Coimbra, Portugal, E-mail: jsacarmo@dec.uc.pt.

information exchange relative to each of these systems are the development pillars underpinning the Hydrographic Basin Plans (HBPs).

Since 1999, pollution has been partially alleviated in Portugal and water quality is slowly improving, now that a different strategy and regulation for water management have been introduced. As a first step towards improvement, water prices and charges for sewage treatment ought to be increased sharply, for all users, and very soon. Also, as a general consequence of the approved HBPs, new strategies and specific programs for safeguarding and improving water resources should be designed and implemented.

A number of environmental problems have been detected during the diagnosis phase of some HBPs, and these need urgent solutions within an integrated management concept. Prominent among them are: dam-reservoir systems, intensive agriculture and pig farming, poor water quality, environmental degradation in cities and flood hazards in the main courses.

This chapter presents and systematizes the main development phases of a general Program already undertaken, and which is currently being implemented in specific case-studies (some Portuguese rivers and estuaries). It is regarded as a highly innovative Program because of the following interrelated aspects: *i*) it adopts the systems approach; *ii*) it is oriented towards new goals of environmental management; *iii*) it involves a new combination of disciplines; *iv*) it indicates new strategies for collecting indicators; *v*) it proposes new, concrete, empirically measurable indices for sustainable development; *vi*) it is likely to develop new indicators, and *vii*) it should provide new tools.

Some results of a one-dimensional model in the horizontal plane (1DH) and preliminary conclusions are presented. Other results, also from a one-dimensional model, but in the vertical plane (1DV), are used in this work to show how pumping water at a reservoir dam influences certain water quality parameters (temperature and dissolved oxygen).

Finally, preliminary results from a third case-study on the impact of eutrophication and river management are given, interpreted within a framework of ecosystem theories.

INTRODUCTION

The economic, social and environmental conditions of various European river basins and estuarine systems have changed dramatically in the last decades as a consequence of anthropogenic effects, and they will go on changing in the years to come due to increasing human pressure. New strategies are needed to develop a multifunctional use structure, which must take into account the many-faceted aims of sustainable development. To support this development, integrated analysis of new information sources is necessary. Furthermore, it should be possible to enact planning scenarios during the decision process, and for that we need to have an instrument that will help decision-makers to find optimised solutions for existing problems.

Recent instruments introduced into EU water policy by the Water Framework Directive create a coherent and comprehensive framework to protect and improve all European Union (EU) waters, through: *i*) an ecological and holistic water status assessment approach; *ii*) river basin planning; *iii*) a strategy to eliminate pollution by dangerous substances; *iv*) public information and consultation, and *v*) financial instruments. The newly-introduced concepts and regulations established provide for the creation of an interface with users, the preparation of reports, and field information about water use proposals and hydrographic network changes.

In Portugal, a Program has been embarked upon to develop and implement tools that incorporate different approaches to fulfilling the following core objectives:

- To develop a computer-based decision support system as a conceptual instrument for evaluating and optimising the sustainability of riverine and estuarine ecosystem management;
- To test and optimise such an instrument by empirical investigations conducted on different scales in chosen target fluvial and estuarine systems;
- To identify the suitability and optimise the design of sets of indicators representing the sustainability of different basin use practices;
- To implement holistic concepts, ecosystem analytical procedures, integrated modelling techniques, and ecological-economic instruments in river basin and estuary management;
- To derive recommendations for environmentally sound, sustainable development of river basins, including estuarine systems, by comparing different uses in a variety of regional conditions;
- To compare, select, and optimise the efficiency of different ecosystem-based concepts, like ecological indicators and ecological orientors, usable as goal functions in the development of structural dynamic ecological models, and cost-benefit analysis;
- To test, through their application to fluvial and estuarine systems, the adaptability of ecosystem protection concepts like ecosystem health, ecological integrity, and ecological functionality.

This Program incorporates different disciplines to attain these objectives, including theoretical analysis, empirical measurement, modelling, optimisation techniques, and economic-ecological considerations.

The whole work is founded on three major components:

- Indication of ecological and economic states: application of an optimal set of indicators that reflects the ecological and economic demands of river and estuarine management and which fits into modern systems-oriented protection strategies. This set of indicators will be tested using data that has been measured on different Portuguese rivers and estuaries.
- Model integration: combining existing models and developing new ones (e. g. structural dynamic model of changes in the trophic chain over time), which can complement the use of indicators carrying out 'predictive monitoring' in riverine and estuarine scenarios for planning purposes.
- GIS based Integration Module: construction of a database in GIS environment that will enable river-estuary system managers to decide on a greatly improved information basis, which will be provided by the indicators and the scenario results.

It is widely appreciated nowadays that an estuary is the consequence of everything that happens along the river basin. So, to achieve sustainable environmental conditions, we cannot separate the river basin from the river estuary. It means, for example, that we must take tidal influence into account. Temperature and salinity can greatly modify riverine ecosystems, and impact on local fauna and flora. The two systems should be inter-connected, studied and managed jointly, within the same hydro-informatics environment.

The first target system is the Portuguese river Lima, which is located in the Northwest of the Iberian Peninsula (UE). It is shared by Portugal and Spain (Figure 1), and occupies an area of 2525 km^2, 1170 km^2 in Portuguese territory, (around 47%). The watershed has a NEE – SWW orientation, with 19 km medium width and an average altitude of 450 m.

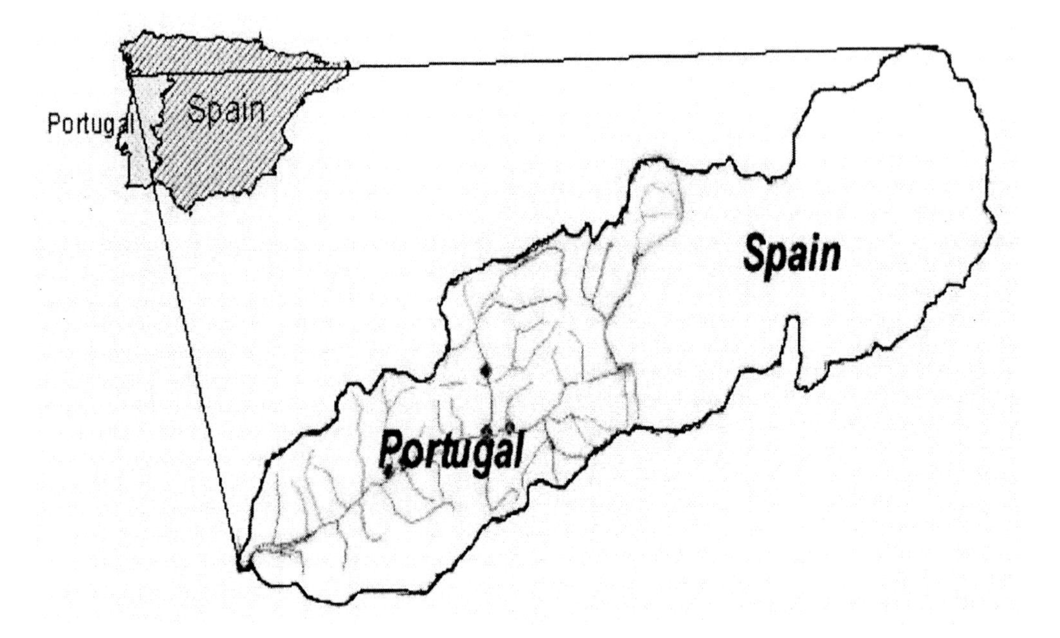

Figure 1. Location of the Lima river watershed (Lopes *et al.*, 2003; Lopes *et al.*, 2004).

The Lima river source is in the mountains of S. Mamede, at approximately 950 *m* altitude, in the province of Orense, Spain. In Portugal, it flows for 67 km before reaching its mouth at Viana do Castelo harbour.

Three other target systems are the river basins and estuaries of the Portuguese Vouga, Mondego and Lis systems, which also appeared to be suitable systems for developing and testing integrated management tools. All of them are of considerable importance to human populations. Figure 2 shows these three river basins, together with the existing 25 water quality measuring points.

The Vouga and Mondego river estuaries support commercial and recreational fishing. Both estuaries contain important harbors, ports and navigation channels.

In these estuaries, too, there are some industries, and their activities often increase the hazards to the existing fragile ecosystems. The various uses of these estuaries and channels place conflicting demands and burdens on water quality. Mathematical models help to assess the effects of these conflicting demands and to develop protective management strategies. The Mondego river-estuary system has undergone several human interventions, notably engineering work to restore a considerable stretch of the channel and margins, which affected the inter-tidal salt marsh. Furthermore, an ongoing eutrophication process, observed in the last 20 years and caused by excessive nutrient release in the estuary, has been causing a shift in primary producers, from macrophytes to macroalgae, as the dominant species. Overall, the low Mondego river and its estuary may be considered to be under severe environmental stress. On the other hand, both the low Vouga and Lis rivers are highly polluted due to intensive agriculture and pig farming. It is a major concern that as much of the existing information as possible is used for the proposed approach, although it has been assumed that this information should be complemented, to a certain extent, with some data collection, for model calibration. This was a major reason underlying the choice of these three target systems, since significant databases already exist in the ambit of the HBP and augmented by a number of studies carried out on these systems over the last two decades.

Figure 2. Vouga (above), Mondego (center) and Lis (below) river basins, showing the existing 25 water quality measuring points (Antunes do Carmo & Marques, 2001).

Consistent sets of data are available on their biological characterisation, biological processes, primary and secondary productivity, pollution studies, sediment characterisation, nutrient loading, and hydraulic regime. Moreover, due to their convenient dimensions, the Vouga, Mondego and Lis river-estuary systems may be considered as field laboratories. These three rivers and estuaries appeared therefore to be suitable systems on which to develop and test integrated management tools.

Brief Description of the Different Modules

Under this Program, research activities are organised in the following four disciplinary modules: Ecology; Sedimentology; Hydraulics and Water Quality, and Landscape and Socio-Economics. An Integration Module is also required to consolidate disciplinary results for management purposes (Figure 3).

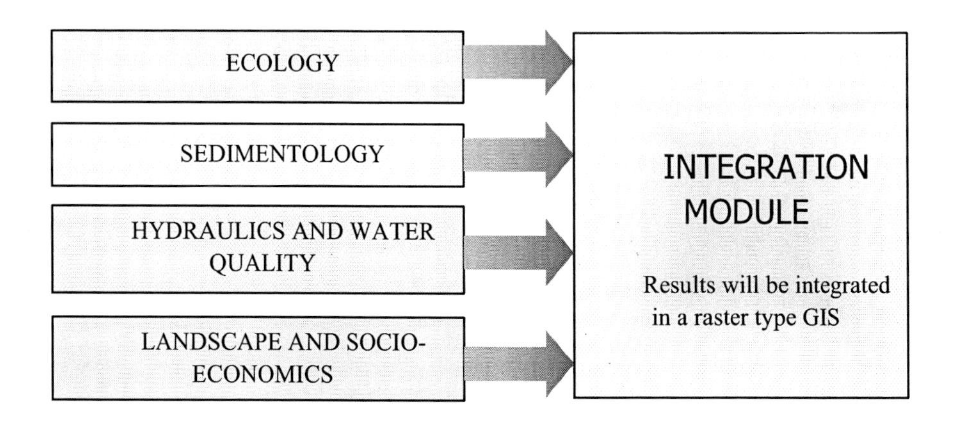

Figure 3. The Program objectives will be attained through an interdisciplinary approach, involving a combination between basic science and socio-economics.

Therefore, each of the disciplinary modules provides input to the Integration Module, where data is used to assist in the development of management scenarios and for the definition of a 'best possible strategy' for the sustainable utilisation of the target ecosystems. It is hoped that the proposed approach, as a more general procedure, might be applicable to river-estuary systems other than the target systems considered.

Ecology Module

All the available information regarding the studies carried out in the target systems will be gathered and then stored in a database.

In order to complement existing data, and any other provided by a literature review, the following two field-work and three modelling activities have been envisaged and incorporated into this module:

- Characterisation of meiofauna, focusing on the study of the meiofauna composition and biomass variation in selected substrata, in all target systems;
- Studies on nutrient mobilisation in the target systems, focusing on the estimation of the nutrients released from sediments. Special attention should be given to phosphorus and nitrogen loading through sediment/water interface processes in tidal pools;
- Balance of the trophic chain. The biomass form in different trophic compartments should be estimated, and fluxes between compartments in different scenarios

identified in the target systems should be evaluated. To accomplish this, it will be necessary to build a steady state model of the trophic chain in the four river-estuary systems, which will be done using the ECOPATH software;

- Analysis of the properties of different goal functions. The properties of three goal functions used as holistic ecological indicators of ecosystem integrity (structure and function), respectively, Exergy, derived from thermodynamics, and Emergy and Ascendancy, derived from network analysis, should be compared. Empirical data on the four river-estuary systems, as well as the output from steady state models (see previous activity) will provide the necessary information on temporal and spatial changes;
- Development of a structural dynamic model, able to simulate the change from non-eutrophicated to eutrophicated communities, using goal functions to drive the model behaviour.

To sum up, these models may be summarized as follows (Figure 4):

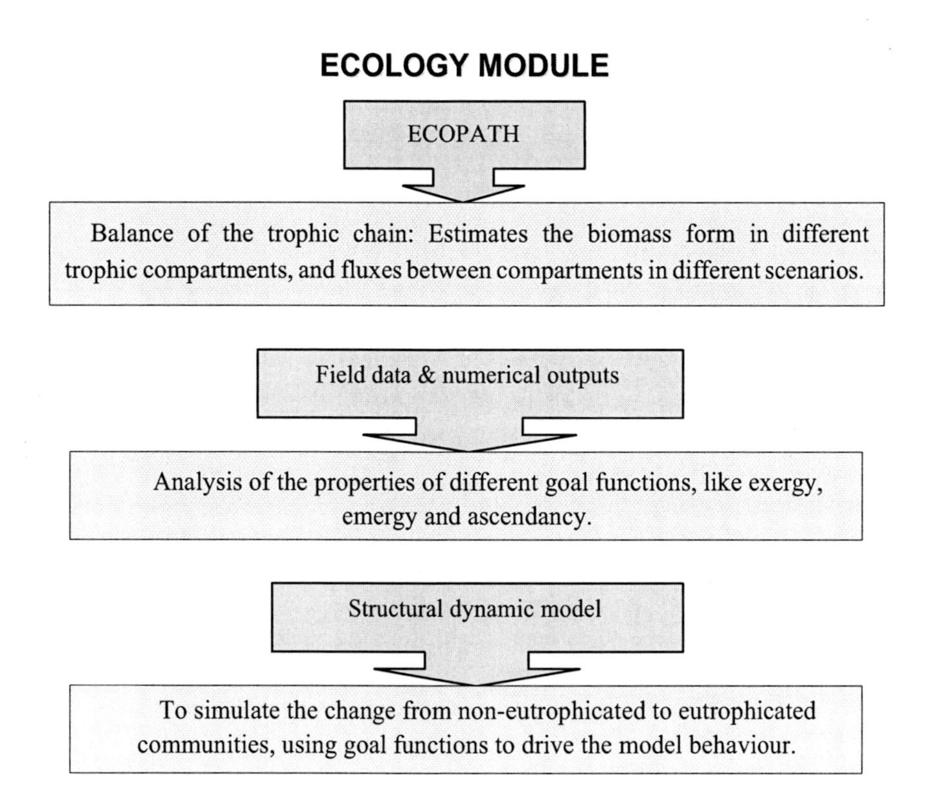

Figure 4. Modelling activities incorporated into the Ecology Module.

The interaction between the different activities of the Ecology Module and the way they provide input to the Integration Module are shown in Figure 5 (Marques *et al.*, 1997; Marques *et al.*, 1998; Nielson *et al.*, 1998; Antunes do Carmo & Marques, 2001).

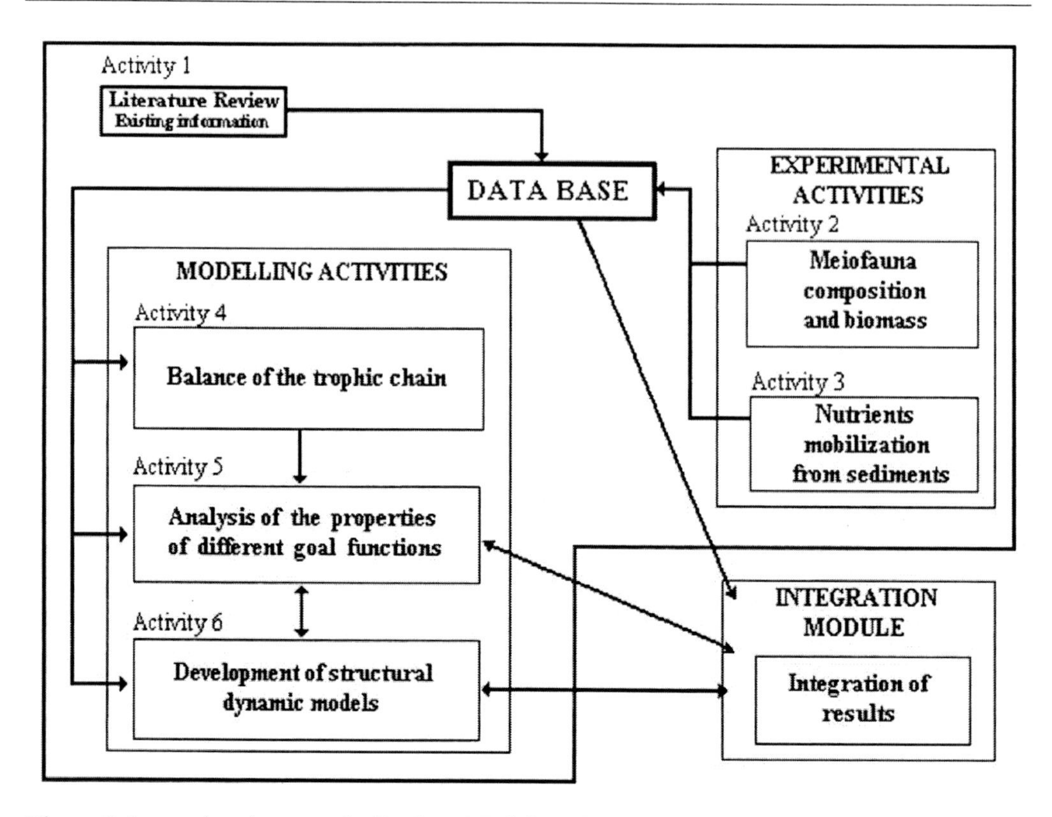

Figure 5. Interactions between the Ecology Module activities and input from this module to the Integration Module (Marques *et al.*, 1997; Marques *et al.*, 1998; Nielson *et al.*, 1998; Antunes do Carmo & Marques, 2001).

Sedimentology Module

The sediment of the four river-estuary targets has been characterised under the HBPs and other research projects, providing a reasonably complete picture of the systems. In the case of the Mondego river basin and estuary, an integrated study of the sedimentary sub-environments has been carried out, and the environmental impact of several human interventions evaluated. In all systems, field reconnaissance, monitoring and sampling, the study of bathymetry charts, and aerial photographs, together with laboratory determinations, have provided a sound set of data. There should therefore be no need to repeat sampling campaigns under this program. Nevertheless, the available data is being re-interpreted, particularly in relation to results from the Hydraulics and Water Quality Module, focusing on the hydrodynamic processes related to the transport, exchange and fate of pollutants, and on the estimation of nutrient release from sediments (see Ecology Module). Moreover, a further step will incorporate data on sediment cartography into a raster type GIS (see Integration Module), which should provide the best possible interface for management purposes.

Hydraulics and Water Quality Module

A sound understanding of the sources, transport mechanisms, storage facilities, and ultimate fate of polluting agents is essential for the proper management of the four target river and estuary ecosystems. After entering the systems, nutrients, trace metals and toxic compounds are transported by the advection motion of water and spread by turbulent mixing and molecular diffusion. During their transport or storage, pollutants can react and change their physical characteristics and chemical composition. The behaviour and transformation of pollutants depend on their physical-chemical parameters, and those of the ambient fluid medium, as well as the hydrodynamic features of the flow.

In this Program, an Eco-hydrodynamic approach is being employed that focuses on the hydrodynamic processes. These are related to the transport, exchange and fate of pollutants, closely linked to a broad range of information provided by the other research modules, that is, the Ecology and Sedimentology Modules. The fate of pollutants and their ultimate impact on biota is affected by the mechanisms of advection, turbulent diffusion and dispersion. Thus, both the dynamics of the flow field and the transport and ultimate fate of aquatic pollutants may be described by means of the Navier-Stokes equations, the equation of continuity and the mass balance equation for pollutants.

It is reasonable to assume that, as with other fluvial systems, the characteristics of the Lima, Vouga, Mondego and Lis systems ought to be controlled by the differences in the chemical and biological characteristics of freshwater and seawater and the degree of mixing that occurs. The degree of mixing is a function of river flows, tidal action, wind and other factors. Differences between the chemical characteristics of freshwater and seawater inflows can affect the relative densities of water layers and the resulting degree of mixing.

The main geometric and flow characteristics of a river include the dimensions of the wet cross-sectional area, bed slope, meandering pattern, flow velocity distribution, water temperature, distribution of sediment particle diameter, concentration of suspended solids and bedload transport. The most important chemical and biological features include dissolved oxygen, pH, bacteria and viruses, fish populations, and aquatic plants and algae.

For most engineering applications, river hydrodynamics are adequately described by the one-dimensional system of continuity and momentum equations (1DH Saint-Venant equations system). For the same kind of applications, estuarine dynamics is quantified by the two-dimensional (2DH) Saint-Venant equations system, in the horizontal plane. For vertically well-mixed estuaries, a one-layer model can adequately describe the flow characteristics, which implicitly assumes a uniform velocity distribution in the water-column. For vertically stratified estuaries, like the Vouga and Mondego, and all existing big reservoirs, particularly Touvedo and Aguieira, respectively in the Lima and Mondego rivers, flow dynamics must be treated by means of a three or more layer system, utilizing a three-dimensional model as presented in Mellor (1998) or Pinho *et al.* (2004). As it is relevant to sedimentary dynamics, at least in some estuaries like the Vouga and Mondego, the local wave propagation over an established current velocity field will be described by means of modified Boussinesq-type equations (Antunes do Carmo & Seabra-Santos, 1996).

A complete water quality computational structure is being developed and implemented (Figure 6). It will be used to make comparisons with field data, and other applications, to analyse different scenarios and to propose recommendations.

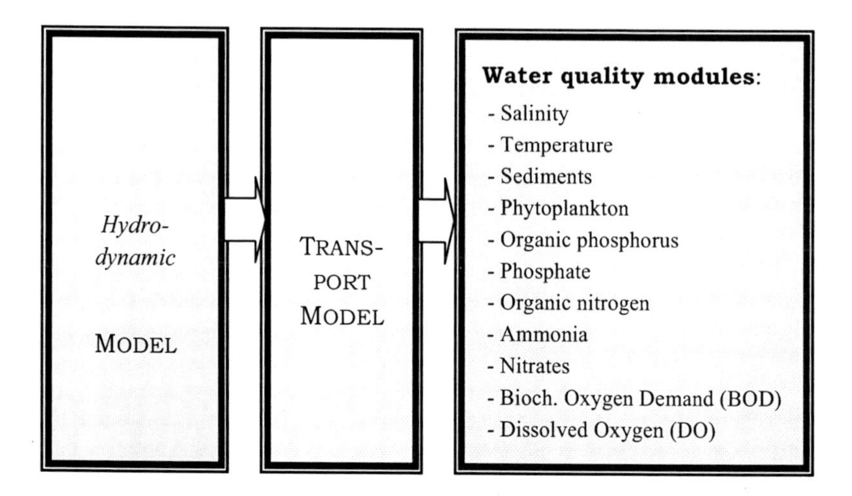

Figure 6. Computational structure proposed for the study of the hydrodynamics and water quality processes in river-estuary systems.

To provide the information needed to calibrate and validate the computational structure, extensive field data should be obtained at various points of the target river systems and estuaries (see Figures 1 and 2). As the effects of water movement, mixing, and transport must be characterized, the hydrodynamic parameters, such as flows, velocities and depths, and measurements of residence time, should be measured. Other parameters will be also measured, including temperature, salinity, suspended sediments, as will other constituents, like phosphorus, ammonia, nitrates, BOD and DO concentrations (Figure 6).

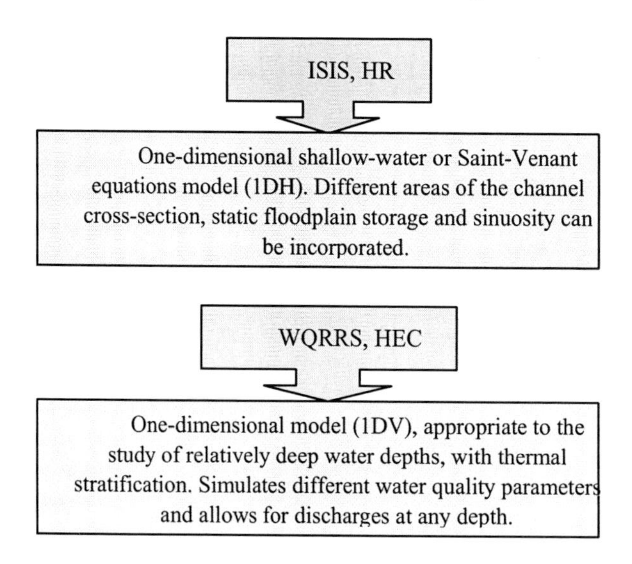

Figure 7. Brief description of the two models incorporated in the Hydraulics and Water Quality Module.

The hydraulics and water quality module is mainly composed of one-dimensional hydrodynamic and water quality models, which are specific for applications along *x*, in the horizontal plane (1DH), or along *z*, in the vertical plane (1DV). Figure 7 briefly describes these two hydrodynamic and water quality models: ISIS Flow & Quality and WQRRS-Water Quality for River-Reservoir Systems.

The first ISIS model was developed and is supported by the Wallingford Software Ltd. and Halcrow Group Ltd., and the second one, WQRRS, was developed by the Hydraulic Engineering Center (HEC). The corresponding mathematical formulations are presented in Figure 8.

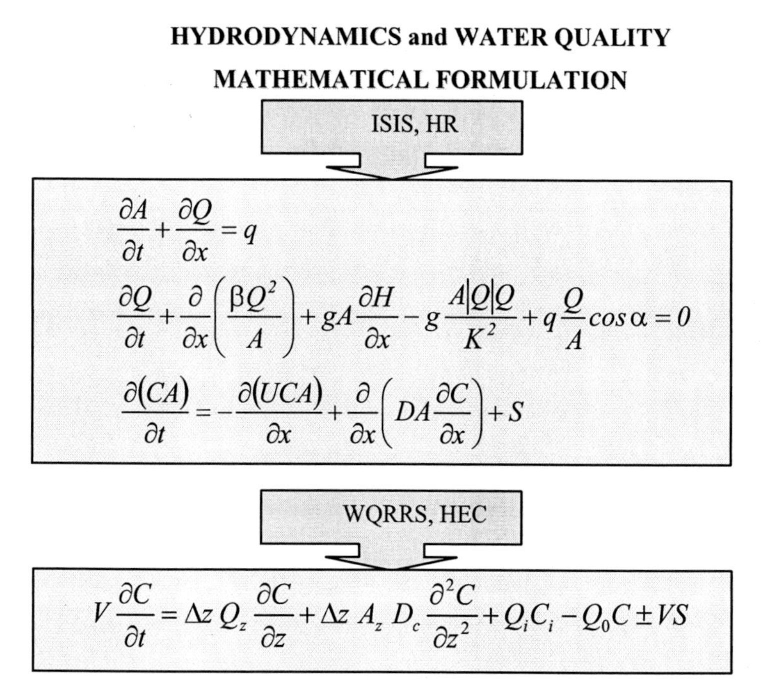

Figure 8. Mathematical formulations of the models incorporated in the Hydraulics and Water Quality Module.

In the ISIS model, A (m^2) represents the cross-sectional flow area; Q (m^3s^{-1}) is the flow; q ($m^3s^{-1}m^{-1}$) is a lateral inflow; ß is the momentum correction coefficient; g ($m\,s^{-2}$) is the gravitational acceleration; *h* (*m*) is the water surface elevation above datum; $S_f = |Q|Q/K^2$ is the friction slope, calculated according to Manning's equation, with $K = A^2R^{4/3}/n^2$ being the channel conveyance; $R = A/P$ (*m*) is the hydraulic radius; *P* (*m*) is the wetted perimeter; n is Manning's roughness coefficient, and α is the angle of inflow; C ($kg\,m^{-3}$) is the pollutant concentration; U ($m\,s^{-1}$) is the cross-sectionally averaged flow velocity; D (m^2s^{-1}) is the diffusion coefficient; S ($kg\,m^{-1}s^{-1}$) is the source/sink term, representing decay, growth, erosion, deposition, etc.

In the HQRRS model, C is the temperature or concentration of any constituent; $V\,(m^3)$ is the volume of each element of the system; Q_z $(m^3 s^{-1})$ is the vertical advection; A_z (m^2) represents the surface area of each element; D_c $(m^3 s^{-1})$ is a diffusion coefficient; C_i represents the thermal energy flux or a concentration of any constituent; Q_i and Q_0 $(m^3 s^{-1})$ are lateral inflows, and S is the source/sink term.

Landscape and Socio-Economics Module

Two complementary actions are needed, which are: the revision of existing policies, programs and plans, and the profiling of the socio-economic baseline of the target systems. The approach to policy programs and plans affecting land-use management and conservation is tackled at several levels (EU, national, regional and local). Such analysis aims to identify the impacts on the land use and on the change of land use patterns driven by legal instruments of policy and planning, at local level (the level of study-areas).

The characterization of the study-areas' socio-economic profiles (population features, settlement, economic activities, landscape characteristics) is designed to identify the framework of anthropogenic-induced 'changes' in the target ecosystems and to identify the relations between the use of the land, the existing natural resources and the conflicts arising from their use. These activities will be carried out by collecting, selecting and systematizing existing data to '*transform data into information*' useful for current management and decision-making processes.

The methodology adopted should mainly involve the review of a range of documents (maps, reports, aerial photographs, monographs, etc) produced by diverse departments, at different times and with different purposes. Following that, field confirmation would be needed. We will also need to identify stakeholders in study areas and establish contact with them at an early stage, to create the social support needed to implement management instruments and to evaluate their accuracy.

Integration Module. Data Storage, Sharing and Dissemination

Results should be integrated in a raster type GIS, which will provide a suitable interface for management purposes. This way, data quality and compatibility will be assured through the construction of a database combining the existing information with that resulting from this Program. This database will also be essential in terms of managing the project, particularly for model calibration and validation. The construction of the database should therefore interact with all the Program Modules.

The development of a flexible data system will allow the storage of heterogeneous data. When applicable, data should be geo-referenced and, in some cases, time referenced, so that the application developed with visual tools, and using languages orientated to the object, can allow the creation of regional maps (raster type). In these maps, not only should the information be registered to permit a more complete spatial analysis, but the data ought to be represented graphically. Besides enabling users to access the charted information, which is one of the main objectives of the application, the database ought to have all the characteristics of a data base system. This means that it should be able to perform complete or partial data listings, using simple and complex filters, as well as to order and process the data included in this management system. The application should have the capacity to store data yielded by future work. Open architecture should allow data to be exported to classic formats, enabling it to be used in some of the commoner existing applications. Besides building up a database, the Integration Module should also address the problem of data acquisition during the project. The most important variables or parameters to be measured should be selected first. Monitoring methodologies and experimental design should then be harmonised.

Finally, analytical techniques should be fine-tuned through collaboration between the researchers involved in the experimental activities. The database compiled by the Integration Module should be shared by the participating institutions and made available to stakeholders and decision-makers.

CASE-STUDIES

The results of some applications of the modelling tools described above are given in the next sections. These are some of the first results obtained in Lima and Mondego rivers, which are now being used to characterize these target systems.

Water Quality Simulation under Different Operational Conditions of the Touvedo Dam

The Touvedo hydroelectric power station is situated on the Lima river, about 47 *km* from its outfall, and is the beginning of the reach to be modelled. When constructing the model the natural channels system (Figure 9) was discretised in internal and external hydraulic units.

As a reference setting for the Touvedo hydroelectric power plant, a discharge of 4.0 $m^3 s^{-1}$ was considered, which is the theoretical summer value of the environmental flow. Water withdrawal takes place close to the surface, at approximately 0.5 *m* depth. The main characteristics of this water mass are temperature 24.3°*C* and dissolved oxygen 7.0 $mg\,l^{-1}$ (these values were observed by at a point close to the upper water intake of the Touvedo dam). The flow contribution of the Tora tributary is 0.5 $m^3 s^{-1}$ and has the following characteristics: temperature 15°*C* and dissolved oxygen 10.1 $mg\,l^{-1}$. The initial conditions in the water mass of the Lima river segment under consideration are shown in Figure 10 (time 0:00:00). The following common values were considered for the whole length of the Touvedo-Ponte da Barca segment: 20°*C* temperature and 9.0 $mg\,l^{-1}$ dissolved oxygen, close to oxygen saturation point.

A second scenario with the same upstream hydrograph was considered, but now with the water intake placed at 20 *m* depth. The water mass temperature decreased from 24.3°*C* to 14.5°*C*, and dissolved oxygen fell from 7.0 $mg\,l^{-1}$ to 5.2 $mg\,l^{-1}$. The initial conditions considered in the Lima river segment, as well for the Tora tributary inflow, are identical to those for the first scenario.

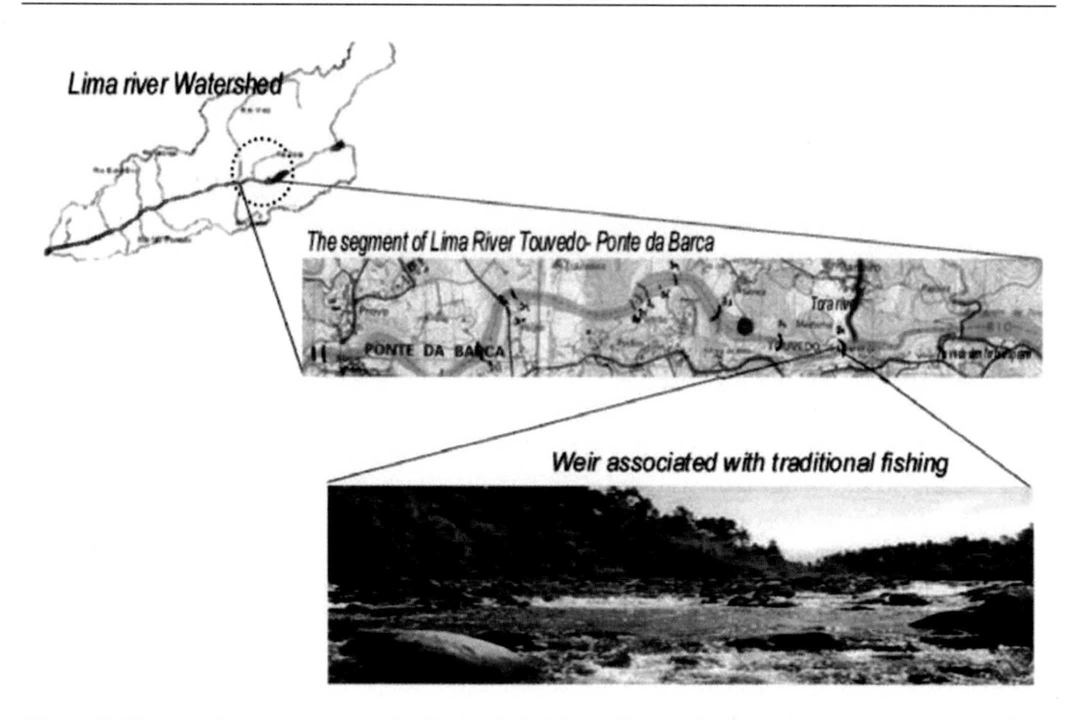

Figure 9. The top-down system results in the definition of several zones: the Lima river watershed with the stream network, the segment of Lima river Touvedo-Ponte da Barca and a weir associated with traditional fishing (Lopes *et al.*, 2003).

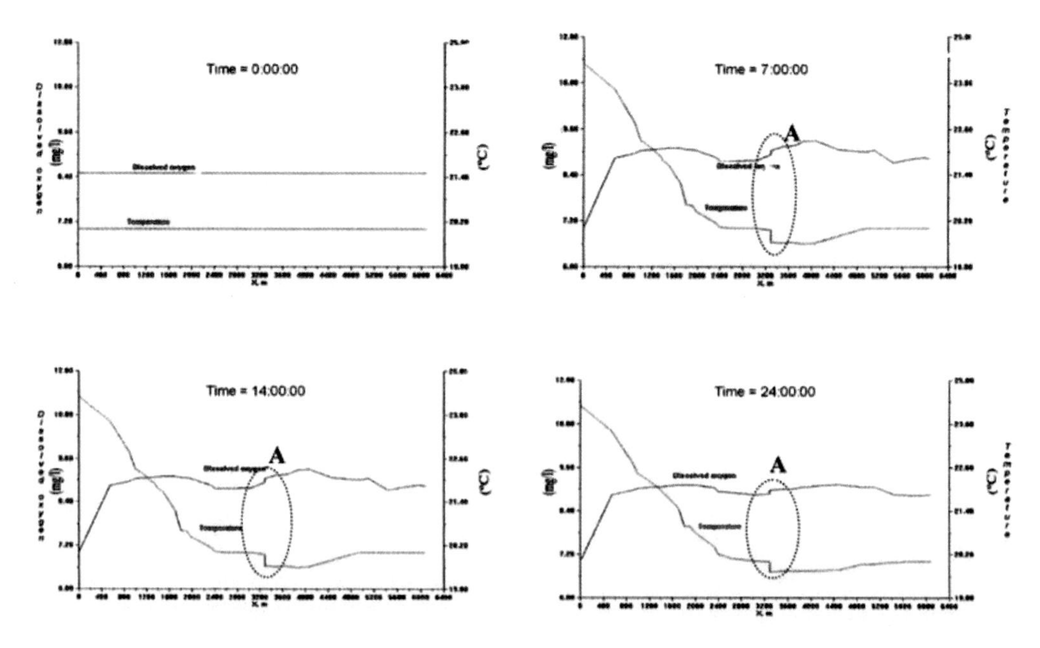

Figure 10. Spatial and temporal variation of the water quality parameters, dissolved oxygen and temperature, for a discharge of 4.0 $m^3 s^{-1}$, and the water intake close to the surface (0.5 *m* depth). A: Effect of Tora river tributary on the water quality parameters of river Lima (Lopes *et al.*, 2004).

Figure 11 shows the dissolved oxygen variation and temperature along the Touvedo-Ponte da Barca segment at different times.

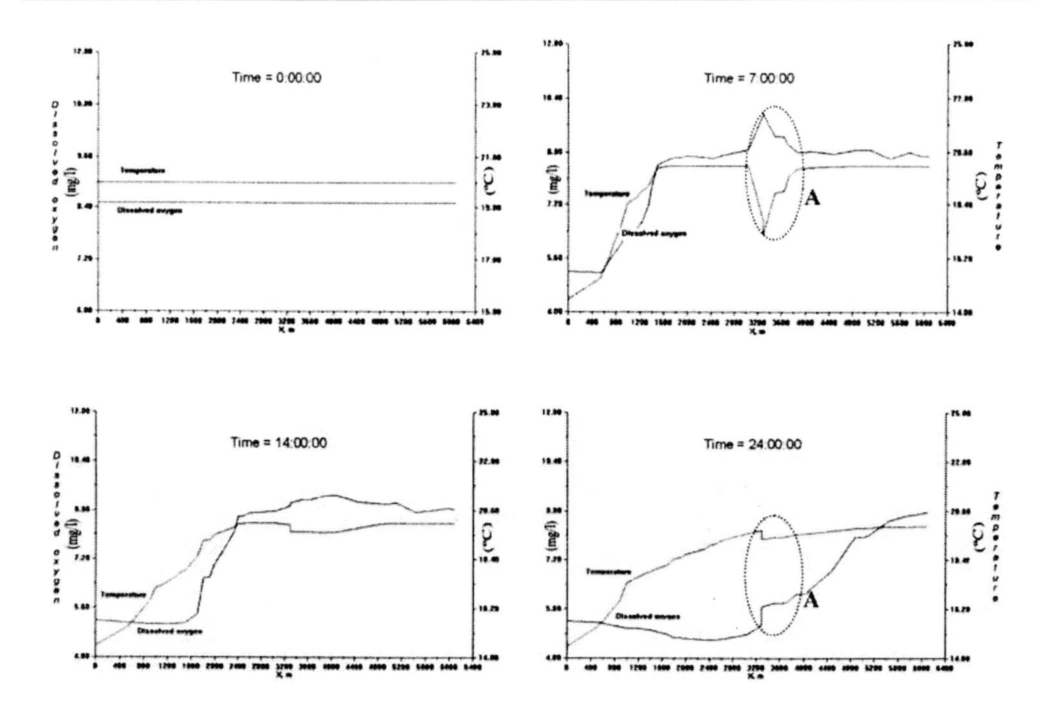

Figure 11. Spatial and temporal variation of the water quality parameters, dissolved oxygen and temperature, for a discharge of 4.0 m^3s^{-1} with the water intake at 20 *m* depth. A: Effect of Tora river tributary on the water quality parameters of river Lima (Lopes *et al.*, 2004).

DISCUSSION

The Touvedo dam reservoir registers a difference of about 10°*C* between the surface and the maximum depth, which has a great influence on the discharged water characteristics downstream of the reservoir. It may be concluded that the discharge level of the water withdrawal is more important than the amount of flow.

Water Quality River-Reservoir System Results in the Aguieira Reservoir

As a first case study in the Mondego River, Figure 12 shows the Aguieira-Raiva system, with the Raiva development situated 10 *km* downstream from Aguieira. The main purposes of the Raiva reservoir are:

- To create a storage reservoir for pumping at Aguieira; and,
- To modulate the flows to be supplied for downstream irrigation.

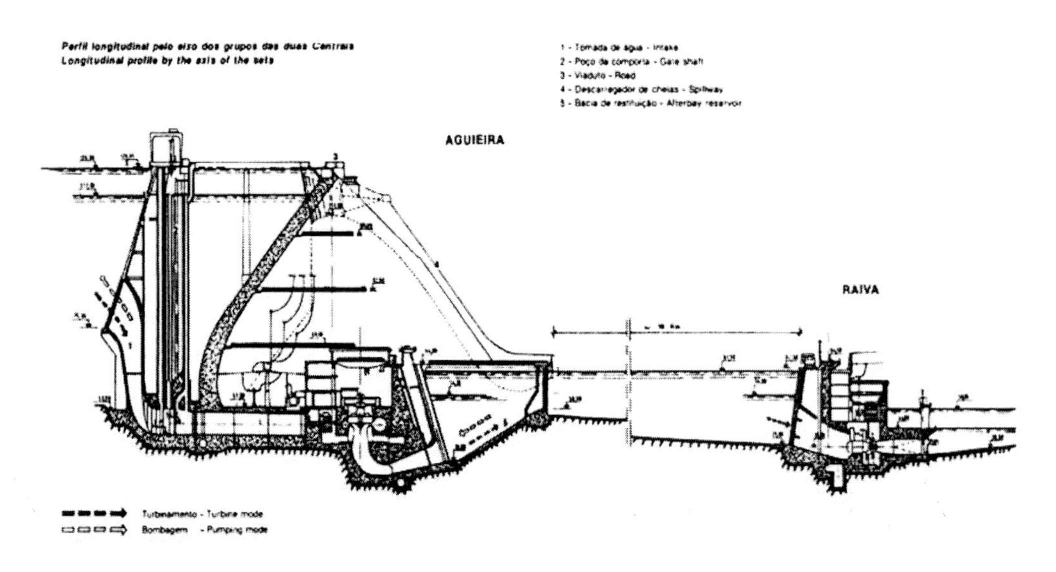

Figure 12. Aguieira-Raiva system (Lopes *et al.*, 2003).

To analyse the pumping mode influence on certain water quality parameters (temperature and dissolved oxygen), the following three scenarios in the Aguieira-Raiva system have been simulated (Coelho *et al.*, 1999):

- Scenario 1 is a normal monthly situation: Pumping of 6 $m^3 s^{-1}$ to the Aguieira reservoir (July 1998, in the case simulation described);
- Scenario 2 is also a normal situation: Without pumping, in the same time period, as scenario 1 (July 1998); and,
- Scenario 3 (unusual situation): Pumping of 35 $m^3 s^{-1}$ to the Aguieira reservoir, considering the same time period.

Simulated temperature and dissolved oxygen results, using the WQRRS program, as described in section *Hydraulics and Water Quality Module*, are shown in Figure 13. It is evident that water temperature has risen about one degree close to and above the intake quota (approximately 40 *m*), in the case of scenario 3 (left graphic, lower line). As in the previous case, dissolved oxygen has been increased about 2.0 $mg\, l^{-1}$ (right graphic, lower line).

Preliminary conclusions:

- Simulation results show good agreement between field data and numerical results;
- Pumping from the Raiva development to the Aguieira reservoir has positive effects on the Aguieira water quality; and,
- For the two parameters (temperature and dissolved oxygen), these study results show the importance of the Raiva development both for the water quality in the Aguieira reservoir and for the flow downstream.

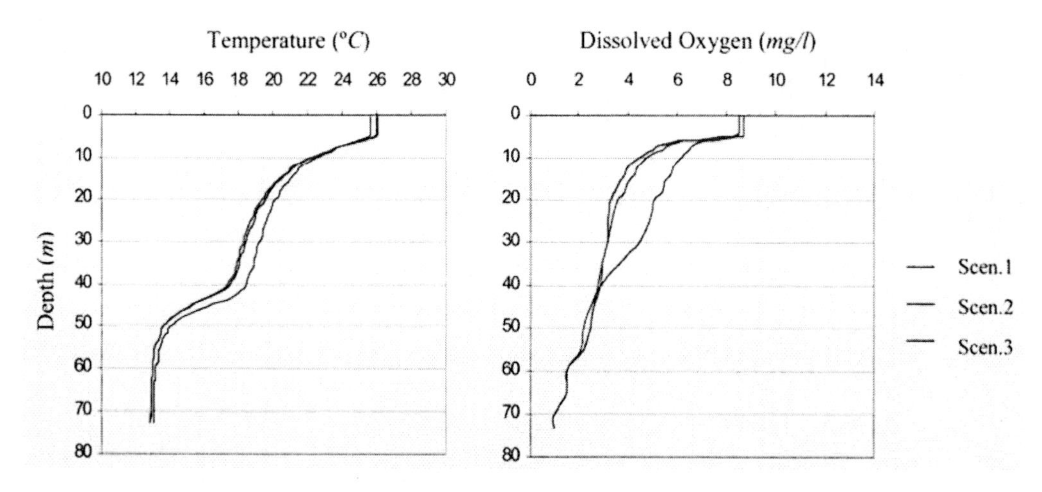

Figure 13. Comparison of model results for the three scenarios under consideration (Scenario 1 -> Scen.1; Scenario 2 -> Scen.2; Scenario 3 -> Scen.3) (Coelho *et al.*, 1999; Lopes *et al.*, 2003).

Discussion

In this case study, it may be concluded that pumping water from the Raiva development to the Aguieira reservoir has generally positive effects on the water quality, which is not only important from the viewpoint of hydroelectric power production but also benefits all other users of the Aguieira-Raiva system.

The Impact of Eutrophication and River Management Interpreted within a Framework of Ecosystem Theories

A second case study is in progress on the Mondego River, and the following are its chief goals:

- To combine past observations and more recent data in an integrated manner;
- To interpret the recent biological evolution of the Mondego estuary within a framework of ecosystem theories;
- To test Exergy, a concept originated in Thermodynamics, and Species Richness as ecological indicators of the state of the ecosystem along the eutrophication gradient; and,
- To compare of the relative efficiency of Exergy and other goal functions in capturing and expressing the state of the ecosystem.

The study site is the Mondego estuary (Figure 14). This estuary is composed of two arms, north and south, which are distinct from a hydraulic point of view: the north arm is deeper than the south arm, and the latter is almost totally silted up in the upstream areas.

This silting up results in the release of the river's fresh water mainly from the north arm, depending on the hydraulic circulation in the tide's south arm, and the release of sweet water from a tributary, the River Pranto, artificially controlled by a floodgate.

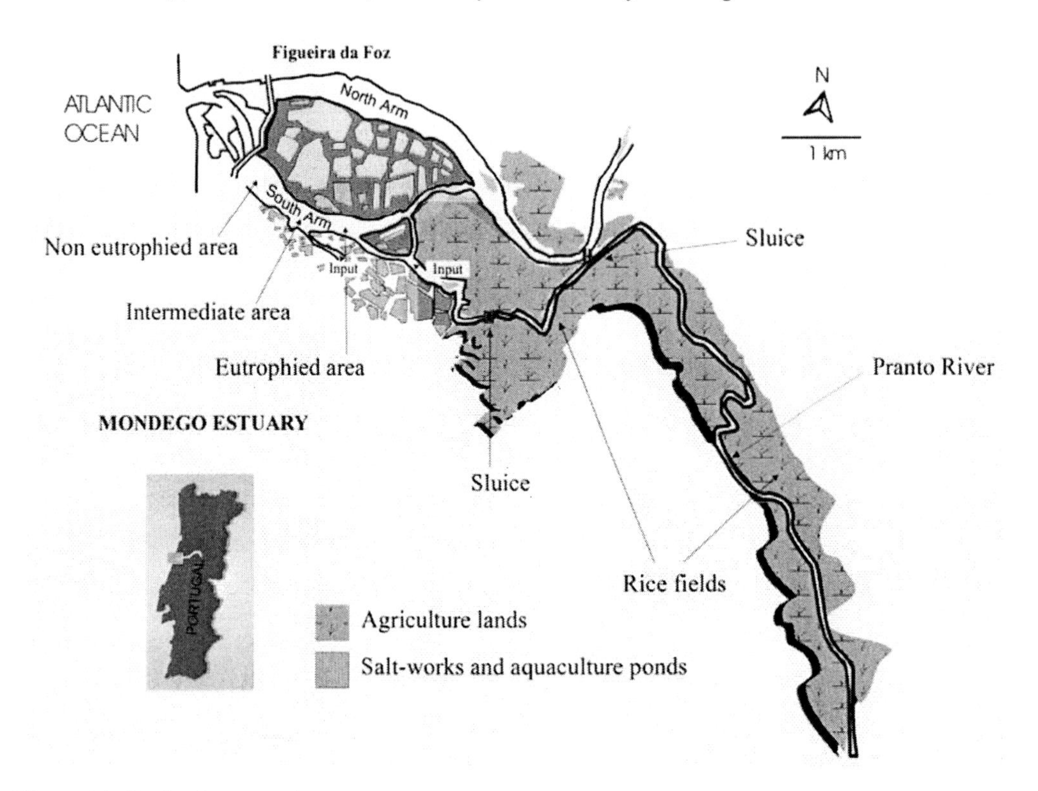

Figure 14. Study site: Mondego river estuary.

Figure 15 shows the circulation pattern along the two arms, considering the characteristic flows of 60.0 m^3s^{-1} and 2.3 m^3s^{-1} from the Mondego and Pranto rivers, respectively. As can be seen, in practical terms there is no connection between the two arms upstream. As a result, the residence time at the south arm is about 48 to 50 hours, which is considered very high to guarantee sustainable conditions for the existing ecosystems.

Approximately 15000 *ha* of the Lower Mondego Valley, the estuary's upstream, are agricultural plots, mostly dedicated to the cultivation of rice. The flow in this area contributes an important release of nutrients (nitrate and phosphorus) and various chemical compounds into the estuary environment, which is in large part achieved through the Pranto River. The release of nutrients into the arm south of the estuary (where the Pranto joins) is therefore important (134 *tons* nitrate annually) and as a function of its greater confinement, has led to a progressive eutrophication process.

Preliminary studies allowed us to identify three distinct areas in the south arm of this estuary:

- Non-eutrophied – *Zostera noltii* beds
- Intermediate eutrophied
- Strongly eutrophied

Figure 15. Hydrodynamic simulation along the two arms of the Mondego river estuary, including the tides' influence zone (just to about 18 *km* from the river mouth), with the objective of obtaining residence times.

The monitoring of biological communities in this estuary has made it possible to construct a database on the space and time variation of the physical-chemical environmental factors and of the characteristics of the benthonic communities (specific composition and respective biomass) along a eutrophication gradient. These data are now being interpreted. Table 1 summarizes the data collected.

The use of this database, along with the design and development of new experimental work based on acquired knowledge, will allow us to make progress in answering a number of questions.

Table 1. Data Collected

Biological	Physicochemical
Specific composition: algae, macrophytes and macroinvertebrates.	**Water factors:** Salinity, Temperature, Dissolved Oxygen, pH, Nitrites, Nitrates, Ammonia, Silica, Phosphates, Chlorophyll_a.
Biomass (*AFDW* m^{-2}) of each species: macrophytes and algae; number of individuals and biomass: macroinvertebrates.	**Sediment factors:** particles size, organic matter contents (*AFDW* m^{-2}).

Coastal systems, by virtue of their characteristics, are more profoundly and drastically affected by disturbances resulting from the agriculture and other human activities, particularly the effects resulting from the unloading of urban and industrial effluents, and particularly at biological community level, that manifest themselves continuously with the passage of time.

How can we predict and evaluate the impact of processes of this nature on the ecosystem? Although the ecological models of aquatic ecosystems developed to date are capable of successfully describing the alterations observed, within certain limits and from a quantitative point of view, as a general rule they are unable to predict qualitative changes (e.g. will there be an alteration in the specific composition of the communities as a result of the increased release of nutrients?). In fact, the methods normally implemented (e.g. production methods) are not effective in evaluating the type of qualitative changes in the ecosystem. More holistic approaches are required, which include the development of dynamic structural models with the application of goal functions.

This new model type embodies the alternative to the more common approaches (inclusion in models with more trophic levels and more organism types to represent each level). It allows the inclusion of qualitative alterations to the ecosystem over time, as a response to external factor changes, while at the same time using mathematical algorithms to direct the ecosystem's selective and adaptive processes, increasing the prediction and generalization capacity of the existing models.

The goal functions can be defined as properties arising from the ecosystems, emerging in the course of its development, and therefore assuming that the ecosystem characteristics resulting from its internal organization processes are the tendencies of its development. Various holistic ecological indicators, capable of acting as goal functions in the ecosystem, have been proposed in the last ten years. These proposals arise from different orientations within ecosystem theory, thus: from thermodynamics, e.g. exergy, and from the analysis of trophic network, e.g. emergy, buffer capacity, ecological efficiency, organization, ascendancy and biomass. These ecosystem properties and others are being optimized in their development.

It is reasonable to assume that in equilibrium situations, ecosystems tend to evolve towards a state of optimal configuration of those properties. As such, the goal functions may not only reflect the natural tendencies of ecosystem evolution, but may also serve as good ecological indicators of their integrity.

Several questions must be answered before the goal functions can be efficiently resorted to, either for modelling purposes or as holistic ecological indicators. Such questions include: *i*) what are the relationships between holistic ecological indicators, such as goal functions, and other more conventional ecological indicators, such as biodiversity ? *ii*) how do we calculate the values of objective functions and render these properties 'operational' for modelling and as holistic ecological indicators ? *iii*) up to what point are the values calculated for the goal functions capable of clarifying the state of the ecosystem ? *iv*) what are the relative efficiencies of the different goal functions at recognizing alterations to the ecosystem ? *v*) how can better estimate values of goal functions be produced ?

A main goal of this research is to help to find an answer to these questions, including the perfection of exergy calculation methods, rendering the estimated values capable of distinguishing between the different states of ecosystem organization, and the comparison of the relative efficiencies of exergy (originating in the field of thermodynamics), emergy and ascendancy (originating in the field of analysis of trophic webs) in capturing and expressing the state of the ecosystem.

The exergy variation in space and time will be calculated on the basis of the organisms' biomass through the use of balancing factors capable of differentiating between different biomass quantities. For this, the organisms' biomass data will be organized as a function of taxonomic levels superior to the species (e.g. biomass of macroalgae, macrophites, Isopodes,

Amphipods, etc.) or of functional groups (e.g. biomass of macroalgae, herbivores, carnivores, etc.).

Biodiversity values will also be calculated, using the biomass data of the species present. Taken into account will be the spatial variation of biodiversity along the estuary eutrophication gradient (diversity) and the annual biodiversity variation. Given that the available data refers to algae, macrophites and macroinvertebrates, we consider it appropriate in this case to use the heterogeneity indexes, which take into consideration the specific richness and regularity of the distribution of species fertility.

In this way, the same set of data will be used to calculate the exergy and biodiversity values. The hypothesis to be tested is that exergy and biodiversity will vary according to the same tendency in space and time. Although the relative subjectivity of the biodiversity methods is a problem, as a qualitative characteristic it is an intuitive concept with respect to the state of the ecosystems (loss of biodiversity = loss of ecosystem integrity). As a consequence, the confirmation of this hypothesis will elucidate up to what point the exergy values calculated are actually capable of clarifying the state of the ecosystem.

In this context, the comparison of the relative effectiveness of exergy, emergy, and ascendancy as holistic ecological indicators in capturing and characterizing the integrity of the ecosystem is essential. The following aspects, advantages and difficulties will be highlighted in relation to the use of emergy and ascendancy as holistic ecological indicators.

Emergy (Odum, 1983) aims to evaluate the exergy necessary for the formation of organisms belonging to different trophic levels. In this case, different types of energy can be converted into solar equivalents.

The calculation can be performed by multiplying the values by an energy transformation ratio (Odum, 1983), assuming that the more steps there are between two types of energy (e.g. trophic levels), the greater the quantity of embodied solar energy, measured in solar equivalents, that will be necessary to produce a unit of this energy type (quality).

Normally, considering a given organism, it is difficult to determine the trophic level **n** to which it belongs, 10^n being the factor used to calculate the emergy from the biomass. In fact, it is frequently difficult to locate, with certainty, a given organism within a determined functional type (many organisms are omnivores) and the energy calculations may thus be incorrect. In addition, the emergy calculations require excellent taxonomic support and a good knowledge of the structure and function of the different parts composing each organism.

Ascendancy (Ulanowicz, 1986) aims to describe and measure the dimension and organization of ecological chains. It can be used to group species into functional groups (without a taxonomic base), keeping in mind their functional and space-time characteristics. In addition, it may also be used to measure structural changes within the ecosystem and could thus serve as an instrument for the evaluation of an ecosystem's evolutionary tendencies.

The ascendancy measures, meanwhile, are instantaneous characterizations of the ecosystem, which requires that they be repeated over a period of time. Furthermore, such as with emergy, ascendancy continues to require strong taxonomic support and a good knowledge of the structure and function of the different parts making up each type of organism.

SELECTED ECOLOGICAL INDICATORS. BRIEF INTRODUCTION

The Exergy concept may be interpreted as a distance between a given state of an ecosystem and what the system would be at thermodynamic equilibrium:

$$Ex = R \cdot T \cdot \sum_{i=0}^{n} \left[C_i \cdot ln\left(\frac{C_i}{C_{i_{eq}}}\right) + \left(C_i - C_{i_{eq}}\right) \right]$$

where R is a gas constant; T is the absolute temperature; C_i is a concentration in the system of component i; $C_{i\,eq}$ is the concentration of component i at thermodynamic equilibrium, and index 0 represents the inorganic compounds of the considered element.

In an ecosystem in equilibrium with the surrounding environment exergy is zero.

During ecological succession exergy is used to build up biomass, which in turn stores exergy. In thermodynamic terms, taking an ecosystem's trophic structure as a whole, there will be a continuous evolution of the structure as a function of changes in the prevailing environmental conditions, during which exergy storage will be optimized.

Ecological succession drives from simpler to more complex ecosystems, which seem, at a given point, to reach a sort of balance between keeping a given structure, emerging for the optimal use of the available resources, and modifying the structure, adapting it to a permanently changing environment.

EXERGY ESTIMATION IN ECOLOGICAL TERMS

The Exergy Index is defined as:

$$Ex = \sum_{i=1}^{n} \beta_i \cdot C_i$$

where C_i is the concentration in the ecosystem of component i (e. g. biomass of a given taxonomic group or functional group), and β_i is a factor able to roughly express the quantity of information embedded in the biomass.

Detritus is chosen as the reference level, i.e. $\beta_i = 1$, and exergy in biomass of different types of organisms are expressed in detritus energy equivalents.

Over time, the variation of exergy in an ecosystem results from the variation of the biomass and of the information built into one unit of biomass (expressing the quality of the biomass):

$$\Delta Ex_{tot} = \Delta Ex_{biom} \cdot \Delta Ex_{inf}$$

If the total biomass ($Biom_{tot}$) in the system remains constant over time, then the variation of exergy (Ex) will be a function of the structural complexity alone. The information embedded in the biomass may be called Specific Exergy ($SpEx$). For each instant, *Specific Exergy* is given by:

$$SpEx = Ex_{tot} / Biom_{tot}$$

Table 2 shows the evolution of *gDNA/cell*, number of genes, and number of cell types for different organisms. The concentration of each organism was multiplied by the proposed weighting factor to estimate Exergy. The weighting factor accounts for the information embodied in the organism in addition to the simple biomass (gm^{-2} *afdw*). For this purpose, it is assumed that detritus (organic matter content in sediments) does not contain relevant structural information. Sources: Li and Grauer (1991) in Jørgensen *et al.* (1995). Values marked with * were taken from Marques *et al.* (1997).

We considered the use of the Margalef index (I) suitable for computing *Species Richness*:

$$I = (n-1)/ln(N)$$

where n is the number of species found, and N is the total number of individuals.

Some of the results obtained are given below (Figures 16 to 20).

Table 2. The Evolution of *g DNA/cell*, Number of Genes, and Number of Cell Types for Different Organisms

Organisms	10^{-12} *g* DNA/cell	Number of genes	Number of cell types	Weighing factor
Detritus	0	0	0	1
Bacteria	0.005	600	1 - 2	2.7
Algae	0.009	850	6 - 8	3.4
Yeast	0.02	2000	5 - 7	5.8
Fungus	0.03	3000	6 - 7	9.5
Sponges	0.1	9000	12 - 15	26.7
Plants, trees	-	10000 - 30000	-	30 to 90 *
Jellyfish	0.9	50000	23	144
Nemertineans	-	-	-	144*
Insects	-	-	-	144*
Crustaceans	-	-	-	144*
Annelid worms	20	100000	60	287
Molluscs	-	-	-	287*
Echinoderms	-	-	-	144*
Fish	20	100000 - 120000	70	287 - 344
Birds	-	120000	-	344
Amphibians	-	120000	-	344
Reptiles	-	120000	-	344
Mammals	50	140000	100	402
Human	90	250000	254	716

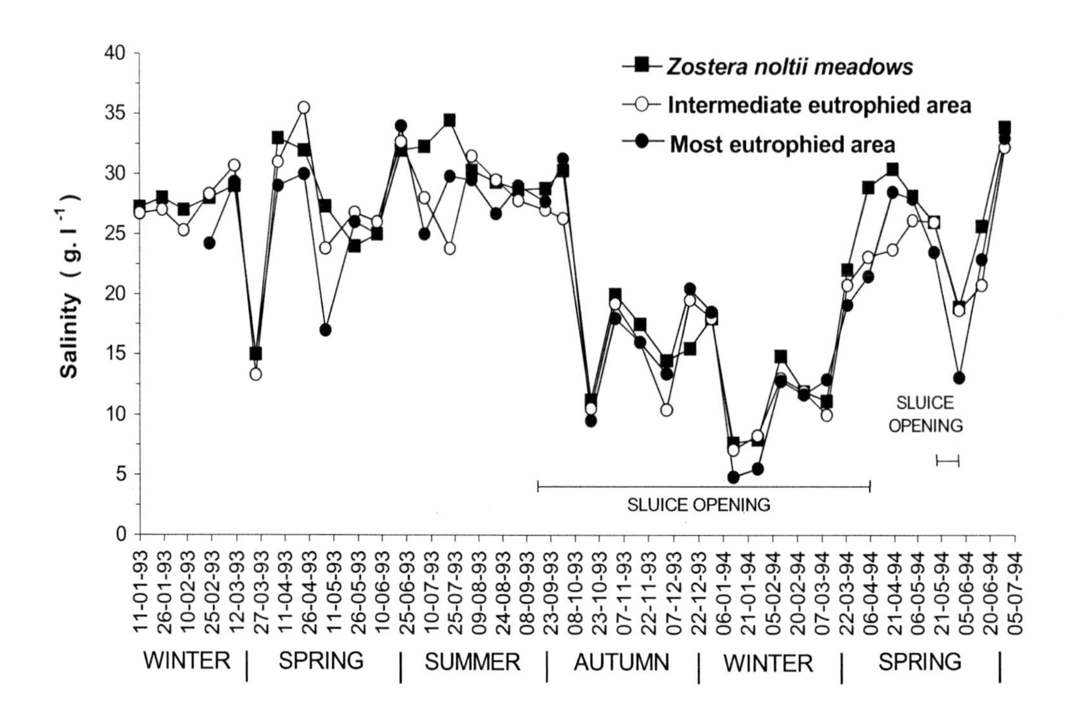

Figure 16. Salinity variation.

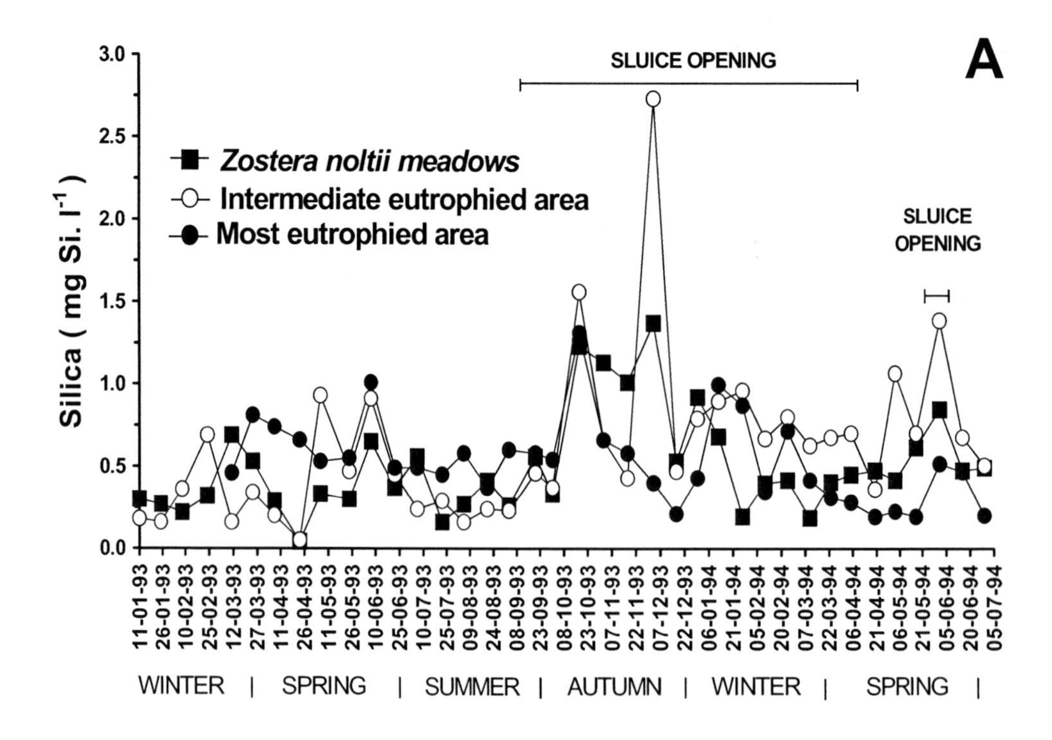

Figure 17. continued on next page.

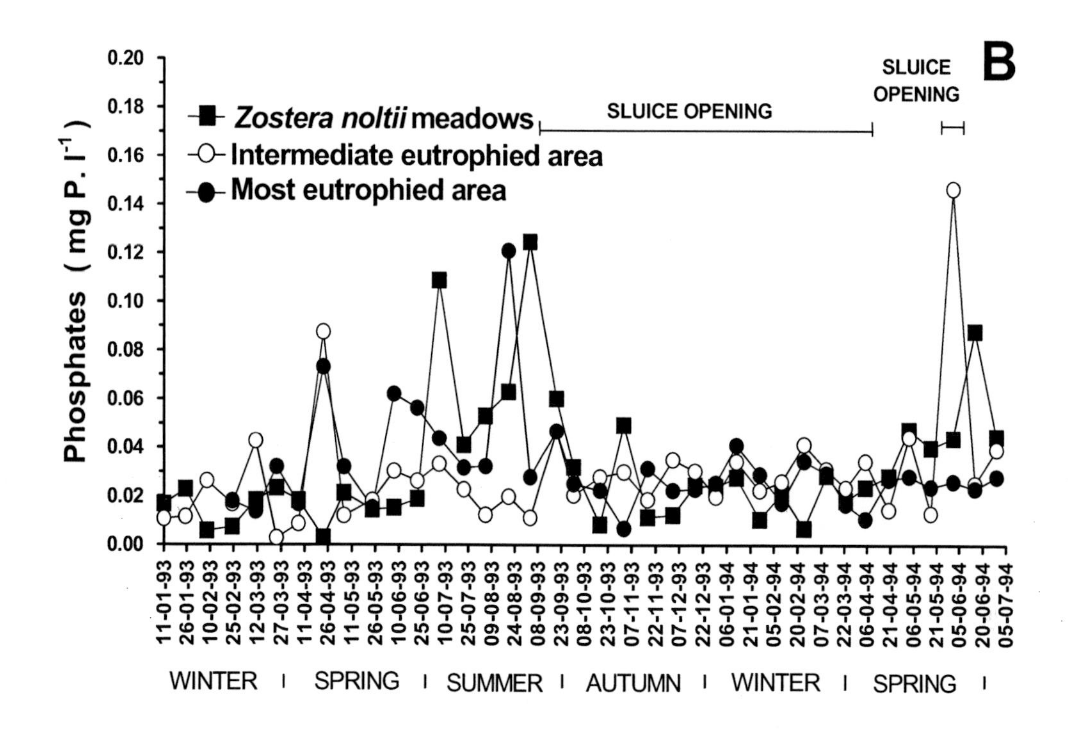

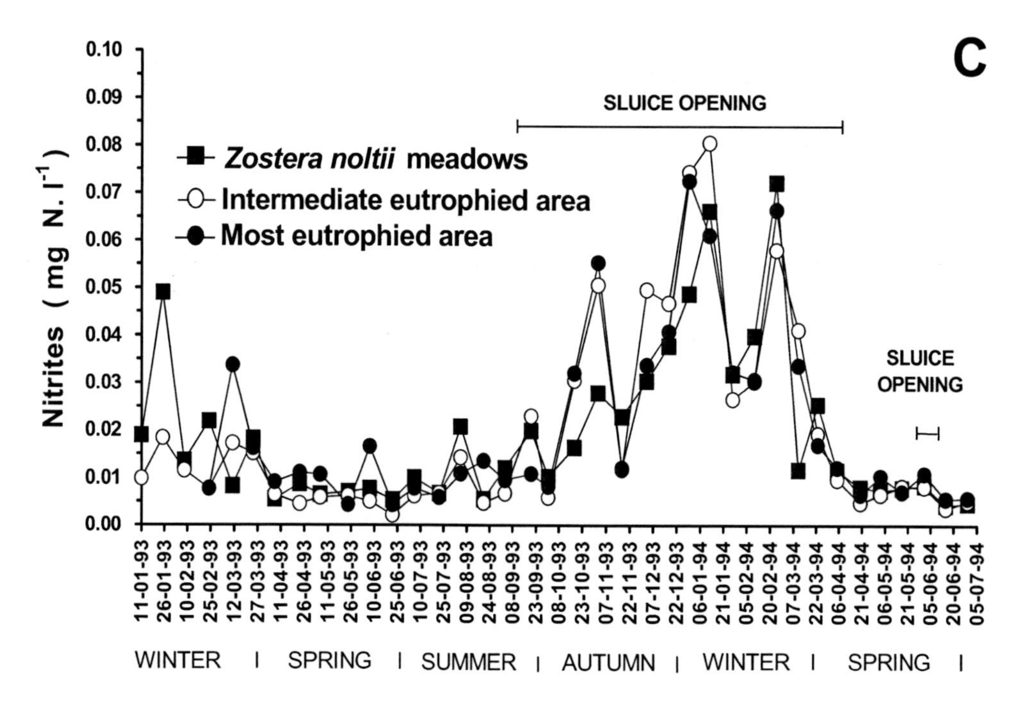

Figure 17. continued on next page.

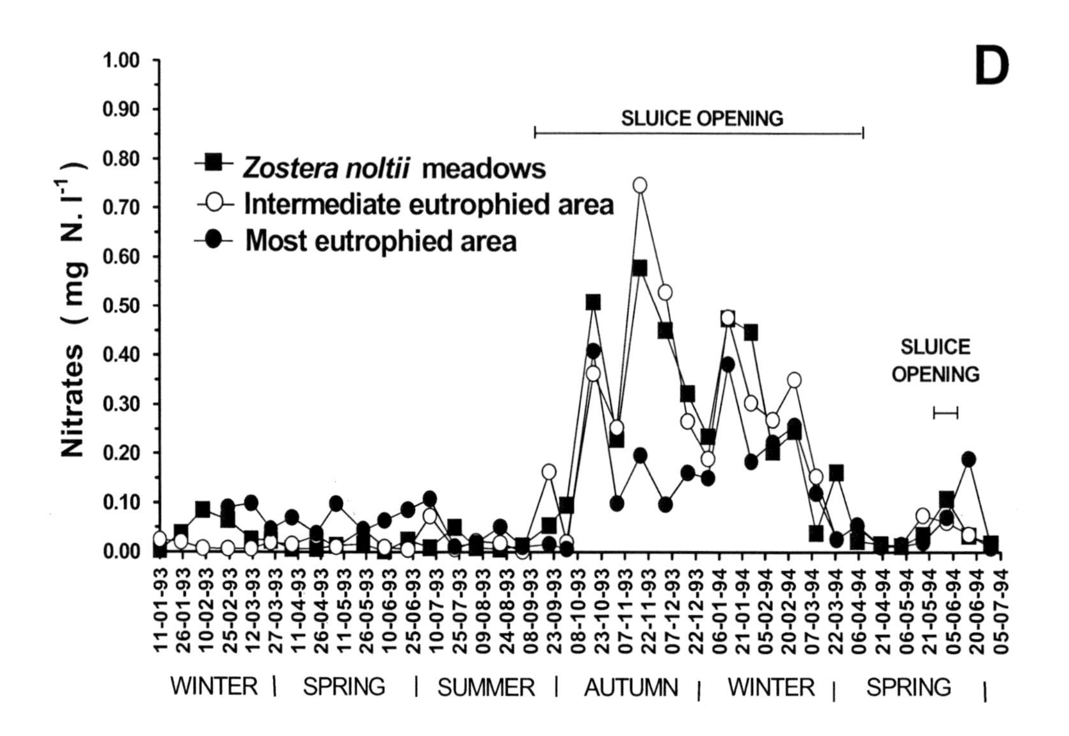

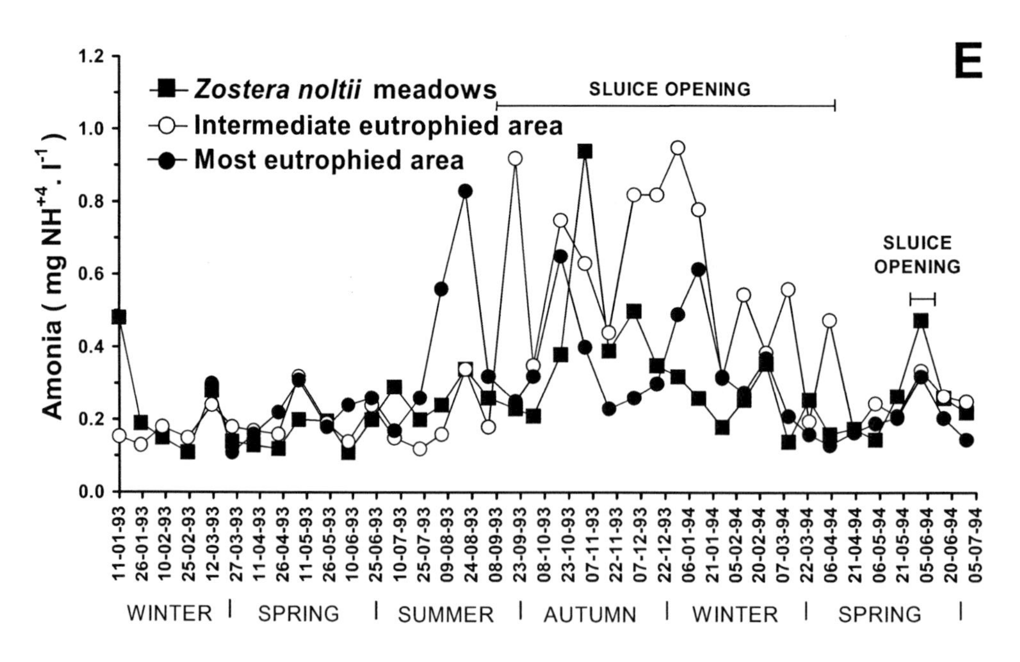

Figure 17. Variation of nutrient concentration in the water column A - Silica; B - Phosphates; C - Nitrites; D – Nitrates, and E - Ammonia

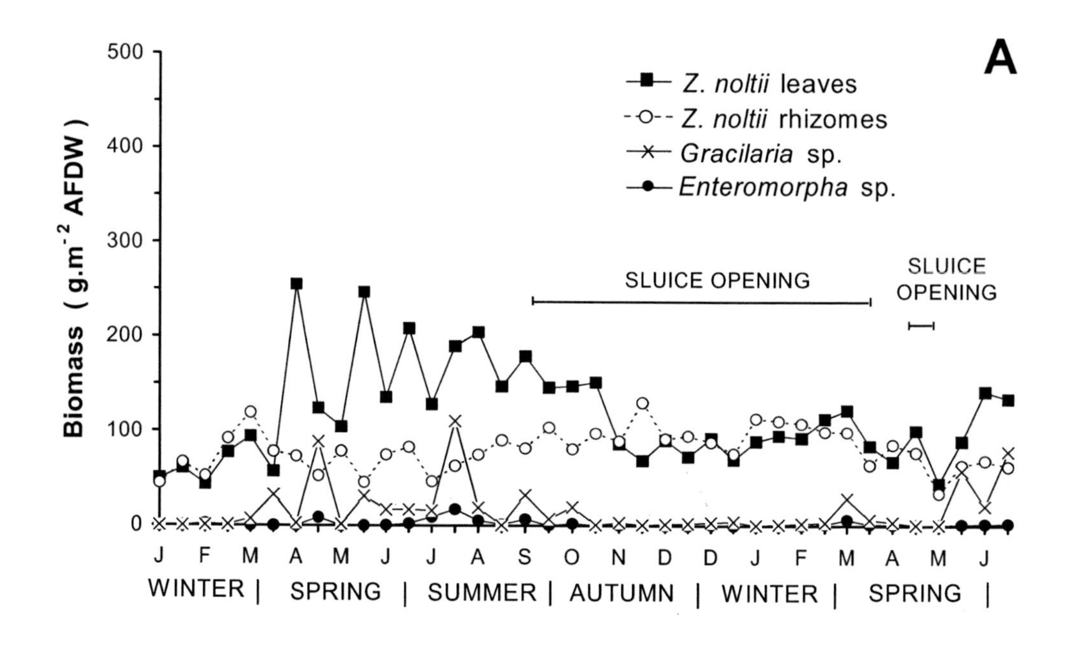

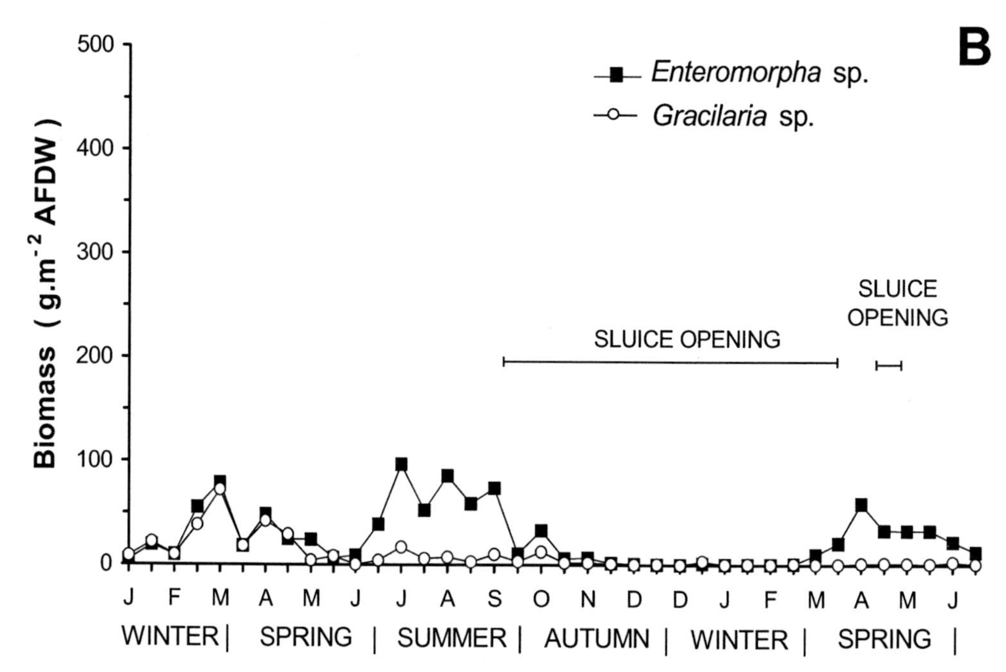

Figure 18. continued on next page.

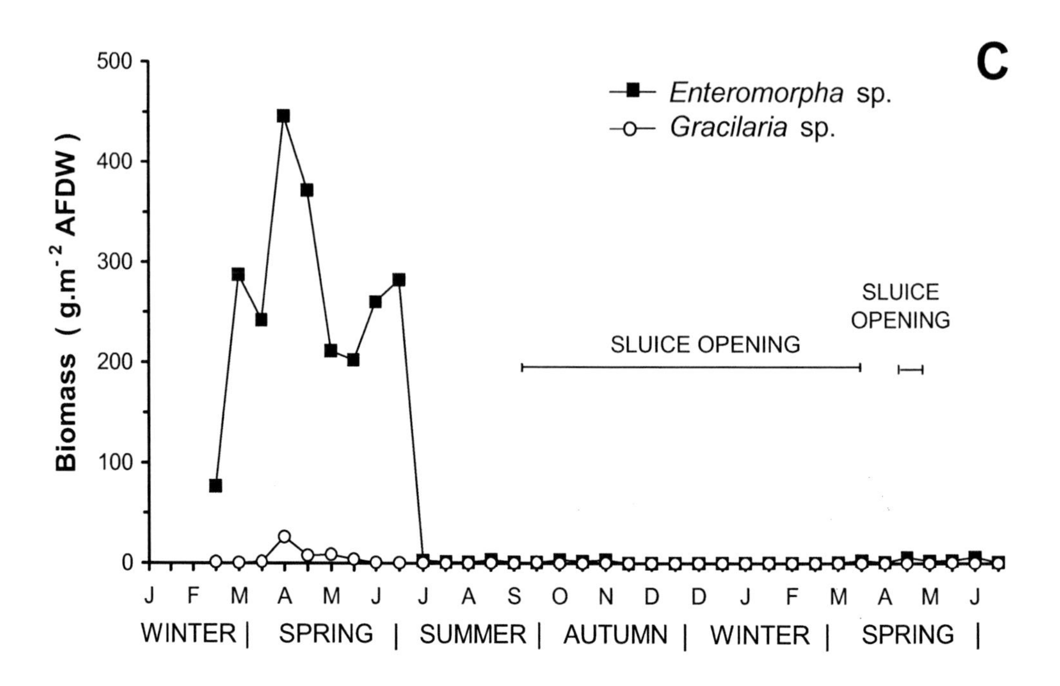

Figure 18. Variation of *Zostera noltii* and macroalgae biomass

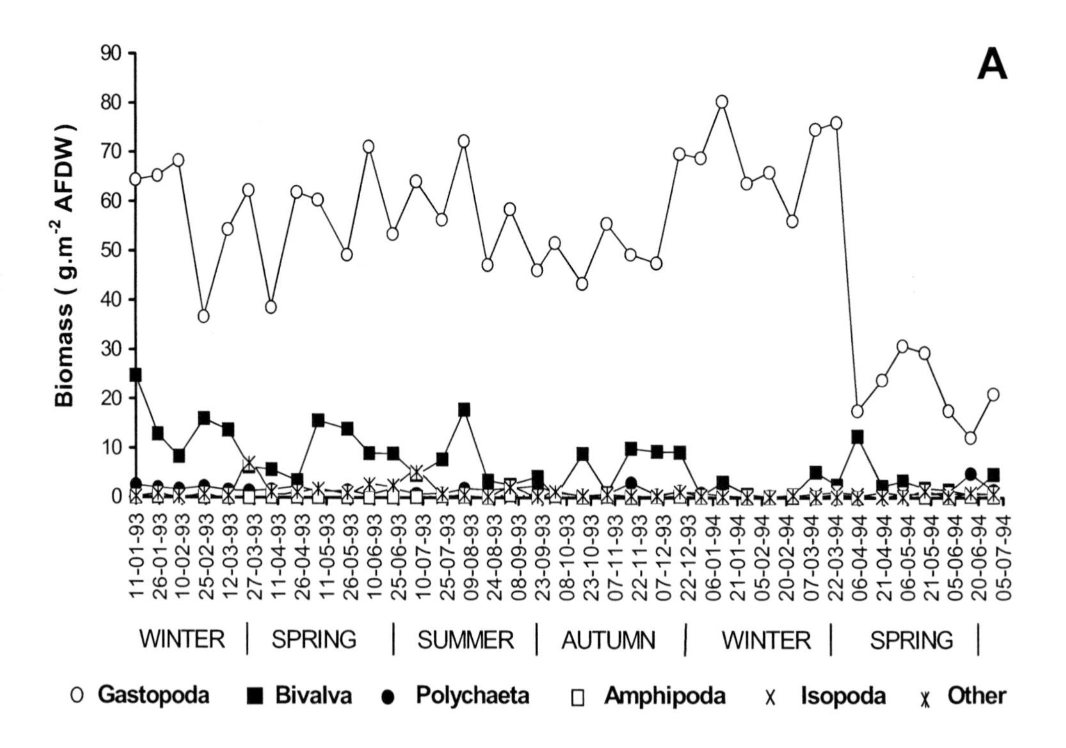

Figure 19. continued on next page.

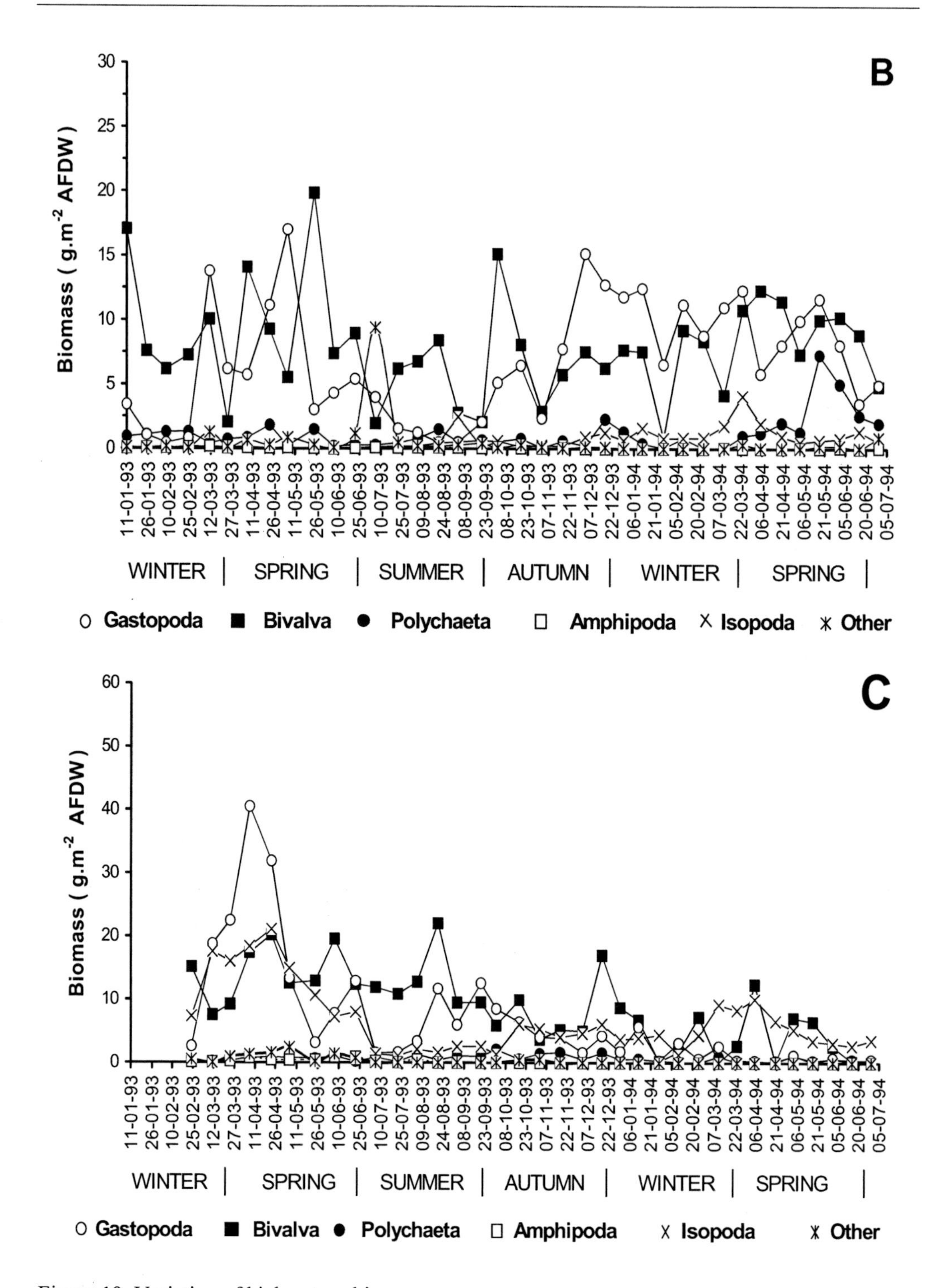

Figure 19. Variation of higher *taxa* biomass

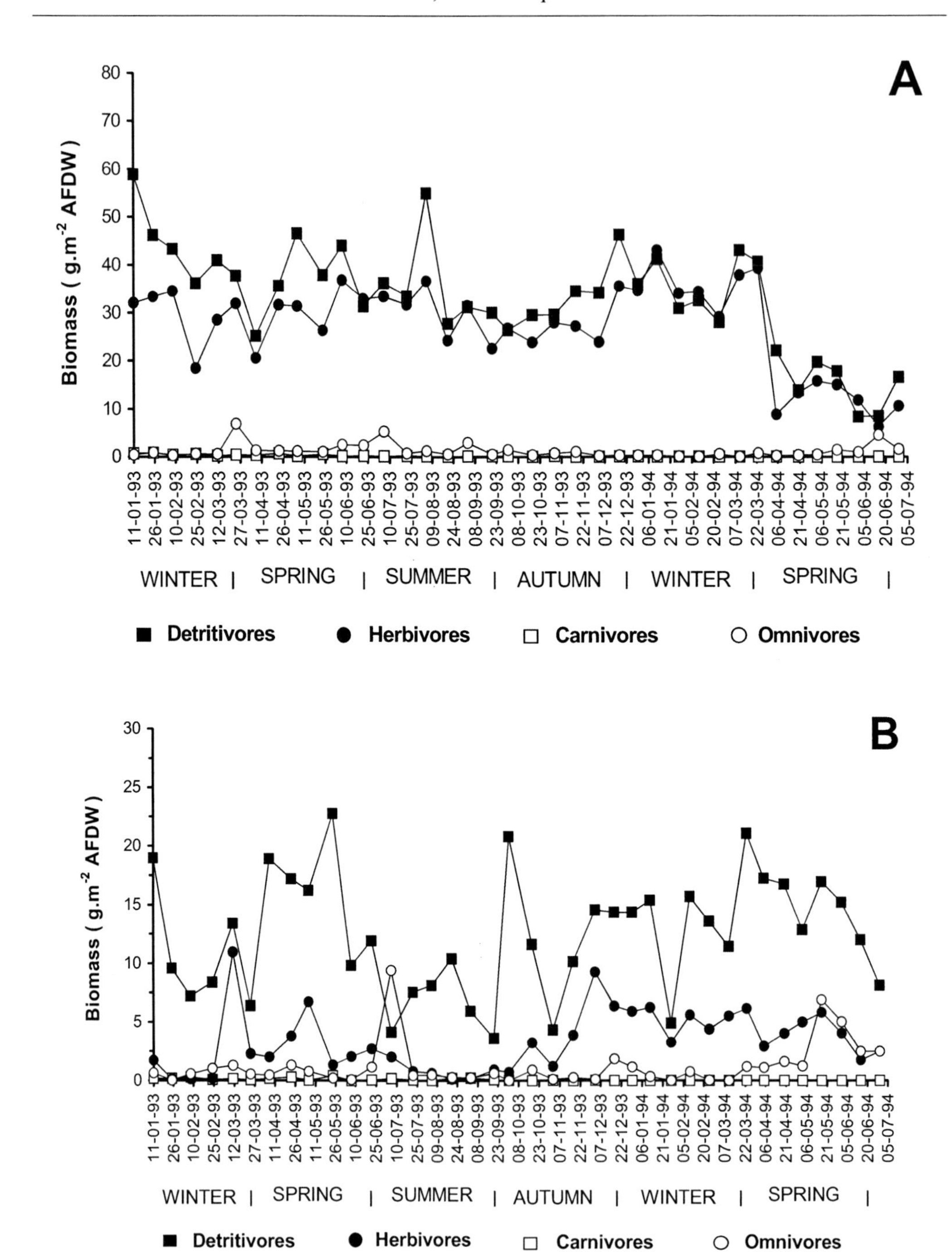

Figure 20. continued on next page.

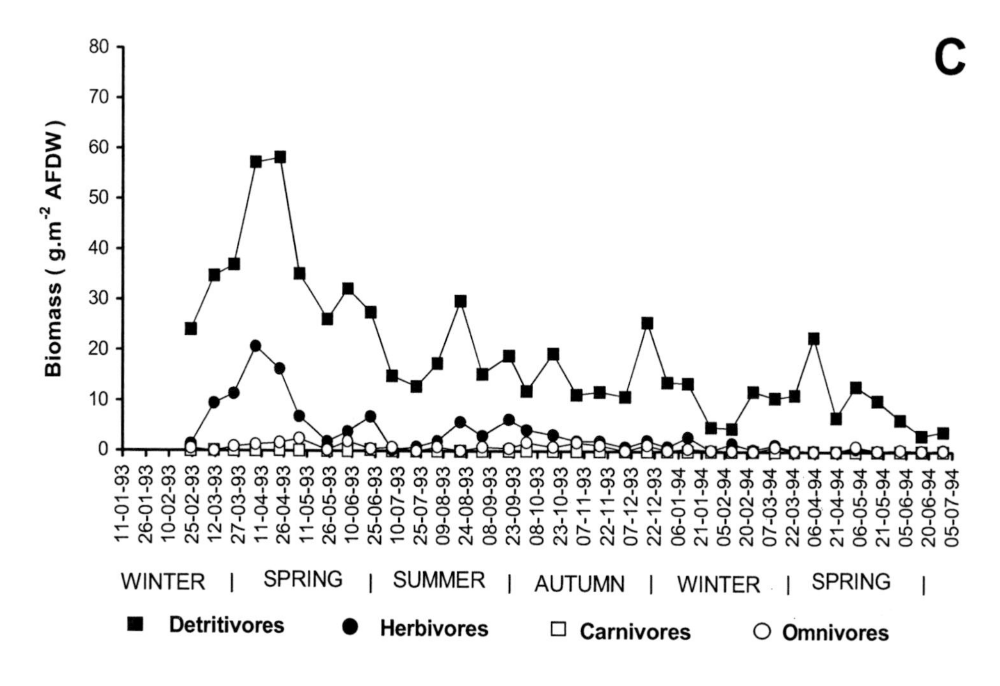

Figure 20. Variation of functional group biomass

Table 3 shows the spatial variation of the Exergy Index, Specific Exergy and Species Richness along the eutrophication gradient. These data was collected from February 24, 1993 to February 24, 1994. Exergy is expressed in gm^{-2} detritus energy equivalents and Specific Exergy is expressed in Exergy per unit of biomass.

Table 3. Exergy Index, Specific Exergy and Species Richness Variation in the Three Scenarios Identified along the Eutrophication Gradient in the South Arm of the Mondego Estuary

Indicator	*Zostera noltii* beds Non eutrophied area			Intermediate eutrophied area			*Enteromorpha* spp dominant Most eutrophied area		
	Average	STD	Range	Average	STD	Range	Average	STD	Range
Exergy Index	25389	±2533 (10%)	20143 to 29530	4802	±2059 (43%)	1803 to 7990	8553	±4485 (52%)	2688 to 19805
Specific Exergy	211	±28 (13%)	120 to 270	151	±132 (87%)	9 to 373	160	±95 (59%)	32 to 255
Species Richness	1.8	±0.36 (20%)	1.2 to 2.4	1.5	±0.6 (40%)	0.5 to 2.5	1.2	±0.4 (33%)	0.4 to 2.0

INTERPRETATION OF THE RESULTS

The acquired behavior of the system and the observations presented here can be summarized as follows:

- In the early eighties – *Zostera noltii* beds widely distributed, extending to the inner areas;
- Increase of nutrient discharges proceeding from mainland;
- Green macroalgae blooms – competition between macroalgae and macrophytes;
- *Zostera noltii* beds decreased – now restricted to downstream areas; and,
- Despite nutrient enrichment of estuarine waters, macroalgae blooms do not occur each year. River management as a function of agricultural practices provides explanation for this.

As a tentative explanation of the observations and results, the following mechanism could be stated:

- Strong freshwater discharges will decrease salinity, and low salinities will inhibit macroalgae growth, despite the increase in nutrients in the water column; and,
- Dissolved nitrogen discharge from mainland with freshwater will increase N/P ratios, since phosphorus is mostly released from estuarine sediments and becomes diluted in the water column. This may determine limitation of phosphorus in the system.

Based on confirmed facts and assuming some hypotheses, a specific interpretation of the system evolution can be achieved.

One fact is that the *Zostera noltii* disappearance in the south arm of the estuary started in the inner areas and progressed downstream.

One basic assumption is that spatial changes in the system may be taken as representing temporal changes. The non eutrophied area, with *Zostera noltii* beds, may represent what the system was twenty years ago, the intermediate eutrophied area may represent changing areas during the process, and the most eutrophied area may represent the most advanced stage in the observed shift between primary producers, where macroalgae have totally ousted macrophytes.

As a consequence, the shift in primary producers in the Mondego estuary and related changes in the food chain may be seen as a structural dynamic shift between species.

As preliminary conclusions we can state that:

- Eutrophication appeared to be the major driving force behind the ongoing changes;
- Two factors play a major role in determining the state(s) of the system: a) Salinity controls the occurrence of macroalgal blooms, and b) nutrient input determines the overall eutrophication state of the system;
- River management appear to be a key question in establishing scenarios, particularly with regard to the impacts of eutrophication;
- Remediation may be achieved by implementing ecological engineering principles in different possible management practices;
- The recent biological evolution in the south arm of the estuary may be interpreted in compliance with a broader ecological theoretical framework; and,
- Further research will be necessary for a more detailed understanding of all the processes involved. The development of a structural dynamic model will be a powerful tool to help achieve this.

Conclusions

It is intended that the results obtained under this Program should lead to a better-integrated and more environmentally correct management approach, especially with respect to protection measures. It could thus be quite an important contribution to Integrated Riverine, Estuarine and Coastal Zones Management, both in Portugal and abroad.

The knowledge acquired from this Program will be useful, for instance, for the following: the definition of floodplain areas; the design of water quality protection measures; correct riverine and costal land-use planning, including the stakeholders in the decision process; the design of beach profile replenishment; a more rational utilisation of shoreline living resources; the definition of set-back lines; the prediction of coastal erosion (shoreline retreat) linked with extreme events, and for the development of a new approach to assessing the integrated river and estuary economic value.

The results already obtained are encouraging enough to proceed with further research. However, much more data, other field information and computational simulations are needed to understand all processes involved and to reach a final correct picture of the systems.

Acknowledgements

This work is a contribution to the project "Transport and dispersion of pollutants in a marine environment: parameterization of the effects and evaluation of the environmental impacts" (PRAXIS/P/EME/12087/1998), supported by the FCT – Fundação para a Ciência e a Tecnologia.

References

Antunes do Carmo, J S & F J Seabra-Santos, 1996. On breaking waves and wave-current interaction in shallow water: A 2DH finite element model. *Int. J. Num. Methods in Fluids*, 22, 429-444.

Antunes do Carmo J S & J C Marques, 2001. River dynamics and restoration. A comprehensive management strategy. Proc. River Basin Managent 2001 Conference. In *River Basin Management*, WIT*PRESS* publisher, 63-72. ISBN 1-85312-876-7.

Coelho P S, A C Rodrigues & P A Diogo, 1999. Water quality modelling in the Aguieira-Raiva multi-purpose system. *Journal of Water Resources*, Vol. 20, 1, 83-91 (in Portuguese).

HEC - Hydrological Engineering Center, 1978. *Water Quality for River-Reservoir Systems*, US Army Corps ofEngineers, USA.

Jørgensen, S E, Halling-Sørensen & S N Nielsen, 1995. *Environmental and Ecological Modelling*. CRC Lewis Publishers.

Lopes, L F G, J S Antunes do Carmo & R M V Cortes, 2003. Influence of dam-reservoirs exploitation on the water quality. Proc. River Basin Managent 2003 Conference. In *River Basin Management II*, WIT*PRESS* publisher, ed. C A Brebia, 221-230. ISBN 1-85312-966-6.

Lopes L F G, J S Antunes do Carmo, R M V Cortes & D Oliveira, 2004. Hydrodynamics and water quality modelling in a regulated river segment: application on the instream flow definition. *Ecological Modelling*, 173, 197-218.

Marques J C, M A Pardal, S N Nielsen & S E Jørgensen, 1997. Analysis of the properties of exergy and biodiversity along an estuarine gradient of eutrophication *Ecological Modelling*, 102:155-167.

Marques J C, S N Nielsen & S E Jørgensen, 1998. Applying thermodynamic orientors: The use of Exergy as an indicator in environmental management. In Müller F. & Leupelt M. (eds) *Ecotargets, goal functions, and orientors. Theoretical concepts and interdisciplinary fundamentals for an integrated, system-based environmental management*. Chapter 5.3: 481-49, Springer-Verlag, Berlim.

Mellor G, 1998. *Users Guide for a Three-Dimensional, Primitive Equation, Numerical Ocean Model*, Princeton University, Princeton, USA.

Nielsen S N, S E Jørgensen & J C Marques, 1998. Case studies: Modelling approaches for the practical application of ecological goal functions. In Müller, F. & Leupelt, M. (eds) *Ecotargets, goal functions, and orientors. Theoretical concepts and interdisciplinary fundamentals for an integrated, system-based environmental management.* Chapter 2.14: 243–254, Springer-Verlag, Berlim.

Odum, H T, 1983. *Systems ecology: An introduction.* John Willey & Sons, Inc., Toronto.

Pinho, J L S, J M Pereira Vieira & J S Antunes do Carmo, 2004. Hydroinformatic environment for coastal waters hydrodynamics and water quality modelling. *Advances in Engineering Software*, Vol. 35, 205-222.

Ulanowicz, R E, 1986. *Growth and development ecosystems phenomenology*. Springer-Verlag, New York.

In: Progress in Aquatic Ecosystems Research
Editor: A. R. Burk, pp. 35-58

ISBN 1-59454-383-6

Chapter 2

ECOLOGICAL INDICATORS AND SOCIETY'S VALUES: MONITORING, RESEARCH AND MANAGEMENT OF WATER QUALITY IN RIVERS

Alonso Aguilar Ibarra*
Universidad Nacional Autónoma de México, México

ABSTRACT

This chapter aims: (i) to give a synthesized yet comprehensive account of the main trends and approaches on monitoring ecological indicators for river research and management, and (ii) to show the importance of the links between society's values and environmental water quality. Human pressure on freshwater ecosystems has increased along with a greater awareness of environmental quality concepts like river health and ecosystem integrity. Several industrialized countries have already integrated these concepts in recent water quality guidelines, giving more emphasis on studying and regenerating biotic communities. Such is the case of the European Water Framework Directive or the Environmental Monitoring and Assessment Program of the U.S. Environmental Protection Agency. These water policies seek to include biological assemblages for defining water quality standards. The most common assemblages used as indicators in running waters are periphyton (i.e. benthic diatoms), macroinvertebrates, and fish. Their ability to signal disturbances vary according to the spatial and temporal scale, collection methods, taxonomic identification and public acceptance. There are different ways to measure these ecological indicators, among which diversity indices, species richness, guilds of species and multi-metric indices have been the most frequently applied. The main considerations which have to be taken into account for implementing effective monitoring programs for research and management comprise: pilot studies and statistical power, cost-effectiveness, selection of reference conditions, and the scale of study and heterogeneity of the landscape. The future development of ecological indicators in aquatic resource management will face challenges such as the prediction of cause-effect

* Universidad Nacional Autónoma de México, Apartado Postal 70-371, Ciudad Universitaria DF 04510, México. Phone: +52 (55) 56 22 49 19 ; Fax : + 52 (55) 56 22 48 28; E-mail: aaibarra@correo.unam.mx

processes, linking society's values and ecological quality concepts, and communicating the message to developing countries.

INTRODUCTION

The economic development of many civilizations has been closely linked to rivers, involving the modification of aquatic habitats world-wide (Dynesius & Nilsson 1994, Everard & Powell 2002). Human influences on river structure and fauna are varied but in general they can be considered as having their sources in four main society's activities: use of water, use of land, species introductions and fisheries (Figure 1). The uses of water include a number of modifications to river morphology, mainly by the construction of dams and channels for irrigation and navigation, resulting in a fragmentation of both biotic communities and their habitats (Poff et al. 1997). Concerning the use of the land in river watersheds, agriculture, urbanization and industrial activities are the main sources of both point and non-point source pollution (Allan et al. 1997, Wang et al. 1997, Harding et al. 1998). River ecosystems are also disturbed by over-exploitation of both commercial and sport fisheries (Arlinghaus et al. 2002), and by the introduction of exotic species (Ross 1991).

These issues have brought the attention of developed countries on the consequences of aquatic ecosystem degradation, leading to the notion of integrity and ecological health as fundamental elements of water quality (Costanza et al. 1997, Walmsley 2002). Introducing such concepts brings into focus the conciliation of economic development and environmental water quality for achieving long-term sustainability (Rapport et al. 1999). Hence, this chapter deals with the relationship of ecological indicators and society's values with a two-fold objective. On the one hand, it describes the main ecological indicators used to assess environmental water quality. It does not pretend to be an extensive review, but to present a synthesized yet comprehensive account of the main trends and approaches on monitoring for research and management. On the other hand, it tries to show the importance of the links between society's values and environmental water quality. Since much work remains to be done in this field, several challenges are described in order to help guiding future research and management efforts.

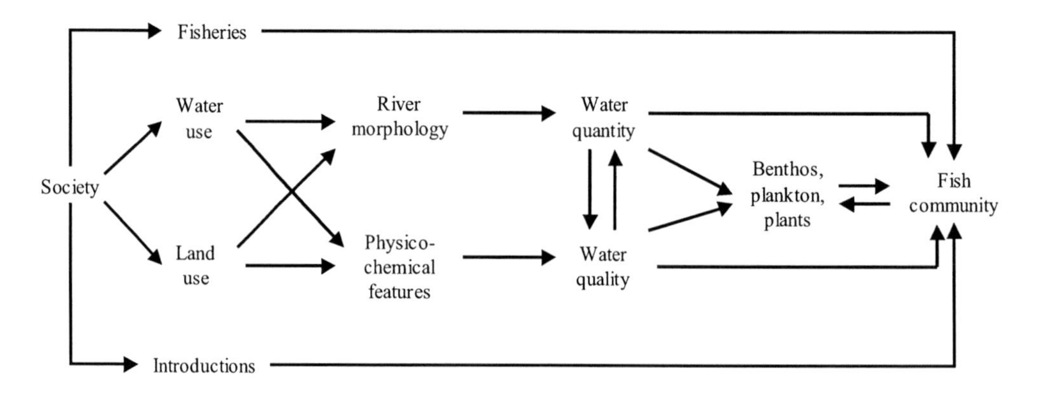

Figure 1. Main human influences on river ecosystems (modified from Welcomme 1983).

SOCIETY'S VALUES TOWARDS ENVIRONMENTAL QUALITY

One of the latest developments on environmental quality is the notion of ecosystem health and integrity. These are, in fact, difficult concepts because each user has a different perception of healthy rivers in terms of water abundance, water quality, or biodiversity conservation. Health and integrity are, nevertheless, useful notions for protecting rivers. Indeed, Boulton (1999) and Karr (1998, 1999) point out that the term 'river health' is useful because it is readily interpreted by the general public and policy makers, and because it evokes societal concern about human impacts on rivers. For example, Karr (1998) explains that rivers and streams may be viewed as a continent's circulatory system, and the study of those rivers, like the study of blood, can diagnose the health not only of the rivers themselves but of their natural populations. A definition of environmental health is given by Costanza et al. (1992) as "an ecological system which is stable and sustainable, that is, it is active and maintains its organization and autonomy, and is resilient to stress". A fish community, for instance, is "healthy" (i.e. has integrity) if its composition and function are comparable with those of natural habitats, presenting minimal human disturbance (Hughes et al. 1998).

A conceptual framework for better understanding the relationship between health, integrity, sustainability and environmental quality is proposed in Figure 2. Environmental quality may be viewed as a continuum of human influence (Karr 1999). At one end of the gradient, severe disturbance eliminates all life. At the other end pristine, or minimally disturbed living systems prevail. In between, a degree of health and sustainability, indicating the level of environmental quality, is defined by a certain threshold of quality. Therefore, in this chapter, the concepts of health, integrity and environmental quality are employed without distinction, knowing that they belong to a continuum of ecological degradation.

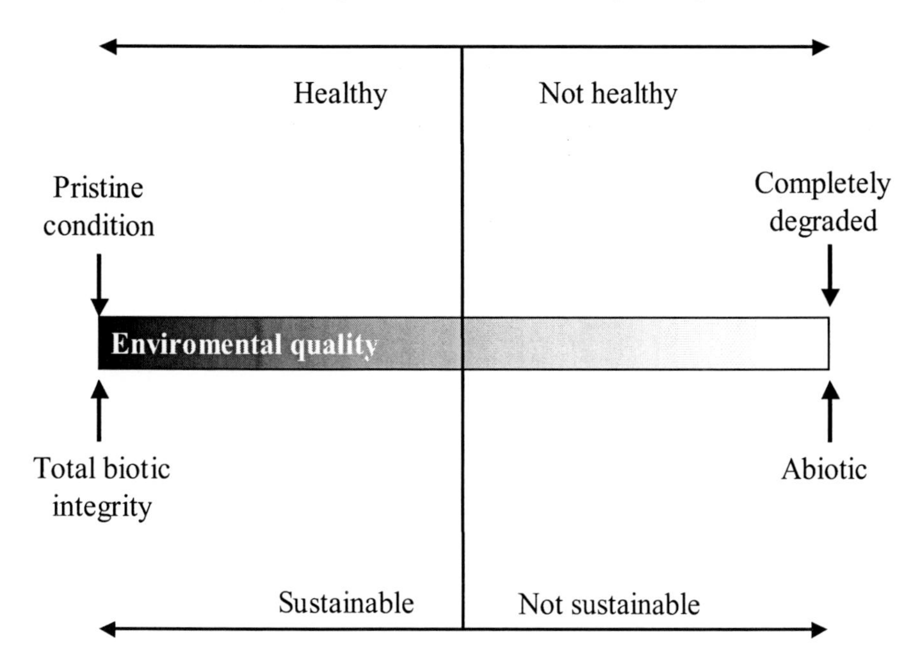

Figure 2. Conceptual framework for environmental quality shown as a continuum (modified from Karr 1999).

Environmental quality has a value only when it is perceived as beneficial to human society (Minns et al. 1996, Boulton 1999, Power 1999, Rapport et al. 1999). In general, for an item of goods or a service to have value, it must meet two conditions: (i) it must provide enjoyment or satisfaction, and (ii) must be scarce (Loomis et al. 1986, Hueting et al. 1998). When these conditions are met, the good or service has an economic value whether or not any financial transaction is made (Loomis et al. 1986, Opschoor 1998). As pristine environments become scarcer, they become more appreciated by society. According to Walmsley (2002), any anthropogenic changes to a natural ecosystem may have direct or indirect economic consequences. This is why environmental quality perception by society is critical in order to put into practice the health or integrity concepts.

Ecosystems with biological integrity maintain the goods and services needed by human society (Karr 1998). In fact, the uses humans have of rivers reflect several types of values (Everard & Powell 2002). For example, Navrud (2001) considers that the total economic value of a marginal change in a freshwater fish stock is what people are willing to pay to get increased fish catches. Total economic value can be divided into several types of values, but for the sake of brevity, only two main types can be considered: use value and non-use value. Use value includes clean drinking water or recreational fisheries. Non-use value is linked to the existence of a resource. Individuals would be willing to pay only to know that biodiversity is preserved in a river, even if they do not use it (i.e. existence value). Besides, they would be able to deliver the existence of biodiversity or pristine rivers to future generations. Both biodiversity protection and aquatic restoration are part of a broader societal goal, that of sustainable development (Rapport et al. 1999).

To achieve sustainability objectives, a shift in thinking and action is needed, and not merely human use of rivers (Everard & Powell 2002). Society's values thus play a central role in assessing ecosystem health. As Wilzbach et al. (1998) point out, cultural values play an important role in driving the impact that humans have on natural resources because they drive political and institutional policy. Hence, the most recent definitions of water quality include aquatic ecosystem health (Hart et al. 1999). Water quality guidelines have been focused over the years on quality for domestic drinking water, and on agricultural, recreational and industrial waters. More recently, the emphasis has been more towards aquatic ecosystem protection under what Hart et al. (1999) call "new generation water quality guidelines". Water policies that have implemented new generation water quality guidelines have been set up in industrialized countries, like the United States and the European Union.

Environmental Water Quality Guidelines in the United States and Europe

Water quality standards in the United States were first incorporated in the Water Quality Act of 1965 but the efforts of integrating a federal water pollution control program continued under the 1972 Clean Water Act (CWA). Biological integrity was first mentioned in the CWA, aiming "to restore and maintain the physical, chemical, and biological integrity of the nation's waters". In practice, however, Karr (1995) points out that its implementation was rather focused on the effectiveness of wastewater treatment technology to control point source pollution and human cancer risk.

The CWA requires states to assess the conditions of their water quality standards according to the water quality criteria set up by the EPA in order to determine any causes of impairment (Davis 1995). Although water quality standards may vary from state to state, the EPA retains the authority to promulgate federal water quality standards for any state that fails to issue standards adequate to achieve the goals of the CWA (Adler 1995). For example, in some states, these biological criteria are formalized into specific regulations (e.g. Delaware, Florida, Ohio), while in others (e.g. Hawaii, Nevada, South Dakota) there is no information on bioassessment criteria (Southerland & Stribling 1995).

In contrast to the environmental water quality policy of the United States, the state members of the European Union have been trying to standardize national water quality guidelines and monitoring methods under the application project of the European Water Framework Directive (EWFD). The EU environmental and water policy development can be divided according to Kallis & Butler (2001) in three periods. During the first period (1973-1986) environmental directives remained on the principles of public health protection and the harmonization of environmental rules to avoid market distortion within the EU. The second period (1987-1992) was marked by the assignment of a European competence for a common environmental policy under its 4th Framework program. This was a period when the emphasis passed from pollution control towards environmental protection. A proposal for a new comprehensive directive on the ecological quality of water coincided with proposals for the revision of the standards for the bathing and drinking water directives during the third phase (1993-onwards) and comprises the 5th and 6th Framework programs.

The WFD sets new goals for the condition of Europe's water and introduces new means and processes for achieving them, being hydrologic basins the management units, instead of political boundaries. Rivers are classified as surface waters for which the goal is to achieve at least a good status of ecological and chemical quality (Article 4 of the EWFD). Surface waters may be classified in five classes of quality defined in Annex V of the Directive: high, good, moderate, poor and bad. Besides, the EWFD requires hydro-morphological, chemical and biological elements to be considered in determining how far a water body departs from a 'reference condition' (Irvine 2004). By the year 2015, all surface and ground waters in Europe should, according to the EWFD objectives, to achieve a good ecological condition (European Commission 2000).

Ecological Indicators of Water Quality in Rivers: Description and Applications

It was noted above that once the human pressure on freshwater ecosystems is recognized, recent water quality guidelines have put more emphasis towards studying and regenerating biotic communities. Under this context, analyzing diversity patterns and distributions has become a focus of aquatic ecology and water quality management (Jenerette et al. 2002), as well as identifying and measuring indicators of aquatic health (Boulton 1999). In this section, therefore, the main ecological indicators of water quality in rivers will be briefly described.

According to Jackson et al. (2000) and Kurtz et al. (2001), ecological indicators are signals which relay a complex message from potentially numerous sources in a simple and useful manner. In fact, this is a critical feature with importance to aquatic resource management. Ecological indicators help to assess environmental trends over time, providing early warnings and helping to diagnose the causes of water quality problems (Dale & Beyeler 2001).

Davis (1995) gives an historical account of ecological indicators of water quality, noting that certain organisms and species were first used in Germany during the early 20th century for signaling water pollution. However, single species indicators were lately dismissed because such an approach ignored complex system-level responses (Barbour et al. 1995), undermining the ability to apply environmental regulations (DeShon 1995). Therefore, it has been widely accepted that biological integrity is best monitored at the community or assemblage level (Hughes et al. 1998). An assemblage is defined by Wootton (1991) as all species in a defined area irrespective of whether they interact or not. Interactions take place in a defined ecological community (Begon et al. 1996).

Periphyton, macroinvertebrates and fish are the most common assemblages used as indicators in rivers and streams. Periphyton refers to the algae, mostly diatoms, linked to substrates. Periphytic diatoms provide food for many primary consumers, such as macroinvertebrates and herbivorous fish (Rosen 1995). Diatoms are very sensitive to chemical alterations of rivers, like for example acidification, eutrophication, organic pollution, nutrient disturbance, sewage discharges, turbidity, and they are especially sensitive to metal toxicity (Rosen 1995, Prygiel et al. 1996, Kelly 1998, Dell'uomo et al. 1999, Wu 1999, Golda et al. 2002). Macroinvertebrates have been one of the most common ecological indicators in freshwater (Rosenberg & Resh 1993). They are used in fact in well established biomonitoring programs of several industrialized countries like Australia (Hart et al. 2001), the United Kingdom (e.g. Wright et al. 1998, Clarke et al. 2003), the United States (e.g. Barbour et al. 1995, Klemm et al. 2002, Butcher et al. 2003), and have been recently applied in developing countries (e.g. Hart et al. 2001, Fenoglio et al. 2002, Ogbeibu & Oribhabor 2002). Fish also have a widespread use around the world as indices of aquatic quality (Karr et al. 1986, Fausch et al. 1990, Soto-Galera et al. 1998, Kestemont et al. 2000, McDowall & Taylor 2000, Oberdorff et al. 2002), and are considered as a good tool for environmental decision-makers (Angermeier & Schlosser 1995).

These three assemblages possess many of the attributes required for suitable ecological indicators (Dale & Beyeler 2001). They present high sensitivity to a broad spectrum of types and degrees of stress, ability to integrate the direct and indirect effects of episodic environmental disturbances, relative constancy over time, they are ecologically diverse and widely distributed. Furthermore, all these assemblages contain many tolerant species which are especially successful in human-disturbed habitats, while others are more sensitive and respond to specific stressors, giving them good diagnostic capabilities.

Since no single indicator can signal all possible human influences under different circumstances, a suite of indicators representative of the structure, function, and composition of ecological systems is thus highly desirable (Dale & Beyeler 2001). However, in practical terms, using several ecological indicators at the same time would be restrained by budgetary conditions. Therefore, it is useful to be aware of the advantages and drawbacks of each assemblage. Their ultimate election would depend on the research and management objectives, and on technical skills and institutional capabilities. Hence, each assemblage will give different responses according to the spatial and temporal scale, collection methods, taxonomic identification and public acceptance. For example, in small streams, both periphyton and macroinvertebrates would be more indicative of local habitat conditions than fish (Rosen 1995, Lammert & Allan 1999, Clarke et al. 2003, Iliopoulou-Georgudaki et al. 2003). Moreover, Kilgour & Barton (1999) reckon that surveys of benthos can be used to make inferences on the condition of fish community composition in wadeable streams. For larger scale studies, however, fish would be a more suitable assemblage since they often present more mobility, range considerable distances and have possibly larger metapopulations (Karr et al. 1986, Lammert & Allan 1999).

There are also differences with respect to the temporal scale. Benthic diatoms have rapid reproduction rates and short life cycles, responding quickly to environmental stress (Rosen 1995). For example, Golda et al. (2002) demonstrated that structural perturbations due to low metal exposure were detected using diatoms in only two weeks. In contrast, both macroinvertebrates and fish present longer life cycles, although they are able to detect short-term catastrophic disturbances (Fausch et al.1990, Scott & Hall 1997, Klemm et al. 2002, Clarke et al. 2003), as well as sustained anthropogenic disturbance induced by agriculture or industrial waste (Dauba et al. 1997, Harding et al. 1998). Nevertheless, macroinvertebrates could be in certain cases more sensitive and quicker to react to momentary perturbations than fish (Kilgour & Barton 1999, Iliopoulou-Georgudaki et al. 2003).

Although diatoms and macroinvertebrates sampling is easier and relatively less expensive than for fish, their taxonomic identification is much more difficult. In fact, Kelly (1998) and Wu (1999) acknowledge that the use of diatom indices should be restricted to specialists. A similar problem is found with macroinvertebrates (Resh 1995), particularly where updated taxonomic keys are not available (Hart et al. 2001). However, identification of higher taxonomic groups in assessment of invertebrates assemblages is indeed very helpful (Barbour et al. 1995). In contrast, fish are relatively easy to identify, staff training is less difficult, and most samples can be identified and sorted at the field site and then released (Karr et al. 1986). Besides, the public acceptance of ecological integrity concepts could be easier achieved by using fish than diatoms or macroinvertebrates (Karr et al. 1986, Rosen 1995). However, society's perception of ecological indicators will depend also on other aspects (see below). Once an indicator (i.e. an assemblage) is chosen, a way to measure it (i.e. by means of an index) has to be found. The following section will thus briefly summarize some of the most common ecological indices for water quality in running waters.

Common Indices Used to Assess Water Quality through Ecological Indicators

One of the first ways to account for the diversity of assemblages was the construction of diversity indices, among which the Shannon-Wiener Index was the most commonly used since the 1960s as a measure of biological response to pollution (Davis 1995, Taylor & Bailey 1997). However, it has been criticized due to its failure in taking into account seasonality, sampling gear, the level of taxonomic resolution, and because it often responds erratically to systematic changes in assemblages (Davis 1995, Karr 1998).

Another simple way to measure an assemblage is to count the number of taxa (i.e. taxa richness) which are contained in it. For example, species richness is, in spite of its limits, a good indicator for environmental management (Walmsey 2002), and is reckoned by Gaston & Spicer (2000) as a "common currency" for the study of biodiversity. Fausch et al. (1990)

mention that the disadvantages of species richness include their dependence on sample size, their limited information about the whole community and that its sensitivity to degradation may vary regionally, seasonally, and with the age of individuals. Indeed, unresponsiveness to degradation is a major problem of using species richness as an index. For example, the total number of species does not always decline with environmental degradation, instead, the replacement of species or changes in relative proportions of species have been observed for invertebrates (Taylor & Bailey 1997) and for fish (Hutagalung et al. 1997). As the number of sensitive species declines with degradation, more tolerant species become more common (Wang et al. 2000, Vila-Gispert et al. 2002).

Grouping species into classes or guilds has eased the problem of differential sensitivity to degradation within an assemblage. For example, the number of aquatic insects of the orders Ephemenoptera (mayflies), Plecoptera (stoneflies) and Trichoptera (caddisflies), commonly known as EPT is an index of environmental condition of streams in France (Compin & Céréghino 2003), the United States (DeShon 1995, Klemm et al. 2002, Butcher et al. 2003), Sweden (Sandin & Johnson 2000a), and recommended by Resh (1995) for developing countries. Another classification, regardless of taxonomy, is based on trophic guilds. Structurally, species are members of guilds as guilds are of communities (Minns et al. 1996). Anthropogenic disturbances in habitat and water quality often result in fluctuations of food supply, which are reflected in the structural changes in trophic composition. Therefore, establishing trophic guilds has proved effective for assessing water quality, river health and ecological integrity. For example, benthivore fish are good indicators of water quality (Scott & Hall 1997), and piscivore fish are of special interest to conservation (Schlosser 1991) and sportfishing (Oberdorff & Hughes 1992). In the case of invertebrates, the proportion of scrapers, the ratio scrapers/shredders, and collectors, are measures of river productivity/respiration ratios (Resh 1995). In fact, both EPT (for invertebrates) and trophic guilds, are commonly used in multi-metric indices.

Multi-metric indices are the most recent approaches to assess environmental quality of rivers. They aim to integrate information of assemblages and environmental variables (i.e. metrics) in order to assess a certain level of impairment by comparing disturbed sites with reference conditions (Karr et al. 1986), providing a quick assessment easy to understand by non-technical resource managers (Hill 2003). The metrics are chosen following diverse criteria, including statistical aspects for eliminating redundant metrics (Butcher et al. 2003). According to Barbour et al. (1995), for a metric to be robust, it must be relevant to the ecological community under study and to the monitoring objectives, be sensitive to stressors, be able to provide a response that can be discriminated from natural variation, be environmental benign to measure in the aquatic environment, and be cost-effective to sample. Barbour et al. (1995) give a detailed account of steps to follow in order to develop and calibrate metrics. Some examples of metrics are species richness, percent and number of tolerant species, percent and number of motile species (e.g. diatoms), percent and number of EPT taxa (e.g. macroinvertebrates), percent and number of top carnivores (e.g. fish). Map-derived variables can also be included, like distance to the river source, surface of agricultural-devoted area in the catchment upstream to the sampling site, or altitude. Ecological and biological traits, such as substratum preference, flow velocity preference, reproductive strategy, dissemination potential, body form, among others, might be also employed on multi-metric approaches (Dolédec et al. 1999).

Multimetric indices have been applied to macroinvertebrates, fish and diatoms. The first effort to integrate information from multiple metrics into one index was the River Invertebrate Prediction and Classification System (RIVPACS). Its development started in the United Kingdom in the late 1970s as a research project "in order to (i) develop a biological classification of unpolluted running water sites in Great Britain, based on their macroinvertebrate fauna, and (ii) to assess whether the type of macroinvertebrate assemblage at a site could be predicted using physical and chemical features" (Wright et al. 1998). The historical development of RIVPACS and recent applications are given by Wright et al. (1998), Chessman (1999) and Jones (2001). This system has been also applied in Australia and Indonesia (Hart et al. 2001). Indices for macroinvertebrates in the United States have not applied the RIVPACS methods but somewhat similar multi-metric approaches. For example, the Ohio Environmental Protection Agency set up in 1987 the Invertebrate Community Index (DeShon 1995) and Butcher et al. (2003) proposed the Benthic Community Index (BCI). More information is given by Rosenberg & Resh (1993).

The Index of Biotic Integrity (IBI), a multi-metric index for fish, was conceived in the United States by Karr et al. (1986). The IBI was originally developed for the mid-western states but has been adapted to other regions (e.g. Fausch et al. 1990, Lyons et al. 2001) and other countries, like Belgium (Kestemont et al. 2000), Canada (Steedman 1988), France (Oberdorff & Hughes 1992), and Mexico (Lyons et al. 2000). The IBI may vary depending on natural variability and the physiographic region (Smogor & Angermeier 2001), the stream location in the basin (Osborne et al. 1992), and the type of river under study (Seegert 2000). A detailed account of the IBI history and functioning can be found in Simon & Lyons (1995). More recently, an index of biotic integrity for diatoms was developed by Hill (2003). Other multimetric indices for diatoms are the Diatom Bioassessment Index (Rosen 1995), the Diatom index (Prygiel et al. 1996, Dell'uomo et al. 1999) and the Trophic Diatom Index (Kelly 1998).

Critiques for multi-metric indices include: (i) the fact of being ill-suited to monitoring the effects of novel stressors, (ii) their categorical scale obscures the magnitude of stressors and effects cannot reflect quantitative alterations of the ecosystem, and (iii) they are frequently reported as absolute values without uncertainty (Taylor & Bailey 1997, Suter 2001, Iliopoulou-Georgudaki et al. 2003). In spite of these criticisms, Davis (1995) and Simon & Lyons (1995) justify their use acknowledging that more research is needed before these indices can be expanded to develop truly ecosystem health indices.

Another index which has proved useful in assessing environmental water quality under human and natural disturbances is the abundance-biomass comparison (ABC) method (Warwick 1986). Although conceived for marine invertebrates, it has been applied for riverine fish (Coeck et al. 1993, Penczak & Kruk 1999). It was Warwick (1986) who gave a first conceptual model using k-dominance curves of species biomass and abundance in three scenarios: unpolluted, moderately polluted and heavily polluted. This led Meire & Dereu (1990) to propose the ABC index using the idea of stress as a more general factor influencing natural populations. The values of the index are negative in heavily stressed conditions, near zero under moderate stress, and positive for no stress. For example, a value of –13.69 was found in a heavily polluted site, whereas an unpolluted and unregulated stretch scored 9.74 in a Belgium river (Coeck et al. 1993). In the case of an overfished river in Poland the index scored -13.23 (Penczak & Kruk 1999). Furthermore, McManus & Pauly (1990) proposed two indices: the Difference in Area by Percent (DAP) and the Shannon-Wiener Evenness

Proportion (SEP) as extensions of the original Warwick's method, however, their application remains to be tested.

MONITORING ECOLOGICAL INDICATORS FOR RESEARCH AND MANAGEMENT

There is an increasing international recognition that monitoring ecological indicators is a critical tool for water resources management. In this way the Ramsar Convention proposed the adoption of early-warning indicators, wetland inventories and frameworks for risk assessment and monitoring (Finlayson 2003). In fact, a number of industrialized countries have regular monitoring programs of ecological indicators. For example, in Europe, consistent water quality biomonitoring started in Great Britain in the late 1970s (Wright et al. 1998, Logan & Furse 2002) and later, Annex V of the EWFD pointed out that biological elements that need to be monitored for assessing water quality (Irvine 2004). In the United States monitoring ecological indicators started about the same time in Ohio state (DeShon 1995), however, not all the states have standardized sampling and processing methods (Adler 1995). This is the reason why the US EPA established the Environmental Monitoring and Assessment Program (EMAP) implemented as a long-term program (Adler 1995, Schiller et al. 2001, Suter 2001). Indeed, effective monitoring of ecological indicators must be designed as a robust tool for assessing river health in the long-term.

The first issue to be addressed in designing a monitoring program is to define clear objectives taking into account realistic financial limitations (Caughlan & Oakley 2001). In the case of ecological indicators, the main objective of a monitoring program should be to provide information to quantify and understand short-term and long-term consequences of management decisions (Southerland & Stribling 1995, Dale & Beyeler 2001, Kurtz et al. 2001). There is an extensive amount of literature referring to sampling methodologies concerning ecological communities or assemblages. Something which is important to bear in mind is that the number of taxa recorded in a site, as well as abundance measures, depend heavily on sampling effort (Angermeier & Smogor 1995, Paller 1995). A weak sampling design invalidates the data and leads to a wrong interpretation, something which may undermine the confidence on the whole program (Hughes 1995, Taylor & Bailey 1997). The main considerations which have to be taken into account for implementing effective monitoring programs for research and management are: (a) pilot studies and statistical power, (b) cost-effectiveness, (c) selection of reference conditions, (d) the scale of study and heterogeneity of the landscape.

Pilot Studies and Statistical Power

A major problem in monitoring is the inherent variability of aquatic ecosystems which may put at risk data validity, implying high costs for the whole program (Chessman 1999, Caughlan & Oakley 2001, Irvine 2004). For example, according to Karr (1998) coefficients of variation for fish samplings are in general very large (from 50 to 300%). Moreover, humans introduce variability along the whole process: from the sampling stage through the interpretation phase, passing by transporting and analyzing samples (Kurtz et al. 2001). A way to determine this variability is by applying prospective power analysis to data from pilot studies. This is done to be sure that sampling methodologies and the quality of data have a high probability of detecting impairment and that the objectives of the study will be fulfilled (Taylor & Bailey 1997, Caughlan & Oakley 2001, Kurtz et al. 2001). Besides, pilot studies can help to define, evaluate, and calibrate metrics (Barbour et al. 1995).

Estimates obtained by the pilot study (and subsequent surveys) should be expressed in statistical terms, that is to say, to measure the uncertainty in the data (i.e. confidence intervals) and to estimate the power of the tests (Sandin & Johnson 2000b, Clarke et al. 2003). Power is the probability of the hypothesis to detect a false null hypothesis and is inversely related to the probability of making a type II error. Thus, low-power pilot studies must increase statistical power by adding more samples to the survey. This is a critical issue in monitoring ecological indicators because it has serious implications for management and decision-making (Hawkins et al. 2000). For example, a type II error would be the incorrect conclusion of no environmental degradation, albeit it has actually occurred (Sandin & Johnson 2000b, Caughlan & Oakley 2001).

Once the monitoring design has been verified by the pilot study, standard procedures for sampling, sorting, identification, and statistical analysis should be set up. These allow repeatability, avoid bias, and impose discipline upon the selection process (Dale & Beyeler 2001). One useful example can be found in the US EPA evaluation guidelines for ecological indicators (Jackson et al. 2000, Kurtz et al. 2001).

Audits to the data can be performed in order to increase the sensitivity of the method to environmental perturbations (Kelly 2001). Data quality controls may include repeating procedures such as re-sampling, or re-identifying samples by independent teams (Klemm et al. 2002, Butcher et al. 2003, Clarke et al. 2003).

Cost-Effectiveness

Designing monitoring programs has always to deal with a trade-off between statistical accuracy, ecological significance and cost-effectiveness. Irvine (2004) warns about ignoring variability in the interests of expediency since repeated sampling is expensive. Although statistical power increases with supplementary sampling collections and independent audits reinforce data quality, cost also increases and the diminishing returns in statistical precision may become marginal (Caughlan & Oakley 2001). Therefore, ways to cope with affordable and significant monitoring programs have to be found. Some examples are: (i) to sample a large area extensively rather than a small area intensively for obtaining a representative number of fish species in streams (Paller 1995), (ii) citizen volunteer monitoring programs, which ease the high cost of professional crews (Lathrop & Markowitz 1995), or (iii) the use of surrogate variables, such as map-based variables, which reduce sampling costs and have powerful predictive capabilities (Davies et al. 2000, Clarke et al. 2003). Long-term ecological monitoring will have a much greater chance of success if cost considerations are explicitly incorporated from the beginning of the program (Caughlan & Oakley 2001).

Selection of Reference Conditions

Reference sites are critical for distinguishing natural from disturbed conditions since they are analogous to a witness control in laboratory experiments. Ideally, reference sites should be pristine and ecologically comparable to the impaired site. However, aquatic ecosystems have been influenced during several centuries by human impacts, therefore, the identification of pristine sites with the natural or the original conditions remains difficult. Instead, the concept of "least-impacted" condition has to be adopted as the reference site (Hughes et al. 1986, Davis & Simon 1995). The selection of the reference conditions consists, according to Hughes (1995), in eight major steps:

1. Define areas of interest on maps
2. Define water body types, sizes, and classes of interest
3. Delineate candidate reference catchments
4. Conduct aerial or photo evaluation
5. Conduct field reconnaissance
6. Subjectively evaluate quality of candidate reference sites
7. Determine number of reference sites desired
8. Evaluate biological health of candidate reference sites

Other criteria include: chemical features, habitat morphology, minimum organism counts (Klemm et al. 2002), detailed consultation with local scientists and experts (Wright et al. 1998), and multivariate analysis (Fausch et al. 1990, Gerritsen et al. 2000).

The Scale of Study and the Heterogeneity of the Landscape

Scale is a fundamental conceptual problem in ecology (Levin 1992), influencing the study of relationships between the landscape and assemblages of macroinvertebrates and fish (Allan & Johnson 1997, Lammert & Allan 1999). According to Smogor & Angermeier (2001) spatial scales range from channel units (e.g. riffles versus pools) to reaches (e.g. upstream versus downstream) to regions (e.g. physiographic regions versus major drainages versus ecoregions) to larger geographical areas. For example, on a large scale, species assemblages may seem predictable, however, local-scale studies reveal high variability (Wiley et al. 1997). In fact, local assemblages are not structured solely by local factors but are closely related to regional and historical processes, implying that a multiple-scale research is needed in order to fully understand the factors determining the distribution and abundance of river ecosystems (Tonn 1990, Allan et al. 1997, Cooper et al. 1998).

Scale is also important for detecting responses to habitat alterations (Minns et al. 1996) and to monitor affected assemblages (Dolédec et al. 1999). Many ecological studies have been focused on small-scale field experiments related to point-source pollution. However, according to Hughes (1995), a problem with site-specific reference conditions is that they possess limited capacity for extrapolation and are no able to detect the severity of non-point source pollution. In fact, little work has been done on different spatial scales to understand aquatic ecosystems under large-scale human (e.g. land-use) and environmental variation (Power 1999, Hawkins et al. 2000). The comparison between reference and impaired sites must be done between similar ecological and geographical conditions (Hughes et al. 1986). In large-scale studies, habitat heterogeneity must be taken into account for designing sound management plans for preserving aquatic resources, assessing biotic integrity and establishing restoration programs (Cowx & Welcomme 1998, Smogor & Angermeier 2001). Hence, deviations from normal community composition caused by human impairment may be better detected by dividing the landscape into regions (Oswood et al. 2000). In some cases, ecoregions and watersheds can be complementary tools for assessing the quality, integrity, and health of ecosystems and their components (Hughes 1995, Omernik & Bailey 1997). However, in other cases, when these ecoregions and watersheds are very large, they may still contain a high diversity of conditions. For example, fish and invertebrate assemblages are different one from another along a river continuum (Vannote et al. 1980). Thus, physiographic provinces (i.e. mountain, piedmont, floodplain) rather than watersheds or ecoregions, account for more natural variation in metrics and best enable biotic integrity indices to detect impairment in rivers (Smogor & Angermeier 2001). This fact has been confirmed for fishes (Lyons 1996, Angermeier & Winston 1998, Oswood et al. 2000, Smogor & Angermeier 2001) and macroinvertebrates (Compin & Céréghino 2003).

Future Challenges in Aquatic Resource Management

Three main challenges for the development of ecological indicators as aquatic resource tools lie ahead:

a. To predict cause-effect processes
b. To link society's values and ecological quality concepts
c. To transmit the message to developing countries

To Predict Cause-Effect Processes

A big challenge is to develop ecological indicators capable of assessing a wider range of stress taking into account the diverse components of variability, but being cost-effective and simple (Hughes 1995, Dale & Beyeler 2001). Such information would be very welcome for management authorities in order to set up reliable environmental standards (Yoder 1995), and to enforce them (Adler 1995). Aquatic resource managers are rather interested in predictive models capable of anticipating environmental problems. Some of the issues needed to be addressed are large-scale processes which cannot be detected by conventional physico-chemical methods, like in the case of aquatic ecosystem restoration, global change, non-point source pollution and biodiversity loss. Although ecological indicators are useful for detecting environmental stress, they do so almost always when damage is already done (Yoder 1995). As Wright et al. (1998) point out: "if such systems are to have broader application, they must be related to physical features of the landscape in order to make inferences on the effects of land-use disturbances". Thus, future ecological indicators should develop predictive capabilities through modeling. In fact, some efforts with new approaches have already been done, like classification systems (Hawkins & Norris 2000), probabilistic models (e.g. Clarke et al. 2003, Oberdorff et al. 2002), and artificial neural networks for diatoms (Gevrey et al.

2004), invertebrates (Walley & Fontama 1998, Schleiter et al. 1999) and fish (Ibarra et al. 2003). However, modeling ecological indicators in its present form is not yet able to discern cause-effect processes. In fact, no ecological indicator by itself can explain cause-effects phenomena based only on observation and correlation (Havens, 1999, Suter 2001). Anyhow, the use of reference sites and long time-series data under carefully designed sampling would be a step towards the understanding of such phenomena.

To Link Society's Values and Ecological Quality Concepts

Another challenge lies in analyzing the relationship between society's values and the implementation of environmental policies. It was noted above that society's values can drive environmental policy. However, this is in fact a feedback process. Therefore, it would be interesting to study whether water policies which take into account ecological health (e.g. EWFD) are capable of inducing changes on society's perception towards the aquatic environment and viceversa. Research on this field implies an integration of biological, physical, cultural and socioeconomic systems (Redman 1999, Barret & Farina 2000). Thus, a stronger focus has to be given to multidisciplinary research and management for aquatic resources and water quality (Gowdy & Ferreri Carbonell 1999, Barret & Farina 2000). Two main ways could be considered to link society's values and ecological indicators: involving stakeholders and estimating the economic value of environmental quality.

Involving Stakeholders through Information

Lathrop & Markowitz (1995) recommend the participation of volunteers in monitoring surveys in order to create a sense of stewardship for aquatic ecosystems. However, a first step in stakeholder involvement for a successful program of ecological indicators is to communicate the results of monitoring in order to demonstrate its significance (Davis 1995, Karr 1999, Irvine 2004). The communication of complex ecological issues has to be easy to understand for a variety of audiences, including the general public and management authorities (Caughlan & Oakley 2001, Dale & Beyeler 2001, Schiller et al. 2001). For example, Schiller et al. (2001) tried to translate scientific terminology into common language regarding diverse USEPA's ecological indicators, finding that the best approach was to describe the kind of information that various combinations of indicators could provide about environmental conditions, rather than to describe what in particular was being measured or how measurements were performed. Indeed, the quality of information received by citizens and authorities will have a direct impact on their perception of environmental quality.

Economic Estimation of Environmental Quality

There are two main arguments for assessing the economic value of environmental quality. On the one hand, there is a need for knowing the optimal level of both quantity and quality of resources in order to preserve them for future generations (Navrud 2001). On the other hand, if the environmental services are not explicitly evaluated in economic terms, then there is a risk of being assumed by decision makers as unimportant (Costanza et al. 1997, Edwards & Abivardi 1998, Daily 2000). Karr (1995) points out that "we remain blissfully ignorant of the actual benefits and costs of regulatory actions ostensibly designed to protect water quality". In fact, an ecological indicator is a difficult asset to value, since no formal markets exists for it. For example, Caughlan & Oakley (2001) explain that non-market monitoring costs include the cost of a type II statistical error (failing to detect a change that has occurred) or the opportunity costs of the dollars spent on monitoring. The main approach to valuing non-market goods is cost-benefit analysis but still, "the contradictions between the rules of market behavior and the rules which govern ecosystems is far from being resolved" (Gowdy & Ferreri Carbonell 1999). The conciliation between ecology and economics is a major challenge for a society in search of a high environmental quality without giving away the comfort of economic progress. Rapport et al. (1999) consider that these two visions should be seen as complementary since society's values reflect both short-term (i.e. economic benefits) and long-term needs (i.e. ecological benefits).

To Transmit the Message to Developing Countries

While the monitoring of ecological indicators is frequently applied in temperate streams, tropical rivers have been somewhat neglected. Only punctual examples exist, like for example, Argentina for diatoms (Gomez 1999), Indonesia (Hart et al. 2001), Nicaragua (Fenoglio et al. 2002) and Nigeria (Ogbeibu & Oribhabor 2002) for benthic macroinvertebrates, and Mexico for fish (Lyons et al. 2000), among others. In fact, most of the concepts on environmental quality come from industrialized countries where public awareness came only after attaining high levels of perturbation in a number of rivers. Why then to wait until environmental degradation arrives to untouched ecosystems in developing countries? This is, nevertheless, no easy task. Several problems constrain the adoption of environmental quality concepts and the implementation of monitoring programs in developing countries. Among these, the lack of trained staff for monitoring, the insufficient knowledge of assemblages taxonomy, and inexistent environmental standards for water quality are the most notorious. Although cooperation with scientists from industrialized countries would certainly help (Resh 1995), it is in developing countries where the main answer remains: a change in current values, attitudes and approaches from both citizens and authorities.

Conclusion

Ecological indicators have gained international recognition as tools for water resources management thanks to an increasing social awareness of environmental quality. There is no silver-bullet for assessing all human disturbances in rivers. Although a combination of species assemblages would be an ideal formula, appropriate information on each group would help to choose the right indicator. For example, benthic diatoms seem to be most suited for small-scale short-term analysis of water quality, preferably related to metal toxicity. Macroinvertebrates would be good indicators for small-scale long-term studies, and fish for large-scale, long term ecological health assessments. To date, the most common form to measure such assemblages is by means of multi-metric indices.

Big challenges lie ahead in the development of ecological indicators and indices, like their use in predicting cause-effect processes on different scales in time and space. Furthermore, designing effective monitoring programs is no easy task since there is always a trade-off between statistical accuracy, ecological significance and limited budgets.

Consequently, the choices that have to be made concerning these issues will heavily depend on both the use and non-use values that society gives to the environmental quality of water, implying the long-term sustainability of freshwater ecosystems.

ACKNOWLEDGEMENTS

The author was partially sponsored by the Conacyt-Sfere Program (No. 131742).

REFERENCES

Adler, R.W. (1995). Filling the gaps in water quality standards: legal perspectives on biocriteria. In: W.S. Davis & T.P. Simon (Eds.), *Biological assessment and criteria: Tools for water resource planning and decision-making* (pp. 345-358). Boca Raton, FL: Lewis Publishers.

Allan, J.D. & Johnson, L.B. (1997). Catchment-scale analysis of aquatic ecosystems. *Freshwater Biology*, 37, 107-111.

Allan, J.D., Erickson, D.L. & Fay, J. (1997). The influence of catchment land use on stream integrity across multiple spatial scales. *Freshwater Biology*, 37, 149-161.

Angermeier, P.L. & Schlosser, I.J. (1995). Conserving aquatic biodiversity: Beyond species and populations. *American Fisheries Society Symposium*, 17, 402-414.

Angermeier, P.L. & Smogor, R.A. (1995). Estimating number of species and relative abundances in stream-fish communities: Effects of sampling effort and discontinuous spatial distributions. *Canadian Journal of Fisheries and Aquatic Sciences*, 52, 936-949.

Angermeier, P.L. & Winston, M.R. (1998). Local vs. regional influences on local diversity in stream fish communities of Virginia. *Ecology*, 79, 911-927.

Arlinghaus, R., Mehner, T. & Cowx, I.G. (2002). Reconciling traditional inland fisheries management and sustainability in industrialized countries, with emphasis on Europe. *Fish and Fisheries*, 3, 261-316.

Barbour, M.T., Stribling, J.B. & Karr, J.R. (1995). Multimetric approach for establishing biocriteria and measuring biological condition. In: W.S. Davis & T.P. Simon (Eds.), *Biological assessment and criteria: Tools for water resource planning and decision-making* (pp. 63-77). Boca Raton, FL: Lewis Publishers.

Barret, G.W. & Farina, A. (2000). Integrating ecology and economics. *BioScience*, 50, 311–312.

Begon, M., Harper, J.L. & Townsend, C.R. (1996). *Ecology* (Third edition). Oxford: Blackwell Science.

Boulton, A.J. (1999). An overview of river health assessment: Philosophies, practice, problems and prognosis. *Freshwater Biology*, 41, 469-479.

Butcher, J.T., Stewart, P.M. & Simon, T.P. (2003). A Benthic Community Index for streams in the Northern Lakes and Forests Ecoregion. *Ecological Indicators*, 3, 181-193.

Caughlan, L. & Oakley, K.L. (2001). Cost considerations for long-term ecological monitoring. *Ecological Indicators*, 1, 123–134.

Chessman, B.C. (1999). Predicting the macroinvertebrate faunas of rivers by multiple regression of biological and environmental differences. *Freshwater Biology*, 41, 747-757.

Clarke, R.T., Wright, J.F. & Furse, M.T. (2003). RIVPACS models for predicting the expected macroinvertebrate fauna and assessing the ecological quality of rivers. *Ecological Modelling*, 160, 219-233.

Coeck, J., Vandelannoote, A. Yseboodt, R. & Verheyen, R.F. (1993). Use of the abundance/biomass method for comparison of fish communities in regulated and unregulated lowland rivers in Belgium. *Regulated Rivers: Research and Management*, 8, 73-82.

Compin, A. & Céréghino, R. (2003). Sensitivity of aquatic insect species richness to disturbance in the Adour–Garonne stream system (France). *Ecological Indicators*, 3, 135–142.

Cooper, S., Diehl, S. Kratz, K. & Sarnelle, O. (1998). Implications of scale for patterns and processes in stream ecology. *Australian Journal of Ecology*, 23, 27-40.

Costanza, R. (1992). Toward an operational definition of ecosystem health. In: R. Costanza, B. Norton & B.J. Haskell (Eds.), *Ecosystem health: New goals for environmental management* (pp. 239-256). Washington, DC: Island Press.

Costanza, R., d'Arge, R., deGroot, R., Farber, S., Grasso, M., Hannon, B., Limburg, K., Naeem, S., O'Neill, R.V.O., Paruelo, J., Raskin, R.G., Sutton, P. & van der Belt, M. (1997). The value of the world's ecosystem services and natural capital. *Nature*, 387, 253-260.

Cowx, I.G. & Welcomme, R.L. (1998). *Rehabilitation of rivers for fish.* Oxford: FAO & Fishing News Books.

Daily, G.C. (2000). Management objectives for the protection of ecosystem services. *Environmental Science and Policy*, 3, 333-339.

Dale, V.H. & Beyeler, S.C. (2001). Challenges in the development and use of ecological indicators. *Ecological Indicators*, 1, 3–10.

Dauba, F., Lek, S., Mastrorillo, S. & Copp, G.H. (1997). Long-term recovery of macrobenthos and fish assemblages after water pollution abatement measures in the river Petite Baïse (France). *Archives of Environmental Contamination and Toxicology*, 33, 277-285.

Davies, N.M., Norris, R.H. & Thoms, M.C. (2000). Prediction and assessment of local stream habitat features using large-scale catchment characteristics. *Freshwater Biology*, 45, 343-369.

Davis, W.S. (1995). Biological assessment and criteria: Building on the past. In: W.S. Davis & T.P. Simon (Eds.), *Biological assessment and criteria: Tools for water resource planning and decision-making* (pp. 15-29). Boca Raton, FL: Lewis Publishers.

Davis, W.S. & Simon, T.P. (1995). Introduction. In: W.S. Davis & T.P. Simon (Eds.), *Biological assessment and criteria: Tools for water resource planning and decision-making* (pp. 3-6). Boca Raton, FL: Lewis Publishers.

Dell'oumo, A., Pensieri, A., & Corradetti, D. (1999). Diatomées épilithiques du fleuve Esino (Italie centrale) et leur utilisation pour l'évaluation de la qualité biologique de l'eau. *Cryptogamie Algologie*, 20, 253-269.

DeShon, J.E. (1995). Development and application of the Invertebrate Community Index (ICI). In: W.S. Davis & T.P. Simon (Eds.), *Biological assessment and criteria: Tools for water resource planning and decision-making* (pp. 217-243). Boca Raton, FL: Lewis Publishers.

Dolédec, S., Statzner, B. & Bournard, M. (1999). Species traits for future biomonitoring across ecoregions: Patterns along a human-impacted river. *Freshwater Biology*, 42, 737-758.

Dynesius, M. & Nilsson, C. (1994). Fragmentation and flow regulation of river systems in the northern third of the world. *Science*, 266, 753-762.

European Commission. (2000). Directive 2000/60/EC of the European Parliament and of the Council of 23 October 2000 establishing a framework for Community action in the field of water policy. *Official Journal of the European Communities*, 22 December 2000.

Edwards, P.J. & Abivardi, C. (1998). The value of biodiversity: Where ecology and economy blend. *Biological Conservation*, 83, 239-246.

Everard, M. & Powell, A. (2002). Rivers as living systems. *Aquatic Conservation: Marine and Freshwater Ecosystems*, 12, 329-337.

Fausch, K.D., Lyons, J., Karr, J.R. & Angermeier, P.L. (1990). Fish communities as indicators of environmental degradation. *American Fisheries Society Symposium*, 8, 123-144.

Fenoglio, S., Badino, G. & Bona, F. (2002). Benthic macroinvertebrate communities as indicators of river environment quality: An experience in Nicaragua. *Revista de Biología Tropical*, 50, 1125-1131.

Finlayson, C.M. (2003). The challenge of integrating wetland inventory, assessment and monitoring. *Aquatic Conservation: Marine and Freshwater Ecosystems*, 13, 281–286.

Gaston, K.J. & Spicer, J.I. (2000). *Biodiversity: An introduction*. Oxford: Blackwell Science.

Gerritsen, J., Barbour, M.T. & King, K. (2000). Apples, oranges, and ecoregions: On determining pattern in aquatic assemblages. *Journal of the North American Benthological Society*, 19, 487-496.

Gevrey, M., Rimet, F., Park, Y.-S., Giraudel, J.L., Ector, L. & Lek, S. (2004). Predictive tool for water quality assessment using diatom assemblages. *Freshwater Biology*, 49, 208-220.

Golda, C., Feurtet-Mazela, A., Coste, M. & Boudou, A. (2002). Field transfer of periphytic diatom communities to assess short-term structural effects of metals (Cd, Zn) in rivers. *Water Research*, 36, 3654–3664.

Gómez, N. (1999). Epipelic diatoms from the Matanza-Riachuelo river (Argentina), a highly polluted basin from the Pampean plain: Biotic indices and multivariate analysis. *Aquatic Ecosystem Health and Management*, 2, 301–309.

Gowdy, J.M. & Ferreri-Carbonell, A. (1999). Toward consilience between biology and economics: The contribution of ecological economics. *Ecological Economics*, 29, 337-348.

Harding, J.S., Benfield, E.F., Bolstad, P.V., Helfman, G.S. & Jones, E.B.D. (1998). Stream biodiversity: The ghost of land use past. *Proceedings of the National Academy of Sciences of the United States*, 95, 14843-14847.

Hart, B.T., Maher, B. & Lawrence, I. (1999). New generation water quality guidelines for ecosystem protection. *Freshwater Biology*, 41, 347-359.

Hart, B.T., Davies, P.E., Humphrey, C.L., Norris, R.N., Sudaryanti, S. & Trihadiningrum, Y. (2001). Application of the Australian river bioassessment system (AUSRIVAS) in the Brantas River, East Java, Indonesia. *Journal of Environmental Management*, 62, 93–100.

Havens, K.E. (1999). Correlation is not causation: A case study of fisheries, trophic state and acidity in Florida (USA) lakes. *Environmental Pollution*, 106, 1-4.

Hawkins, C.P. & Norris, R.H. (2000). Performance of different landscape classifications for aquatic bioassessments: Introduction to the series. *Journal of the North American Benthological Society*, 19, 367-369.

Hawkins, C.P., Norris, R.H., Gerritsen, J., Hughes, R.M., Jackson, S.K., Johnson, R.K. & Stevenson, R.J. (2000). Evaluation of the use of landscape classifications for the prediction of freshwater biota: Synthesis and recommendations. *Journal of the North American Benthological Society,* 19, 541-556.

Hill, B.H., Herlihy, A.T., Kaufmann, P.R., DeCelles, S.J. & Vander Borgh, M.A. (2003). Assessment of streams of the eastern United States using a periphyton index of biotic integrity. *Ecological Indicators*, 2, 325–338.

Hueting, R., Reijnders, L., de Boer, B., Lambooy, J. & Jansen, H. (1998). The concept of environmental function and its valuation. *Ecological Economics*, 25, 31-35.

Hughes, R.M. (1995). Defining acceptable biological status by comparing with reference conditions. In: W.S. Davis & T.P. Simon (Eds.), *Biological assessment and criteria: Tools for water resource planning and decision-making* (pp. 31-47). Boca Raton, FL: Lewis Publishers.

Hughes, R.M., Larsen, D.P. & Omernik, J.M. (1986). Regional reference sites: A method for assessing stream potentials. *Environmental Management*, 10, 629-635.

Hughes, R.M., Kaufman, P.R., Herlihy, A.T., Kincaid, T.M., Reynolds, L. & Larsen, D.P. (1998). A process for developing and evaluating indices of fish assemblage integrity. *Canadian Journal of Fisheries and Aquatic Sciences*, 55, 1618-1631.

Hutagalung, R.A., Lim, P. Belaud A. & Lagarrigue, T. (1997). Effets globaux d'une agglomération sur la typologie ichtyenne d'un fleuve: Cas de la Garonne à Toulouse (France). *Annales de Limnologie*, 33, 263-279.

Ibarra, A.A., Gevrey, M., Park, Y.-S., Lim, P. & Lek, S. (2003). Modelling the factors that influence fish guilds composition using a back-propagation network: Assessment of metrics for indices of biotic integrity. *Ecological Modelling*, 160, 281-290.

Iliopoulou-Georgudaki, J., Kantzaris, V., Katharios, P., Kaspiris, P., Georgiadis, Th. & Montesantou, B. (2003). An application of different bioindicators for assessing water quality: A case study in the rivers Alfeios and Pineios (Peloponnisos, Greece). *Ecological Indicators*, 2, 345–360.

Irvine, K. (2004). Classifying ecological status under the European Water Framework Directive: The need for monitoring to account for natural variability. *Aquatic Conservation: Marine and Freshwater Ecosystems,* 14, 107–112.

Jackson, L.E., Kurtz, J.C. & Fisher, W.S. (2000). *Evaluation guidelines for ecological indicators*. EPA/620/R-99/005. Research Triangle Park, NC: U.S. Environmental Protection Agency, Office of Research and Development.

Jenerette, G.D., Lee, J., Waller, D.W. & Carlson, R.E. (2002). Multivariate analysis of the ecoregion delineation for aquatic systems. *Environmental Management*, 29, 67-75.

Jones, J.G. (2001) Freshwater ecosystems-structure and response. *Ecotoxicology and Environmental Safety*, 50, 107-113.

Kallis, G. & Butler, D. (2001). The EU Water Framework Directive: Measures and implications. *Water Policy*, 3, 125-142.

Karr, J.P. (1995). Protecting aquatic ecosystems: Clean water is not enough. In: W.S. Davis & T.P. Simon (Eds.), *Biological assessment and criteria: Tools for water resource planning and decision-making* (pp. 7-13). Boca Raton, FL: Lewis Publishers.

Karr, J.R. (1998). Rivers as sentinels: Using the biology of rivers to guide landscape management. In: R.J. Naiman & R.E. Bilby (Eds.), *River ecology and management: Lessons from the Pacific coastal ecoregion* (pp. 502-528). New York: Springer-Verlag.

Karr, J.R. (1999). Defining and measuring river health. *Freshwater Biology*, 41, 221-234.

Karr, J.R., Fausch, K.D., Angermeier, P.L., Yant P.R. & Schlosser, I.J. (1986). Assessing biological integrity in running waters: A method and its rationale. *Illinois Natural History Survey Special Publication*, 5, 1-28.

Kelly, M.G. (1998). Use of the Trophic Diatom Index to monitor eutrophication in rivers. *Water Research*, 32, 236-242.

Kelly, M.G. (2001). Use of similarity measures for quality control of benthic diatom samples. *Water Research,* 35, 2784–2788.

Kestemont, P., Didier, J., Depiereux, E. & Micha, J.C. (2000). Selecting ichthyological metrics to assess river basin ecological quality. *Archives für Hydrobiologie Supplement (Monographic Studies)*, 121, 321-348.

Kilgour, B.W. & Barton, D.R. (1999). Associations between stream fish and benthos across environmental gradients in southern Ontario, Canada. *Freshwater Biology*, 41, 553-566.

Klemm, D.J., Blocksom, K.A., Thoeny, W.T., Fulk, F.A., Herlihy, A.T., Kaufmann, P.R. & Cormier, S.M. (2002). Methods development and use of macroinvertebrates as indicators of ecological conditions for streams in the Mid-Atlantic Highlands region. *Environmental Monitoring and Assessment*, 78, 169-212.

Kurtz, J.C., Jackson, L.E. & Fisher, W.S. (2001). Strategies for evaluating indicators based on guidelines from the Environmental Protection Agency's Office of Research and Development. *Ecological Indicators*, 1, 49–60.

Lammert, M. & Allan, J.D. (1999). Assessing biotic integrity of streams: Effects of scale in measuring the influence of land use/cover and habitat structure on fish and macroinvertebrates. *Environmental Management*, 23, 257-270.

Lathrop, J.E., & Markowitz, A. (1995). Monitoring water resource quality using volunteers. In: W.S. Davis & T.P. Simon (Eds.), *Biological assessment and criteria: Tools for water resource planning and decision-making* (pp. 303-314). Boca Raton, FL: Lewis Publishers.

Levin, S.A. (1992). The problem of pattern and scale in ecology. *Ecology*, 73, 1943-1967.

Logan, P. & Furse, M. (2002). Preparing for the European Water Framework Directive: Making the links between habitat and aquatic biota. *Aquatic Conservation: Marine and Freshwater Ecosystems*, 12, 425–437.

Loomis, J.B., Sorg, C. & Donnelly, D. (1986). Economic losses to recreational fisheries due to small-head hydro-power development: A case study of Henrys Fork in Idaho. *Journal of Environmental Management*, 22, 85-94.

Lyons, J. (1996). Patterns in the species composition of fish assemblages among Wisconsin streams. *Environmental Biology of Fishes*, 45, 329-346.

Lyons, J., Gutiérrez-Hernandez, A., Diaz-Pardo, E., Soto-Galera, E., Medina-Nava M. & Pineda-Lopez. R. (2000). Development of a preliminary index of biotic integrity (IBI) based on fish assemblages to assess ecosystem condition in the lakes of central Mexico. *Hydrobiologia*, 418, 57-72.

Lyons, J., Piette, R.R. & Niermeyer, K.W. (2001). Development, validation, and application of a fish-based index of biotic integrity for Wisconsin's large warmwater rivers. *Transactions of the American Fisheries Society*, 130, 1077-1094.

McDowall, R.M. & Taylor, M.J. (2000). Environmental indicators of habitat quality in a migratory freshwater fish fauna. *Environmental Management*, 25, 357-374.

McManus, J.W. & Pauly, D. (1990). Measuring stress: Variations on a theme by R.M. Warwick. *Marine Biology*, 106, 305-308.

Meire, P.M. & Dereu, J. (1990). Use of the abundance/biomass comparison method for detecting environmental stress: Some considerations based on intertidal macrozoobenthos and bird communities. *Journal of Applied Ecology*, 27, 210-223.

Minns, C.K., Kelso, J.R.M. & Randall, R.G. (1996). Detecting the response of fish to habitat alterations in freshwater ecosystems. *Canadian Journal of Fisheries and Aquatic Sciences,* 53 (Suppl. 1), 403-414.

Navrud, S. (2001). Economic valuation of inland recreational fisheries: Empirical studies and their policy use in Norway. *Fisheries Management and Ecology*, 8, 369-382.

Oberdorff, T. & Hughes, R.M. (1992). Modification of an index of biotic integrity based on fish assemblages to characterize rivers of the Seine basin, France. *Hydrobiologia*, 228, 117-130.

Oberdorff, T., Pont, D., Hugueny, B. & Porcher, J.P. (2002). Development and validation of a fish-based index for the assessment of 'river health' in France. *Freshwater Biology*, 47, 1720-1734.

Ogbeibu, A.E. & Oribhabor, B.J. (2002). Ecological impact of river impoundment using benthic macro-invertebrates as indicators. *Water Research*, 36, 2427–2436.

Omernik, J.M. & Bailey, R.G. (1997). Distinguishing between watersheds and ecoregions. *Journal of the American Water Research Association*, 33, 935-949.

Opschoor, J.B. (1998). The value of ecosystem services: Whose values? *Ecological Economics*, 25, 41-43.

Osborne, L.L., Kohler, S.L., Bayley, P.B., Day, D.M., Bertrand, W.A., Wiley M.J. & Sauer, R. (1992). Influence of stream location in a drainage network on the index of biotic integrity. *Transactions of the American Fisheries Society*, 121, 635-643.

Oswood, M.W., Reynolds, J.B., Irons, J.G., Milner, A.M., Rabeni, C.F. & Doisy, K.E. (2000). Distributions of freshwater fishes in ecoregions and hydroregions of Alaska. *Journal of the North American Benthological Society*, 19, 405-418.

Paller, M.H. (1995). Relationships among number of fish species sampled, reach length surveyed, and sampling effort in South Carolina coastal plain streams. *North American Journal of Fisheries Management*, 15, 110-120.

Penczak, T. & Kruk, A. (1999). Applicability of the abundance/biomass comparison method for detecting human impacts on fish populations in the Pilca River, Poland. *Fisheries Research*, 39, 229-240.

Poff, N.L., Allan, D., Bain, M.B., Karr, J.R., Prestegaard, K.L., Richter, B.D., Sparks R.E. & Stromberg, J.C. (1997). The natural flow regime: A paradigm for river conservation and restoration. *BioScience*, 47, 769-784.

Power, M. (1999). Recovery in aquatic ecosystems: An overview of knowledge and needs. *Journal of Aquatic Ecosystem Stress and Recovery*, 6, 253-257.

Prygiel, J., Lévêque, L. & Iserentant, R. (1996). Un nouvel Indice Diatomique Pratique pour l'évaluation de la qualité des eaux en réseau de surveillance. *Revue des Sciences de l'Eau,* 9, 97-113.

Rapport, D.J., Bohm, G., Buckingham, D., Cairns, J., Costanza, R., Karr, J.R., DeKruijf, H.A.M., Levins, R., McMichael, A.J., Nielsen, N.O. & Whitford, W.G. (1999).

Ecosystem health: The concept, the ISEH, and the important tasks ahead. *Ecosystem Health*, 5, 82-90.

Redman, C.L. (1999). Human dimensions of ecosystem studies. *Ecosystems*, 2, 296-298.

Resh, V.H. (1995). Freshwater benthic macroinvertebrates and rapid assessment procedures for water quality monitoring in developing and newly industrialized countries. In: W.S. Davis & T.P. Simon (Eds.), *Biological assessment and criteria: Tools for water resource planning and decision-making* (pp. 167-177). Boca Raton, FL: Lewis Publishers.

Rosen, B.H. (1995). Use of periphyton in the development of biocriteria. In: W.S. Davis, & T.P. Simon (Eds.), *Biological assessment and criteria: Tools for water resource planning and decision-making* (pp. 209-215). Boca Raton, FL: Lewis Publishers.

Rosenberg, D.M. & Resh, V.H. (Eds.). (1993). *Freshwater biomonitoring and benthic macroinvertebrates*. New York: Chapman & Hall.

Ross, S.T. (1991). Mechanisms structuring stream fish assemblages: Are there lessons from introduced species? *Environmental Biology of Fishes*, 30, 359-368.

Sandin, L. & Johnson, R.K. (2000a). Ecoregions and benthic macroinvertebrate assemblages of Swedish streams. *Journal of the North American Benthological Society*, 19, 462-474.

Sandin, L. & Johnson, R.K. (2000b). The statistical power of selected indicator metrics using macroinvertebrates for assessing acidification and eutrophication of running waters. *Hydrobiologia*, 422/423, 233-243.

Schiller, A., Hunsaker, C.T., Kane, M.A., Wolfe, A.K., Dale, V.H., Suter, G.W., Russell, C.S., Pion, G., Jensen, M.H. & Konar, V.C. (2001). Communicating ecological indicators to decision makers and the public. Conservation Ecology, 5, available from: URL: *http://www.consecol.org/vol5/iss1/art19*.

Schleiter, I.M., Borchardt, D., Wagner, R., Dapper, T., Schmidt, K.D., Schmidt H.H. & Werner, H. (1999). Modelling water quality, bioindication and population dynamics in lotic ecosystems using neural networks. *Ecological Modelling*, 120, 271-286.

Schlosser, I.J. (1991). Stream fish ecology: A landscape perspective. *BioScience*, 41, 704-712.

Scott, M.C. & Hall, L.W. (1997). Fish assemblages as indicators of environmental degradation in Maryland coastal plain streams. *Transactions of the American Fisheries Society*, 126, 349-360.

Seegert, G. (2000). Considerations regarding development of index of biotic integrity metrics for large rivers. *Environmental Science and Policy*, 3, S99-S106.

Simon, T.P. & Lyons, J. (1995). Application of the Index of Biotic Integrity to evaluate water resource integrity in freshwater ecosystems. In: W.S. Davis & T.P. Simon (Eds.), *Biological assessment and criteria: Tools for water resource planning and decision-making* (pp. 245-262). Boca Raton, FL: Lewis Publishers.

Smogor, R.A. & Angermeier, P.L. (2001). Determining a regional framework for assessing biotic integrity of Virginia streams. *Transactions of the American Fisheries Society*, 130, 18-35.

Soto-Galera, E., Diaz-Pardo, E., Lopez-Lopez, E. & Lyons, J. (1998). Fish as indicators of environmental quality in the Rio Lerma Basin, Mexico. *Aquatic Ecosystem Health and Management*, 1, 267-276.

Southerland, M.T. & Stribling, J.B. (1995). Status of biological criteria development and implementation. In: W.S. Davis & T.P. Simon (Eds.), *Biological assessment and criteria:*

Tools for water resource planning and decision-making (pp. 81-96). Boca Raton, FL: Lewis Publishers.

Steedman, R.J. (1988). Modification and assessment of an index of biotic integrity to quantify stream quality in southern Ontario. *Canadian Journal of Fisheries and Aquatic Sciences*, 45, 492-501.

Suter, G.W. (2001). Applicability of indicator monitoring to ecological risk assessment. *Ecological Indicators,* 1, 101–112.

Taylor, B.R. & Bailey, R.C. (1997). *Technical evaluation on methods for benthic invertebrate data analysis and interpretation.* Final Report. Ottawa: Aquatic Effects Technology Evaluation (AETE) Program. Project 2.1.3.

Tonn, W.M. (1990). Climate change and fish communities: A conceptual framework. *Transactions of the American Fisheries Society*, 119, 337-352.

Vannote, R.L., Minshall, G.W., Cummins, K.W., Sedell, J.R. & Cushing, C.E. (1980). The river continuum concept. *Canadian Journal of Fisheries and Aquatic Sciences*, 37, 130-137.

Vila-Gispert, A., García-Berthou, E. & Moreno-Amich, R. (2002). Fish zonation in a Mediterranean stream: Effects of human disturbances. *Aquatic Sciences*, 64, 163–170.

Walley, W.J. & Fontama. V.N. (1998). Neural network predictors of average score per taxon and number of families at unpolluted river sites in Great Britain. *Water Research*, 32, 613-622.

Walmsley, J.J. (2002). Framework for measuring sustainable development in catchment systems. *Environmental Management*, 29, 195-206.

Wang, L., Lyons, J., Kanehl, P.D. & Gatti, R. (1997). Influences of watershed land use on habitat quality and biotic integrity in Wisconsin streams. *Fisheries*, 7, 6-12.

Wang, L., Lyons, J., Kanehl, P., Bannerman, R. & Emmons, E. (2000). Watershed urbanization and changes in fish communities in southeastern Wisconsin streams. *Journal of the American Water Research*, 36, 1173-1175.

Warwick, R.M. (1986). A new method for detecting pollution effects on marine macrobenthic communities. *Marine Biology*, 92, 557-562.

Welcomme, R.L. (1983). River basins. *FAO Fisheries Technical Paper*, 202, 1-60.

Wiley, M.J., Kohler, S.L. & Seelbach, P.W. (1997). Reconciling landscape and local views of aquatic communities: Lessons from Michigan trout streams. *Freshwater Biology*, 37, 133-148.

Wilzbach, M.A., Mather, M.E., Folt, C.L., Moore, A., Naiman, R.J., Youngson, A.F. & McMenemy, J. (1998). Proactive responses to human impacts that balance development and Atlantic salmon (*Salmo salar)* conservation: An integrative model. *Canadian Journal of Fisheries and Aquatic Sciences,* 55 (Suppl. 1), 288-302.

Wootton, R.J. (1991). *Ecology of teleost fishes.* New York: Chapman & Hall.

Wright, J.F., Furse, M.T. & Moss, D. (1998). River classification using invertebrates: RIVPACS applications. *Aquatic Conservation: Marine and Freshwater Ecosystems*, 8, 617–631.

Wu, J.-T. (1999). A generic index of diatom assemblages as bioindicator of pollution in the Keelung river of Taiwan. *Hydrobiologia*, 397, 79-87.

Yoder, C.O. (1995). Policy issues and management applications of biological criteria. In: W.S. Davis & T.P. Simon (Eds.), *Biological assessment and criteria: Tools for water resource planning and decision-making* (pp. 327-343). Boca Raton, FL: Lewis Publishers.

In: Progress in Aquatic Ecosystems Research
Editor: A. R. Burk, pp. 59-82
ISBN 1-59454-383-6

Chapter 3

THE ROLE OF BIOMANIPULATION IN AQUATIC ECOSYSTEM RESTORATION

Douglas J. Spieles
Denison University

ABSTRACT

In response to a perceived decline in the quality of aquatic resources and associated changes in public policy, a great deal of recent ecological research has investigated the feasibility of restoring aquatic ecosystems. In 1992, the National Research Council (NRC) on Aquatic Ecosystem Restoration called for a more systematic and holistic approach to aquatic restoration that addresses not only structural aspects of morphology and hydrology but also attributes of ecosystem function. In this chapter I review the status of lake, river, and wetland restoration since the NRC recommendations. In particular, I focus on the restoration of trophic functionality in these three ecosystems and the degree to which it has been achieved by structural modification and by biomanipulation. Trophic restoration of lakes, while typically featuring the reduction of external and internal nutrient loads, has successfully incorporated biomanipulation of fish populations to manage phytoplankton through trophic cascade. River and wetland restoration efforts have included trophic function to a lesser degree. In these ecosystems, hydrologic regime and landscape connectivity have necessarily been given far more attention. This body of research shows, however, that structural modification alone does not guarantee functional success. Thus both river and wetland restorations have begun to address the management of biota to achieve system-level effects through trophic interaction. Biomanipulation is taking a broader form in river and wetland systems and includes habitat restoration to facilitate the establishment of a variety of feeding guilds. Though trophic manipulation has some serious constraints and clearly cannot be achieved without management of the abiotic regime, it can be an important link between the structural aspects of the restored ecosystem and the emergent properties of a functional ecosystem.

INTRODUCTION

Human management of ecosystems is much older than the concept of ecosystem management. For thousands of years, human cultures have altered their environments through harvest, manipulation, and selection—presumably with little regard for the ecological relationships being altered. Only in the last few centuries has there been a call for conservation and preservation of ecological resources and an accompanying notion that an ecosystem damaged by human influence can be restored to what it once was. This field of ecological restoration is the process of modifying a degraded ecosystem in an attempt to reestablish some pre-disturbance state or condition—a return to "the diversity and dynamics of the indigenous ecosystem" (Throop 2000). The science of restoration ecology is still rather young, for while restoration activities have long been a part of natural resource management, the systematic study of restoration processes and outcomes has blossomed only in the last quarter century. Restoration ecology also evolves rather slowly, as results from particular management methods may not be apparent for years or even decades. This combination of youth and slow development means that we are only now reaching a position to evaluate the goals, objectives, and methods of restoration ecology. This chapter is a comparative review of restoration research and progress in three types of aquatic ecosystems.

Freshwater ecosystems have received particular attention from restoration ecologists. In part, this is a response to the disproportionate ecological importance and highly modified state of aquatic systems. Lakes, rivers and wetlands have been substantially altered around the world by activities of the burgeoning human population. In some cases, the alteration constitutes a complete loss of an ecosystem, as in the case of wetland drainage. The aquatic ecosystems that remain are nearly all subjected to more subtle changes in physical, chemical, and biological characteristics. Aquatic ecosystems are often physically affected by modifications to the adjacent uplands—including agricultural practices, erosion, creation of impermeable ground cover, or loss of vegetative cover. More direct physical changes in the form of dams, flood control structures, and channel or basin engineering are ubiquitous in developed regions of the world. Water chemistry is inextricably linked to changes in the physical environment, as reduced light penetration, restructured hydrologic regime and even variation in water temperature affect dissolved oxygen levels, pH, and chemical speciation. Industrial, agricultural and municipal pollutants add to the chemical alteration of aquatic ecosystems. The final scourge—biological degradation—involves the extirpation of native biota and the intentional or accidental introduction of exotic species, some of which prove to be formidable foes to indigenous species. In all of these ways, humans have manipulated aquatic ecosystems very differently and at a far greater rate than natural succession, leaving the restoration ecologist with a daunting task indeed.

It has been recognized that no restoration can ever be perfect; that is, no restored ecosystem will ever truly replicate the ecosystem prior to disturbance. Instead, the goal of restoration ecology is "to emulate a natural, functioning, self-regulating system that is integrated with the ecological landscape in which it occurs (NRC 1992)." Emulation is much more achievable than replication, but even this reasonable goal raises questions. Emulating a natural system implies that for each restoration there should be either a clear concept of the predisturbance ecosystem or an undisturbed reference ecosystem with which to compare. Unfortunately, the predisturbance structure of an ecosystem is often difficult to discern, and predisturbance functions even more so. Reference ecosystems, while undeniably important components of restoration research, are not always available and are seldom free from anthropogenic disturbance themselves, making any emulation of natural systems somewhat subjective. The goal of restoring a functioning ecosystem is similarly problematic. Which functions should be restored? Some functions, like retention of potential floodwaters, are decidedly anthropocentric and may be in direct conflict with other functions, like oxidation of organic matter during hydrologic drawdown. Thus the goals of ecological restoration are a bit nebulous. In fact, even the goal of designing self-regulating systems suggests that we are not certain of the endpoint—a successful restoration is one that allows for changes in functions we desire within a template we have defined.

Despite these philosophical shortcomings with the concept of ecological restoration, the fact remains that many aquatic ecosystems have been destroyed or severely degraded by human activity. In attempt to reverse the tide (sometimes literally), we have turned to restoration—nebulous goals and all. In 1992, the National Research Council issued a review of aquatic restoration and identified a need for a more systematic and strategic restoration of lakes, rivers, and wetlands. The report calls for more clearly defined and prioritized goals, objectives, and performance indicators, improved monitoring and assessment programs, and a more holistic approach to restoration and evaluation that considers attributes of structure and function at the population, community, ecosystem, and landscape levels. While acknowledging our inability to recreate pristine ecosystems, the authors offer compelling reasons why restoration efforts must nevertheless be undertaken, including meeting the growing human demand for freshwater, tempering the effects of global climate change, and supporting the biota upon which both human and biological communities depend. In short, the NRC articulated the urgency of restoring ecosystems to a "superior ecological condition that far surpasses the degraded one, so that valuable ecosystem services will not be lost."

A dozen years have passed since the NRC recommendations on the restoration of aquatic ecosystems. In that time, a great many ecological restoration projects have been undertaken in a variety of aquatic environments with a plethora of methods. Through both success and failure, we have learned much about ecosystem structure and function. But these efforts have not been concerted, and the necessary focus on site-level detail has offered too few glimpses of the big picture. My objective here is to assess the status of lake, river, and wetland restoration since the NRC recommendations. In particular, I focus on the restoration of trophic state and the management of nutrient metabolism in these three ecosystems. I compare techniques used to manage the abiotic aspects of ecosystem structure and consider efforts to restore ecosystem function through the emerging technique of biomanipulation—management of biota to achieve system-level effects through trophic interaction. Strategies of restoring lakes, rivers and streams have developed somewhat in isolation from one another, and much may be gained by cross-system analysis. As theory becomes applied and experimentation yields understanding, such analyses will be an important part of the next decade of restoration ecology.

LAKE RESTORATION

Lakes have historically been degraded in a number of different ways, but the most pervasive form of degradation is cultural eutrophication. Excessive nutrient additions—

particularly nitrogen and phosphorus—from both point and nonpoint sources have changed the physicochemical nature and biological complement of lakes around the world. Hypereutrophication by intentional discharge or incidental runoff of nutrients or organic matter ultimately have the same effects on lake function: excessive growth, followed by decreased light penetration, extreme microbial decomposition and oxygen depletion. Other changes in lake structure and function may be related to eutrophication, including inordinate macrophyte growth, invasive species, food web simplification, and loss of functional redundancy and resilience. Accompanying changes in lake aesthetics, recreational value, and water quality make hypereutrophic lakes less desirable for human use.

At first glance, hypereutrophication is the simplest stress on lakes and the easiest to address—one must simply remove the source of allochthonous nutrients and allow the lake to return to a more mesotrophic state. In fact, much progress has been made on the control of point source loadings in the past few decades, particularly by limiting the amount of highly degradable organic matter and readily labile nutrients released into lakes. Unfortunately, this has not been a panacea for lake eutrophication for three basic reasons (Carpenter and Lathrop 1999, NRC 1992). First, control of point source pollutants like municipal wastewater requires extensive infrastructure that is not available in many regions of the world. Second, even in developed nations that do regulate point source effluent, the focus has been on reduction of biochemical oxygen demand and pathogens, not on reduction of limiting nutrients like nitrogen and phosphorus. Finally, the problem of nonpoint source pollutants has not been addressed to nearly the same extent, and many lakes remain in a hypereutrophic state due to diffuse runoff from agricultural and urban areas (Carpenter et al. 1998). Thus, eutrophication remains a lake degradation problem and is one of the most common targets of lake restoration efforts.

Restoration of lake trophic state has been undertaken in two basic forms (Table 1). The first is reduction of nutrient load on the lake, or "bottom-up" management (McQueen et al. 1986). There are many approaches to this goal, including diversion of a waste stream, treatment prior to lake discharge, source reduction or modification, and land use changes. Some rather famous examples of restoration efforts on Lakes Erie, Washington, and Mendota illustrate the effectiveness of load reduction. Lake Erie represents both the great improvement possible with nutrient diversion and the complications of lake recovery from hypereutrophication. Lake Erie received excessive anthropogenic phosphorus for many decades, dramatically altering the physicochemical and biological character of the lake (Ludsin et al. 2001). Eutrophication-induced anoxia eliminated some important benthic invertebrate species and limited the habitat for vertebrates throughout the lake. Phosphorus abatement programs in the 1970's successfully reduced bioavailable P and phytoplankton biomass, but differences in morphology across the basin and the invasion of the zebra mussel obscure the overall effect of nutrient diversion. Lake Washington similarly received heavy nutrient loads from wastewater treatment plants until diversion was completed in 1968. This was followed by a period of algal biomass decline and increased transparency (Lehman 1988), but without an accompanying change in the planktonic species that were present during eutrophication (Edmondson et al. 2003). Lake Mendota was enriched with sewage until diversion in 1971, when the biologically available P load was reduced by about 30% (Lathrop 1992). A significant nutrient contribution from non-point sources—particularly agriculture—has limited the improvement of the lake's water quality since diversion of point sources (Carpenter and Lathrop 1999).

Table 1. Selected Physical, Chemical, and Biomanipulation Techniques for the Trophic Restoration of Lakes

Technique	Mechanism	References
Physical and Chemical Manipulation		
Nutrient Diversion	Reduce external nutrient load	Ludsin et al. 2001 Carpenter and Lathrop 1999
Phosphorus Immobilization	P sequestration in sediment	
Aluminum		Welch and Cooke 1999
Iron		Deppe and Benndorf 2002
Calcium sulfate		Varjo et al. 2003
Oxidation	Prevent anoxic hypolimnion	Gachter and Wehrli 1998
Dredging	Physical removal of P-rich sediment	Annadotter et al. 1999
Hypolimnetic withdrawal	Physical removal of P-rich water	Macdonald et al. 2004
Biomanipulation		
Fish Elimination	Remove planktivore and benthivore effects	Lammens 1999
Piscivore Stocking	Predator-mediated reduction of planktivores	Berg et al. 1997 Lathrop et al. 2002
Planktivore Removal	Increase zooplankton population	Annadotter et al. 1999
Benthivore Removal/Exclusion	Reduce bioturbation	Karjalainen et al. 1999
Habitat Modification	Maintain fish population after stocking	Christiansen et al. 1996 Scheffer et al. 1992

Long term research sites such as these show that nutrient diversion generally reduces available P, chlorophyll-*a*, and turbidity of the receiving lake. The eutrophic state can be quite stable, however, and load reduction alone does not guarantee a return to a pre-stress condition. Phosphorus from internal loading continues to be an issue after diversion and may be significant enough to trigger recurring algal blooms (Scharf 1999), particularly in lakes with anoxic hypolimnia. Internal phosphorus loading occurs as sediment phosphorus becomes re-dissolved and mixed with surface waters during turnover or as buoyant organic matter and algae transport P to the epilimnion (Welch and Weiher 2004). The plankton community does not necessarily revert to pre-eutrophication conditions after diversion either; Villena and Romo (2003) found only a slight decrease of cyanobacteria dominance after nutrient load reduction in Lake Albufera in Spain. Carvalho et al. (1995) speculate that diversion of external P loads alone may never lead to P limitation of phytoplankton biomass. For these reasons, many lakes tend to remain in a eutrophic state after diversion, leading some managers to use post-discharge nutrient reduction methods aimed at reducing or inactivating nutrient concentrations in the lake. Most commonly, this has involved chemical additions to modify the quality of lake water either by precipitation or by manipulation of the reduction-oxidation potential.

Artificial nutrient immobilization has been aimed almost exclusively at phosphorus, considered the limiting nutrient in many lacustrine systems. Alum (aluminum sulfate) has long been used to induce phosphorus precipitation. Welch and Cooke (1999) evaluated alum treatment in 21 lakes in the United States and found effective long term control of internal P loading—on the order of 10 years for polymictic lakes and 15 years for dimictic lakes. Aluminum addition at a molar ratio of 4:1 with labile phosphorus has been found to effectively reduce internal loading (Reitzel et al. 2003), and may continue to reduce mobile P at lower molar ratios (Lewandowski et al. 2003). Other researchers have experimented with ferrous iron (Deppe and Benndorf 2002), nitrate (Sondergaard et al. 2000), and gypsum (Varjo et al. 2003) additions to bind phosphorus. While all of these chemical treatments provide some decrease in the availability of P, chemical immobilization as the sole restoration tool is limited for two reasons. First, the threshold for P driven eutrophication is quite low—perhaps lower than 0.05 mg/L total P. This is a level very difficult to achieve and maintain (Villena and Romo 2003). Second, sediment sequestration of P may not have the desired effect on the lake biota. Cyanobacteria can tolerate a wide range of P availability and may persist in lakes with reduced P levels (Gulati and Van Donk 2002); this may delay the onset of increased transparency. Altered P and light availability affects macrophyte growth, but this appears to be a double edged sword. Muhammetoglu et al. (2002) found excessive macrophyte growth and decay to be a significant contributor to post-diversion eutrophication and advocate macrophyte harvesting. Robertson et al. (2000) similarly cite excessive macrophyte growth as a restoration roadblock in Delavan Lake in Wisconsin. Conversely, Gulati and Van Donk (2002) note positive effects of macrophytes as they compete with phytoplankton for nutrients, provide habitat for zooplankton and fish, and reduce bioturbation of sediment. Generally speaking, in-lake nutrient immobilization measures, used alone or in conjunction with nutrient diversion, have not consistently improved the trophic state of lakes or achieved successful lake restoration. Schauser et al. (2003) contend that these approaches are not always applied systematically according the lake characteristics, and hold that the success rate of in-lake nutrient immobilization would greatly increase with appropriate decision criteria. Even with successful nutrient immobilization, many lake managers have turned to a trophic food web approach to complement nutrient management in the restoration process.

BIOMANIPULATION IN LAKES

Biomanipulation in its broadest sense is the attempt to alter the trophic state of an ecosystem by altering its food web (Shapiro et al. 1975). For lakes, this generally means introduction of some organisms and elimination of others, particularly fish (Kasprzak et al. 2002). Based on the trophic cascade model, the intended mechanism is that a managed predator exerts a "top-down" control on lower trophic levels and ultimately on algae, as opposed to the "bottom-up" control of nutrient management (McQueen et al. 1986, Carpenter et al. 1985). A common biomanipulation practice in lakes is the introduction of piscivorous fish—intended to reduce planktivores, increase zooplankton populations, and decrease phytoplankton—and removal of benthivorous fish, which increase turbidity by suspending benthic organic matter and sediments. Lake biomanipulation techniques range from the drastic removal of all fish followed by selective restocking to gradual piscivore, planktivore, and benthivore supplementation and removal. Biomanipulation of fish has some clear advantages, including that it may be much easier than direct management of nutrients or plankton (Lammens 1999). Trophic relationships are not quite so simple, however, and there are many confounding factors for lake biomanipulation (Perrow et al. 2002). Feeding preferences of stocked fish can change as they age, resulting in a seasonally variable predation pressure or omnivory. Zooplankton populations, too, undergo seasonal fluctuations and are not always easily maintained at high density. The presence and abundance of macrophytes can play a role in control of fish and plankton populations by providing refuge sites for zooplankton (Perrow et al. 1999a and 1999b), but also can load a lake with oxygen-demanding detritus. And, of course, all of these food web interactions are occurring under the umbrella of lake nutrient availability and physicochemical state, which can further complicate management and sometimes render biomanipulation ineffective.

Despite the potential pitfalls, biomanipulation has been used with considerable success in lakes and must be considered as a significant component of lake restoration. In a comprehensive review of lake biomanipulation literature, Drenner and Hambright (1999) found that 61% of biomanipulation projects successfully improved water quality parameters, at least in the short term. Piscivore stocking was much more successful when combined with partial or total fish removal prior to stocking. When combined with benthivore removal, biomanipulation has been shown to lead to increased biomass and density of macroinvertebrates (Leppa et al. 2003) and macrophytes (Strand and Weisner 2001) and a decrease in cyanobacteria (Robertson et al. 2000). Recovered macrophyte assemblages, along with associated periphyton, may further act to compete with phytoplankton and maintain the clear water state. Macrophytes also provide habitat for fish, and further littoral habitat modification is being explored to augment predator populations (Skov and Berg 1999). For all of these reasons Mehner et al. (2002), while noting the complexity of using biomanipulation as a tool across seasonal and morphological scales and population structures, give biomanipulation a positive assessment as a fundamental tool for lake restoration.

In short, biomanipulation has become an important technique for lake restoration and has significantly improved the trophic restoration success rate beyond the capacity of nutrient diversion or immobilization alone. Aside from the general caution that biomanipulation be applied specifically to the needs of the individual lake, there appear to be three caveats. First, biomanipulation cannot stand on its own. Bottom-up nutrient dynamics cannot be ignored—biomanipulation is most successful when nutrient levels are reduced to the point where piscivory may play a role (Perrow 1999a). Second, fish population management must be revisited regularly, particularly for benthivores which may return to pre-management levels within several years (Leppa et al. 2003) or in cases of significant overfishing (Carpenter and Lathrop 1999). Third, biomanipulation will be more successful in some geomorphic settings than others. Shallow lakes that mix frequently will respond better to biomanipulation than deep stratified lakes that develop an anoxic hypolimnion (Mehner et al. 2002).

As a general observation, we may note that simple management of the abiotic—nutrient loading, in this case—has in many cases been insufficient for successful lake restoration, primarily because of recurring internal phosphorus loading. Even in-lake nutrient immobilization, while perhaps restoring the physicochemical structure of the lake, has not been as successful in restoring trophic function—a task for which biomanipulation is increasingly being employed. With a combination of all three approaches, lake restoration efforts are attempting to address both structure and function. Though different in form, the same issues of structure and function must be addressed in the trophic restoration of rivers, to which we now turn.

RIVER RESTORATION

Lotic systems function very differently than lentic systems, and are consequently the subject of very different ecological research. Like lakes, however, rivers have been affected by intense anthropogenic alterations of trophic structure. Curiously, while great attention has been given to the effects of primary nutrients like phosphorus in lakes, these nutrients have been virtually ignored as drivers of degradation in rivers that are similarly enriched. The consequences of excessive organic contamination in rivers has comparatively—and justifiably—been given more attention. Rapid nutrient spiraling, catchment and landscape-level effects, and complexities of trophic interactions have all limited research of top-down and bottom-up restoration strategies in streams. It is clear, however, that primary nutrients have a measurable effect on stream fish communities (Miltner and Rankin 1998), and that both nutrient sources and top predators influence other trophic levels in streams. We now consider structural modification and the role of biomanipulation in the restoration of rivers.

The most prevalent techniques of river restoration aim to remove impediments of the natural hydrologic regime, to rehabilitate the channel, and to improve the catchment or watershed through which the river runs (Table 2). Poff et al. (1997) make a clear case that it is the flow regime that governs all other river characteristics, including water quality, energy exchange, physical habitat, and biotic interactions, and that it should be the foremost consideration of any restoration. Attempts at flow regime restoration often include dam removal, channel restoration, and floodplain enhancement. Dam removal has immediate effects on the temperature, sediment load, and velocity of water flow downstream and can profoundly alter the biota of the river system. Channel restoration involves the creation of meanders in artificially straightened rivers and often includes banks stabilization in areas prone to erosion. Channel morphology may also be restored, by adding riffles and vanes to direct water flow, protect banks from erosion, and create habitat. Landscape level restoration may include reclamation of riparian buffers, modification of impervious cover or stormwater runoff, and alteration of landuse practices—particularly those that contribute to rapid runoff, soil loss, and elevated pollutant loads on the river.

Table 2. Selected Physical, Chemical, and Biomanipulation Techniques for the Trophic Restoration of Rivers

Technique	Mechanism	References
Physical and Chemical Manipulation		
Nutrient Diversion	Reduce external nutrient load	Miltner and Rankin 1998
Dam Removal	Restore hydrologic regime	Pizzuto 2002, Stanley and Doyle 2002
Re-meandering	Slower, more heterogeneous flow	Moerke and Lamberti 2003
Meso-habitat Restoration	Riffle and pool installation	Kemp et al. 1999
Obstruction Installation	Bank protection, pool restoration	Collins and Montgomery 2002
Corridor Rehabilitation	Re-connect floodplain with river	Buijse et al. 2002, Zalewski et al. 1998
Biomanipulation		
Fish Elimination	Remove competitor or nuisance fish	Shepard et al. 2002
Fish Stocking	Enhance predator population	De Groot 2002 Dannewitz et al. 2004
Habitat Modification	Maintain fish population after stocking	Fjellheim et al. 2003, Hendry et al. 2003

All of these techniques are ostensibly aimed at restoring the channel morphology and hydrologic regime, although the degrees to which they are successful depend upon the criteria by which river restorations are judged. Often, standards of water quality and biotic indices based on regional high-quality streams are used to evaluate the state of a river. We may ask, then, if these structural restoration processes successfully bring about an improvement in the water quality and trophic dynamics in streams.

The effects of an increased nutrient load on biota are more complex in streams than in lakes. In the simplest terms, streams act as transporters of nutrients and the bulk of the nutrient load affects the receiving body of water, not the stream itself. It has become more apparent, however, that rivers are transformers of nutrients, not just transporters. The degree to which nutrients are metabolized and transformed in rivers is a function of multiple factors, including the form of nutrient inputs, stream morphology and flow rates, material and energy exchanges between river subsystems, and disturbance regime (Fisher et al. 1998). Consequently, nutrient dynamics vary greatly from system to system. Generally speaking, human "improvements" to rivers—simplification of morphology and flow regime—increase the capacity of rivers to transport nutrients and decrease nutrient retention and transformation (Royer et al. 2001). Major eutrophication has been a concern not for the river, then, but for the lake or estuary into which the river flows. Interestingly, the very act of river restoration changes this relationship. The removal of dams, revegetation of banks and riparian corridors, restoration of a flood regime, and reconstruction of meanders, pools and backwaters all increase the spatial and temporal complexity and allow for increased nutrient retention and metabolism. For instance, dam removal may have profound effects on incision, erosion, and floodplain processes both above and below the dam (Pizzuto 2002). The result is a new stream morphology, which alters flow rates, sediment transport, particle size distribution, the extent of sediment-water contact, and the trophic dynamics of the system (Stanley and Doyle 2002). In some ways, then, restoration of natural fluvial processes without reduction of the nutrient load may make the distant problem of eutrophication a much more local concern.

The biotic effects of morphological river restoration and accompanying changes in nutrient retention are not entirely clear. In the short term, the restoration activity likely will increase turbidity and alter sediment load and distribution (Koebel et al. 1999), which may be detrimental for some benthic organisms through both the physical effect of burial and toxicity of released pollutants. After the initial upheaval of restoration, fluvial processes begin to take effect and the tropic structure of the river may begin to change. Hein et al. (1999) found increased nutrient metabolism in the Danube system as connectivity with side channels was restored. They noted increased phytoplankton productivity, strongly correlated with soluble phosphorus levels in the post-restoration side channels. Such backwater algal blooms are common in river systems (Wen 1992; Heiler et al. 1995), but not nearly as prevalent in straightened, channelized rivers. By restoring backwaters, meanders, and low-turbidity pools we are effectively changing the trophic state of the river.

This shift from a heterotrophic toward an autotrophic state may well be spatially and temporally variable—heavily dependent on the geomorphology and flow regime—in a different way than the autotrophic response to eutrophication varies in lakes. Even so, there can be distinct food web responses to increased phytoplankton production in river systems following restoration. Stanley et al. (2002) found a dramatic shift in the macroinvertebrates assemblage flowing restoration of the Baraboo River in Wisconsin; previously impounded reaches supported invertebrate biota indistinguishable from unimpounded reaches within one year. Friberg et al. (1998) similarly found rapid macroinvertebrate density and diversity response to restoration in the River Gelsa in Denmark, which the authors attribute to increased substrate and macrophyte-related habitat. The fish community may respond as well.

Hill et al. (1993) note increased fish diversity following impoundment removal on the Chipola River in Florida.

Despite the obvious differences in river and lake ecology, there are some parallels emerging here. Both systems, in areas highly disturbed by anthropogenic processes, are overburdened with primary nutrients, and trophic effects are apparent in each. In both systems, trophic restoration processes have predominantly focused on the abiotic—nutrient diversion and immobilization in lakes, and fluvial geomorphic restoration in rivers. In both systems, however, eutrophication is not so easily dismissed. As we have seen, lake restoration has successfully employed biomanipulation as a top-down regulation of trophic state. In the river restoration literature there is almost no reference to biomanipulation for restoration of trophic state. Given the obvious connections between nutrient load, river geomorphology, and biota, biomanipulation as a tool for river restoration seems at least worthy of consideration.

Biomanipulation in Rivers

The ecological status of a river is most frequently judged by an index of biotic integrity (Karr 1991; modified by Emery et al. 2003), and the fish community is commonly the subject. As one might expect, anthopogenic influences to river systems have greatly affected fish populations, particularly migratory and intolerant specialist species. Many physical, chemical, and biological factors can negatively affect a river fish assemblage, and eutrophication is certainly among them (Growns et al. 1998). River restoration strategies have, in some cases, been shown to allow spontaneous recoveries of key fish populations (particularly of salmonid dominated rivers; see Pretty et al. 2003). Restoration activities based solely on geomorphology and flow regime do not always result in restored fish communities, however, particularly in reach-scale projects that are still subjected to landscape level processes (Pretty et al. 2003; Moerke and Lamberti 2003).

In general, the effects of morphological river restoration on heterotrophic dynamics are poorly understood. In part this is due to longitudinal variability. Aarts and Nienhuis (2003) demonstrate the longitudinal changes in feeding guild, with zoobenthivorous and periphytivorous species giving way to zooplanktivorous and phytivorous species as one moves downstream. Osmundson et al. (2002) consider longitudinal distribution of a Colorado River piscivore, the pikeminnow. They found this top predator to be concentrated in upstream reaches, where prey fish, macroinvertebrates, and periphyton were also the most concentrated—a distribution strongly linked to flow rate and sedimentation. There are also significant issues of transversal variability (perpendicular to main flow direction) that influence piscivore establishment, including current zonation, spawning substrate, and riffle-pool arrangement (Aarts et al. 2004, Fisher et al. 1998).

Given all these physical and morphological variables of the river environment, the general approach to biotic restoration has been to do the best job possible of recreating the proper physical environment and allow the fish community to re-establish itself. But piscivore introduction in rivers is not uncommon—including impressive efforts to stock millions of brown trout, Atlantic salmon, sturgeon, and shad (Fjellheim et al. 2003; Reinartz et al. 2003; Brown et al. 2000). This so-called propagation assistance, though generally with extremely high mortality, has resulted in some successful reintroductions. In some cases, stocking has been attempted in the absence of geomorphic or water quality restoration, and resulting failures have made it clear that physical and chemical restoration are necessary precursors to any sort of biomanipulation (Fjellheim et al. 2003). The focus of fish introduction projects has typically been on re-establishing the populations of these economically valuable species, however, and not on trophic consequences. An exception, presented by Deegan et al. (1997), manipulated the top predator of an arctic stream and monitored the macroinvertebrate and algal response in both high and low nutrient environments. Their results showed little top-down regulation, as neither fish nor grazers had a significant effect on their respective food source. This suggests far greater trophic control by the availability of primary nutrients. Unfortunately, we lack similar studies on trophic cascade in rivers of other orders and in other ecoregions.

The most viable attempts at biotic restoration of rivers are currently in the area of habitat modification. These are physical means of improving the likelihood that fish populations will recover. Such approaches are necessarily specific to stream order. In restored headwater streams, the retention of allochthonous leaf litter has been explored as a key to trophic restoration (Muotka and Laasonen 2002). Numerous researchers are exploring the role of coarse woody debris in providing fish habitat in low- to mid-order streams (e.g. Collins and Montgomery 2002; Crook and Robertson 1999). In lowland rivers, these small scale in-stream structures may be less important that restoration of connectivity with lateral channels and backwater habitats (Pretty et al. 2003). Biomanipulation in rivers may not be quite as feasible as it is in lakes, but these habitat modification approaches deserve more attention; for as Ormerod (2003) notes, restoring past physical or chemical conditions is not always sufficient for ecosystem recovery. Habitat manipulation is an important link between structural restoration and functional recovery in lotic and lentic systems as well as in wetlands, which exhibit some characteristics of both.

WETLAND RESTORATION

Wetlands occur in many geomorphic settings with a wide range of hydrologic regimes, and they support different ecological communities than either lakes or streams. The task of wetland restoration is further complicated by historic hydrologic alteration—meaning many wetlands to be restored are not currently in existence. The necessary precursor to trophic restoration, then, is hydrologic restoration. These factors make wetland restoration a very different challenge for restoration ecologists than lake or stream restoration. And yet, there are some striking similarities. As with lakes and streams, wetland restoration efforts have thus far been concentrated on restoration of abiotic structure and processes. And, while plant establishment has been studied in some detail, the trophic arrangement of and role of plankton and higher trophic levels in restored wetlands have not been given the same attention. As emphasis shifts from structural to functional replacement, restoration of wetland trophic interactions should be increasingly scrutinized.

Wetlands are generally regarded as nutrient-rich environments and are thus valued for their role in nutrient transformation or retention, thereby reducing the nutrient load on adjacent bodies of water. Indeed, many natural wetlands are high nutrient, highly productive systems. But wetland ecosystems have a wide variation in fertility, and many wetlands—in unaltered state—are chronically limited by nutrients and support biota that are adapted to nutrient poor conditions (Keddy 2000). As we have seen, humans tend to overload aquatic ecosystems with primary nutrients, and wetlands are no exception. The result is that wetland ecosystems subjected to anthropogenic disturbance are often hypereutrophic, and restored wetlands are disproportionately nutrient enriched. Hypereutrophic emergent wetlands often exhibit a shift in the plant community, with increased biomass and decreased species richness (Weiher and Keddy 1995). In open water areas, the opposite may occur, with a dramatic increase in phytoplankton biomass that excludes macrophytes (Moss 1983).

More driven by regulatory mandates than lake or river restoration, wetland restoration has been largely focused on area-based replacement of the structural qualities of natural wetlands. The foremost consideration is placement within the landscape as dictated by regional hydrology, for wetland restorations are by definition the re-establishment of wetland conditions where they formerly existed (Table 3). Wetlands established where none formerly existed are classified as created wetlands. In either case, establishment requires a source of water—some combination of a groundwater spring or seep, river, lake or tide inundation, surface runoff, tile effluent, and precipitation. The source of water is of critical importance, for it will define the biogeochemical character of the wetland (Hunt et al. 1999). In this sense wetland restoration is similar to stream restoration, for the seasonal water level fluctuations, called the wetland's hydroperiod, are analogous to the natural flow regime of rivers. The hydroperiod determines the amount and duration of oxygen exposure in the wetland substrate and is therefore the regulator of many chemical and biological processes, including reduction and oxidation reactions, seed germination, and colonization (Mitsch and Jorgensen 2004).

Table 3. Selected Physical, Chemical, and Biomanipulation Techniques for the Trophic Restoration of Wetlands

Technique	Mechanism	References
Physical and Chemical Manipulation		
Hydrologic Restoration	Establish hydroperiod	Loucks 1990, Hunt et al. 1999
Landscape Connectivity	Enhance disturbance regime and dispersal	Middleton 1999, Bedford 1999, Seabloom and van der Valk 2003
Soil Amendment	Establish substrate and seed bank	Middleton 2003, van der Valk and Pederson 1989
Peat Restoration	Re-establish conditions for peat accumulation	Wind-Mulder et al. 1996, Gorham and Rochefort 2003
Biomanipulation		
Plant Introduction	Enhance diversity, reduce invasion	Callaway et al. 2003 Mitsch and Jorgensen 2004
Plant Removal	Artificial selection for desired species	Clark and Wilson 2001 Wilson et al. 2004
Herbivore Exclusion	Allow growth of desired plant species	Llewellyn and Shaffer 1993 Sherfy and Kirkpatrick 2003
Fish Elimination/Exclusion	Reduce bioturbation and zooplanktivory	Schrage and Downing 2004, Zimmer et al. 2002, Lougheed et al. 1998
Fish Introduction	Predator-mediated reduction of planktivores	Walker and Applegate 1976, French et al. 1999
Habitat Modification	Enhance diversity of biotic community	Levin and Talley 2002, Brady et al. 2002

Middleton (1999) makes a strong case that restoration of a water source is not enough; the disturbance regime is an essential component to wetland functional development. Natural disturbances may include flooding, drought (or periods of drawdown), storm activity, physical effects of organisms like beaver, and even fire. A great many wetland restorations have been completed without natural disturbance dynamics, resulting in relatively static systems that do not mimic natural wetlands. In some cases restored wetlands are too wet, without seasonal drawdown; in other cases they lack flood subsidies and are chronically too dry (Gallihugh and Rogner 1998; Erwin 1991). Restoration of a natural disturbance regime is a fine goal, but landscape level changes throughout the history of human perturbations may have rendered the original regime unattainable. Perhaps the best wetland restoration goal is to design systems that are at least open to environmental fluctuations, thereby allowing for a dynamic ecosystem.

Soil and Seed

Closely associated with hydrologic restoration is the establishment of wetland soil (Table 3). Ideally, restoration takes place on hydric soils—those that have been saturated or inundated long enough to develop anaerobic characteristics. In most cases, restoration occurs where hydric soils once existed but have since been altered by drainage or other hydrologic modification. An advantage to restoration on existing wetland soils is that these soils are more likely to contain a complement of viable wetland plant seeds, called a seed bank. Alternatively, some restoration ecologists have used seedbanks transplanted from other wetlands to accelerate plant development at the restored site; this appears to be a particularly useful technique for revegetation of herbaceous wetland plants (Middleton 2003). Wetland restoration without a remnant or donor seedbank is possible (Leck 2003), but it may lengthen the time required to develop the hydric soil and macrophyte community characteristic of natural wetlands—perhaps by decades.

After hydrologic and soil restoration, the strongest emphasis has been placed on revegetation. The most passive approach to wetland plant restoration is reliance upon natural recruitment. This has the potential advantage of recruiting only those species that are well suited to the restored environment, including those that may be present in a remnant seed bank. While natural revegetation has been successful in a number of different settings, it appears to be a particularly viable approach in systems that are open to dispersal from a nearby seed source (Primack 1996). A disadvantage, of course, is that quick dispersing r-strategists may dominate the colonization, and these may not include the most desired plant species (Odum 1987). For this reason, many wetland restoration projects use active revegetation to broaden the cohort. One technique, described above, is the use of a transplanted seed bank (van der Valk and Pederson 1989). Though labor intensive and impractical for large areas, this has the advantage of accelerating both soil and plant community development. Alternatively, plants can be introduced directly by dissemination of seed, vegetative fragments, or mature plants. Though mortality can be quite high, direct introduction is commonly used to accelerate development of diverse plant community (Mitsch and Jorgensen 2004). Ultimately, the plant community will develop by succession that is governed in part by the abiotic features of the restored system. The macrophyte community itself is itself a structural feature of the wetland—it affects the velocity of water flow, reduces light penetration, influences the redoximorphic soil environment, and provides a site for microbial attachment. These structural qualities of restored vegetation are often seen as important as any trophic function, if not more so. Thus, as we have seen for lakes and streams, wetland restoration has predominantly been based upon restoration of hydrologic, geomorphic and physicochemical structure.

Biomanipulation in Restored Wetlands

Comparatively little research has been done on the restoration of trophic function in wetlands. We do, however, have an understanding of the effects of hypereutrophication on wetland trophic systems, primarily from study of wetlands used for wastewater treatment. In treatment wetlands, high nutrient loads clearly change the biota, including a shift toward phytoplankton dominance over submerged macrophytes, altered role of herbivores, and changes in the macroinvertebrate community structure (Spieles and Mitsch 2000). Hypereutrophication in open water habitats has been associated with a loss of piscivores and domination by planktivores and benthivores (Whillans 1996). In restored systems that are also subjected to eutrophication, we thus have the dual uncertainty of successional trajectory and nutrient load pressures on trophic processes. And, though we know something about the assembly rules of restored wetland communities, we know far less about their assembly in nutrient rich conditions. McKenna (2003) addresses just this question in a study of production and respiration in nutrient-rich restored and natural reference wetlands in New York. He reports a rapid development of metabolic processes in the restored wetland, with gross production and respiration rates statistically similar to the reference site. Abundant populations of zooplankton and macroinvertebrates inhabited both wetlands, although fish were present only in the reference wetland. Results from this study suggest that top-down regulation of trophic processes have little effect in high-nutrient, detritus-dominated wetlands.

Angeler et al. (2003) note the resemblance of nutrient load effects in wetlands to those seen in shallow eutrophic lakes and propose that biomanipulation may play a similar management role in certain conditions. In particular, high sediment phosphorus loads, dense algal biomass and a fish community dominated by zooplanktivores and benthivores occur in both eutrophic wetlands and lakes. Documented attempts at using biomanipulation of fish for remediation of these effects in wetlands are few, and those that exist take two forms (Table 3). The first is fish removal or exclusion. Zimmer et al. (2002) report that the presence of zooplantivorous fathead minnows in prairie pothole wetlands increased total phosphorus, turbidity and chlorophyll-*a* in the water column; removal of minnows reversed these effects. Similar results have been noted by the exclusion of benthivorous fish like carp (Schrage and Downing 2004, Lougheed et al. 1998, Loughheed and Chow-Fraser 2001). A second approach is the introduction or encouragement of piscivores. Walker and Applegate (1976) found strong effects of walleye predation on minnows in a semi-permanent wetland in South Dakota, and French et al. (1999) have experimented with structures to allow northern pike in Lake Erie coastal wetlands while excluding carp. Less studied are the potential effects of macroinvertebrate grazers on the amount and form of primary production, although recent studies indicate that macroinvertebrate recruitment can be facilitated through inoculation and stocking (Brady et al. 2002) and that grazers can significantly influence the wetland primary production standing crop (Spieles and Mitsch 2003).

As the potential of wetland biomanipulation compares with that of shallow lakes, so too do the constraints. Systems that are continuously nutrient rich due to sustained external loading are not likely to respond to biomanipulation (McKenna 2003). The omnivorous nature of some wetland zooplankton is also a confounding factor, as they consume both phytoplankton and other grazers, negating any control of primary production (Ortega-Mayagoita et al. 2002). Third, many wetland systems are not hydrologically suitable for fish communities, and thus classical biomanipulation is not a possibility. Finally, wetlands with a high frequency of disturbance (particularly of inundation and drawdown) may be poor candidates for top-down control of primary production, as consistent predatory pressure may be difficult to maintain. In some restored wetlands, then, traditional biomanipulation may not be a useful tool. However it does appear that trophic development is a valuable but understudied measure of wetland function, and consideration of trophic cascades may be an important next step for wetland restoration.

Synthesis and Conclusions

Considering these three ecosystems in parallel illustrates the differential progress and challenges of restoration in dissimilar aquatic environments. In some ways, restoration of these systems is following the same path (Figure 1). In all three, there are prevailing abiotic forces that govern community assembly and that must be the foundation of the restoration effort. Much of the aquatic restoration research to date has been focused on these forces: reducing bioavailable nutrients in lakes, restoring the flow regime and channel morphology in rivers, and restoring the seasonal hydroperiod in wetlands. Though entirely warranted, this body of research on abiotic structure has yielded techniques that have not always resulted in functional restoration. We see this, too encountered in all three aquatic environments, with phytoplankton dominance, a shift in macrophyte community composition, and simplified food webs dominated by benthivores and planktivores in the restored system. Having encountered the limitations of structural modification, lake, river, and wetland restoration are thus poised on the new frontier of functional rehabilitation.

And what functions should we restore? From a human perspective, aquatic ecosystems may function by performing services for humanity, by transforming or transporting nutrients, retaining floodwaters, recharging aquifers, or supporting economically valuable species. These functions are all potentially measurable in restored aquatic ecosystems. But there are also ecological functions to consider—processes that make some anthropocentric ecosystem services possible. These include development of metapopulations and patch dynamics, successional pathways, community metabolism, and trophic interactions. Trophic restoration, as this review demonstrates, is emerging as an important method of functional restoration. In lakes this is not such a new frontier; in fact lake restoration has moved beyond structure to experiment with trophic dynamics to a greater extent than either stream or wetland restoration. Lakes are simply better suited to analyses of trophic cascades—there are fewer disturbances, more predictable niches, and less spatial heterogeneity. Trophic restoration in rivers and wetlands faces significant obstacles, mostly due to the fluctuating hydrologic regime that shapes the biogeochemical environment. Any yet, recent research in stream and wetland restoration has begun to follow the lead of lake restoration; more research is being focused on what occurs after geomorphic and hydrologic restoration. Habitat enhancement for particular species is emerging as an important link between structural restoration and trophic function (Figure 1).

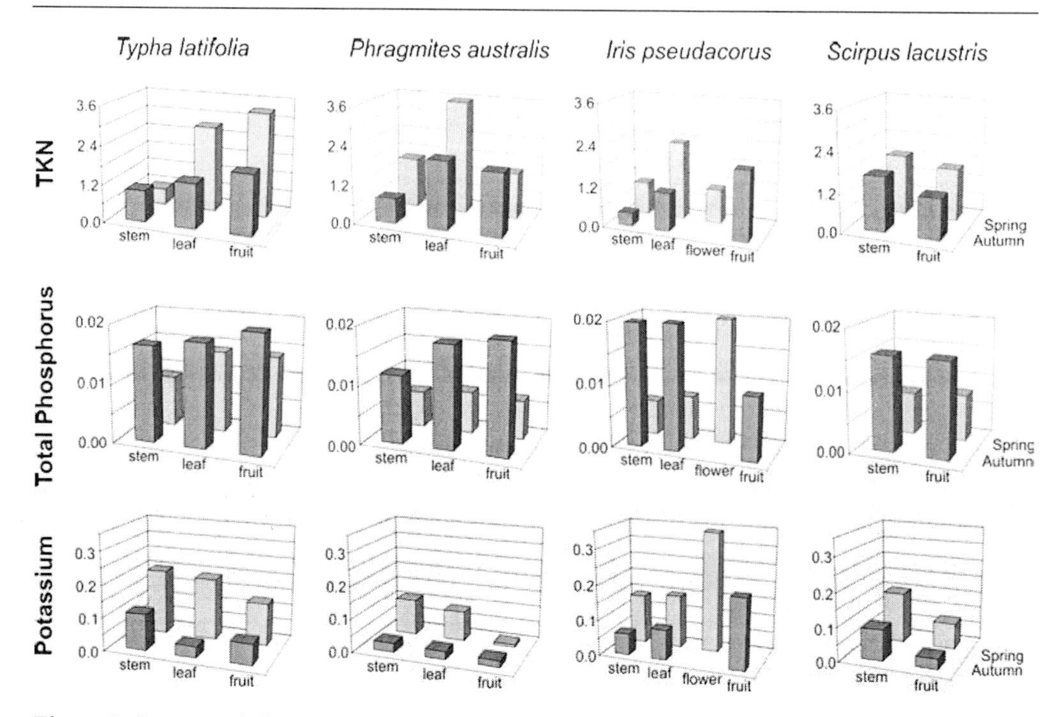

Figure 1. Conceptual diagram of a restored aquatic ecosystem with emphasis on trophic structure and function. Habitat modification is shown as a link between structural restoration and biomanipulation.

As a management tool, biomanipulation will probably never be as important in streams and wetlands as it is in lakes, but an understanding of the restored trophic environment is just as essential. This review yields several conclusions about the current state of and possible future directions of trophic restoration:

1. Any control of primary production by trophic cascade is unlikely in the face of overwhelming external nutrient load. The lake restoration experience has made it clear that the system must approach nutrient limitation for top-down control to be effective. For functional restoration of trophic state, then, aquatic ecosystems must first be protected from hypereutrophication.
2. Appropriate habitat is a key to trophic restoration. The process of natural recruitment can be accelerated and stocking may be enhanced with careful attention to habitat requirements. In lakes this has taken the form of macrophyte structure, spawning substrate and passage, and littoral woody debris. In rivers—depending on order—leaf litter, coarse woody debris, and backwater connectivity may all play a role in the reintroduction of higher trophic levels. In wetlands, the effects of exclusion have been better studied than predator reintroduction. Much research has been devoted to plant establishment and plant community assembly in restored wetlands; faunal establishment deserves the same attention.
3. Consideration of trophic restoration reaffirms what restoration ecologists have long known—ecosystems do not occur in a vacuum, and any attempt at restoration must include landscape level processes. The catchment, the riparian floodplain, the upstream condition, the hydrologic source: all are intricately linked with the goals of

nutrient regulation and habitat formation and can either be great aids to restoration or, if improperly incorporated, the cause of a swift failure.

4. Indices of biotic integrity continue to be important tools for monitoring the development of restored aquatic ecosystems. Indices based on pollution tolerance, nativity, and deformities are certainly useful indicators of ecosystem health, but those that incorporate feeding guild are particularly useful for assessing trophic restoration.
5. Finally, this review makes it clear that trophic restoration—and perhaps any functional restoration—must be periodically revisited. Again, the lake experience is instructive. Benthivores return after removal, piscivores are overharvested after introduction, and new sources of external nutrients emerge after diversion. The goal of designing systems for minimal maintenance is a fine one, but restored aquatic ecosystems exist in a very human-dominated landscape and may lack features that make natural ecosystems more resilient; thus continuing management may be necessary. Vigilance, it seems, is one of the prices of functional restoration.

Biomanipulation in the narrow sense, that is, management of fish populations to effect a change in primary production, may indeed be a useful tool in some aquatic restorations. But biomanipulation in the broad sense—whole-ecosystem rehabilitation of biota and associated habitats—is much more than a tool; it represents a paradigm of aquatic restoration. The NRC recognized trophic state as a component of functional restoration in 1992, but only now are we beginning to experimentally approach trophic restoration in rivers and wetlands as we have in lakes. As we gain the ability to restore trophic dynamics, we may find that greater ecosystem properties like stability, response to disturbance, and resilience follow suit—or we may discover how little we know about ecosystem emergent properties. In either case, we will be one step closer to emulating natural aquatic ecosystems.

REFERENCES

Aarts BGW, Nienhuis PH. 2003. Fish zonations and guilds as the basis for assessment of ecological integrity of large rivers. *Hydrobiologia* 500:157-178.

Aarts BGW, Van Den Brink FWB, Nienhuis PH. 2004. Habitat loss as the main cause of the slow recovery of fish faunas of regulated large rivers in Europe: the transversal floodplain gradient. *River Research and Applications* 20(1):3-23.

Angeler DG, Chow-Fraser P, Hanson MA, Sanchez-Carrillo S, Zimmer KD. 2003. Biomanipulation: a useful tool for freshwater wetland mitigation? *Freshwater Biology* 48:2203-2213.

Annadotter H, Cronberg G, Aagren R, Lundstedt B, Nilsson PA, Strobeck S. 1999. Multiple techniques for lake restoration. *Hydrobiologia* 395/396:77-85.

Bedford BL. 1999. Cumulative effects on wetland landscapes: links to wetland restoration in the United States and southern Canada. *Wetlands* 19(4):775-788.

Berg S, Jeppesen E, Sondergaard M. 1997. Pike (*Esox lucius* L.) stocking as a biomanipulation tool: 1. effects on the fish population in Lake Lyng, Denmark. *Hydrobiologia* 342/343:311-318.

Brady VJ, Cardinale BJ, Gatham JP, Burton TM. 2002. Does facilitation of faunal recruitment benefit ecosystem restoration? An experimental study of invertebrate assemblages in wetland mesocosms. *Restoration Ecology* 10(4):617-626.

Brown BL, Gunter TP, Waters JM, Epifanio JM. 2000. Evaluating genetic diversity associated with propagation-assisted restoration of American shad. *Conservation Biology* 14(1):292-303.

Buijse AD, Coops H, Staras M, Lans LH, Van Geest GJ, Grift RE, Ibelings BW, Oosterberg W, Roozen FCJM. 2002. Restoration strategies for river floodplains along large lowland rivers in Europe. *Freshwater Biology* 47(4):889-907.

Callaway JC, Sullivan G, Zedler JB. 2003. Species-rich plantings increase biomass and nitrogen accumulation in a wetland restoration experiment. *Ecological Applications* 13(6):1626-1639.

Carpenter SR, Bolgrien D, Lathrop RC, Stow CA, Red T, Wilson MA. 1998. Ecological and economic analysis of lake eutrophication by nonpoint pollution. *Australian Journal of Ecology* 23:68-79.

Carpenter SR, Kitchell JF, Hodgson JF. 1985. Cascading trophic interactions and lake productivity. *Bioscience* 35:634-639.

Carpenter SR, Lathrop RC. 1999. Lake restoration: capabilities and needs. *Hydrobiologia* 395/396:19-28.

Carvalho L, Beklioglu M, Moss B. 1995. Changes in a deep lake following sewage diversion: a challenge to the orthodoxy of external phosphorus control as a restoration strategy? *Freshwater Biology* 34(2):399-410.

Christensen DL, Herwig BJ, Schindler DE, Carpenter SR. 1996. Impacts of lakeshore residential development on coarse woody debris in northern temperate lakes. *Ecological Applications* 6:1143-1149.

Clark DL, Wilson MV. 2001. Fire, mowing, and hand-removal of woody species in restoring a native prairie wetland in the Willamette Valley of Oregon. *Wetlands* 21(1):135-144.

Collins BD, Montgomery DR. 2002. Forest development, wood jams, and restoration of floodplain rivers in the Puget Lowland, Washington. *Restoration Ecology* 10(2):237-247.

Crook DA, Robinson AI. 1999. Relationships between riverine fish and woody debris: implications for lowland rivers. *Marine and Freshwater Research* 50(8):941-953.

Dannewitz J, Petersson E, Dahl J, Prestegaard T, Lof AC, Jarvi T. 2004. Reproductive success of hatchery-produced and wild-born brown trout in an experimental stream. *Journal of Applied Ecology* 41(2):355-364.

Deegan LA, Peterson BJ, Golden H, McIvor C, Miller MC. 1997. Effects of fish density and river fertilization on algal standing stocks, invertebrate communites, and fish production in an arctic river. *Canadian Journal of Fisheries and Aquatic Sciences* 54(2):269-283.

De Groot SJ. 2002. A review of the past and present status of anadromous fish species in the Netherlands: is restocking the Rhine feasible? *Hydrobiologia* 478:205-218.

Deppe T, Benndorf J. 2002. Phosphorus reduction in a shallow hypereutrophic reservoir by in-lake dosage of ferrous iron. *Water Research* 36(18):4525-4534.

Drenner RW, Hambright KD. 1999. Biomanipulation of fish assemblages as a lake restoration technique. *Archiv fuer Hydrobiologie* 146(2):129-165.

Edmonson WT, Abella SEB, Lehman JT. 2003. Phytoplankton in Lake Washington: long term changes 1950-1999. *Archiv fuer hydrobiology Supplement* 139(3):275-326.

Emery EB, Simon TP, McCormick FH, Angermeier PL, Deshon JE, Yoder CO, Sanders RE, Pearson WD, Hickman GD. Reash, RJ, Thomas JA. 2003. Development of a multimetric index for assessing the biological condition of the Ohio River. *Transactions of the American Fisheries Society* 132(4):791-808.

Erwin KL. 1991. *An evaluation of wetland mitigation in the South Florida Water Management District, Vol. I.* Final report to South Florida Water Management District, West Palm Beach, FL, USA.

Fisher SG, Grimm NB, Marti E, Holmes RM, Jones JB Jr. 1998. Material spiraling in stream corridors: a telescoping model. *Ecosystems* 1(1): 19-34.

Fjellheim A, Barlaup BT, Gabrielsen SE, Raddum GG. 2003. Restoring fish habitat as an alternative to stocking in a river with strongly reduced flow. *Ecohydrology and Hydrobiology* 3(1):17-26.

French JRP III, Wilcox DA, Nichols SJ. 1999. Passing of northern pike and common carp through experimental barriers designed for use in wetland restoration. *Wetlands* 19:883-888.

Friberg N, Kronvang B, Hansen HO, Svendsen LM. 1998. Long-term, habitat-specific response of a macroinvertebrate community to river restoration. *Aquatic Conservation* 8(1):87-99.

Gachter R, Wehrli B. 1998. Ten years of artificial mixing and oxygenation: no effect on the internal phosphorus loading of two eutrophic lakes. *Environmental Science and Technology* 32:3659-3665.

Gallihugh JL, Rogner JD. 1998. *Wetland mitigation and 404 permit compliance study, Vol. I. United States Fish and Wildlife Service*, Region III, Burlington, IL, USA and United States Environmental Protection Agency, Region V, Chicago, IL, USA.

Gorham E, Rochefort L. 2003. Peatland restoration: a brief assessment with special reference to *Sphagnum* bogs. *Wetlands Ecology and Management* 11(1-2):109-119.

Growns IO, Pollard DA, Gehrke PC. 1998. Changes in river fish assemblages associated with vegetated and degraded banks, upstream of and within nutrient enriched zones. *Fisheries Management and Ecology* 5(1):55-69.

Gulati RD, Van Donk E. 2002. Lakes in the Netherlands, their origin, eutrophication and restoration: state of the art review. *Hydrobiologia* 478:73-106.

Heiler G, Hein T, Schiemer F, and Bornette G. 1995. Hydrological connectivity and flood pulses as the central aspects for the integrity of river-floodplain systems. *Regulated Rivers Research and Management* 11:351-362.

Hein T, Heiler G, Pennetzdorfer D, Riedler P, Schagerl M, Schiemer F. 1999. The Danube River Restoration Project: functional aspects and planktonic productivity in the floodplain system. *Regulated Rivers Research and Management* 15:259-270.

Hendry K, Cragg-Hine D, O'Grady M, Sambrook H, Stephen A. 2003. Management of habitat for rehabilitation and enhancement of salmonid stocks. *Fisheries Research* 62(2):171-192.

Hill MJ, Long EA, Hardin S. 1993. Effects of dam removal on Dead Lake, Chipola River, Florida. Apalachicola River Watershed Investigations, Florida Game and Freshwater Fish Commission. *Wallop-Breaux Project F-39-R*. 12 p.

Hunt RJ, Walker JF, Krabbenhoft DP. 1999. Characterizing hydrology and the importance of groundwater discharge in natural and constructed wetlands. *Wetlands* 19(2):458-472.

Karjalainen J, Leppa M, Rahkola M, Tolonen K. 1999. The role of benthivorous and planktivorous fish in a mesotrophic lake ecosystem. *Hydrobiologia* 408/409:73-84.

Karr JR. 1991. Biological integrity: a long-neglected aspect of water resource management. *Ecological Applications* 1:66-84.

Kasprzak P, Benndorf J, Mehner T, Koschel R. 2002. Biomanipulation of lake ecosystems: an introduction. *Freshwater Biology* 47:2277-2281.

Keddy PA. 2000. *Wetland ecology: principles and conservation*. Cambridge: Cambridge University Press. 614 p.

Kemp JL, Harper DM, Crosa GA. 1999. Use of "functional habitats" to link ecology with morphology and hydrology in river rehabilitation. *Aquatic Conservation* 9(1):159-178.

Koebel JW Jr., Jones BL, Arrington DA. 1999. Restoration of the Kissimmee River, Florida: water quality impacts from canal backfilling. *Environmental Monitoring and Assessment* 57(1):85-107.

Lammens EHRR. 1999. The central role of fish in lake restoration and management. *Hydrobiologia* 395/396:191-198.

Lathrop RC. 1992. Nutrient loadings, lake nutrients, and water clarity. In: J.F. Kitchell editor. *Food web management: a case study of Lake Mendota*. Springer-Verlag, New York. p. 69-96.

Lathrop RC, Johnson BM, Johnson TB, Vogelsang MT, Carpenter SR, Hrabik TR, Kitchell JF, Magnuson JJ, Rudstam LG, Stewart RS. 2002. Stocking piscivores to improve fishing and water clarity: a synthesis of the Lake Mendota biomanipulation project. *Freshwater Biology* 47(12):2410-2424.

Leck MA. 2003. Seed bank and vegetation development in a created tidal freshwater wetland on the Delaware River, Trenton, New Jersey USA. *Wetlands* 23(2):310-343.

Lehman JT. 1988. Hypolimnetic metabolism in Lake Washington, Washington USA: relative effects of nutrient load and food web structure on lake productivity. *Limnology and Oceanography* 33(6):1334-1347.

Leppa M, Hamalainen H, Karjalainen, J. 2003. The response of benthic macroinvertebrates to whole-lake biomanipulation. *Hydrobiologia* 498:97-105

Levin LA, Talley TS. 2002. Natural and manipulated sources of heterogeneity controlling early faunal development of a salt marsh. *Ecological Applications* 12(6):1785-1802.

Lewandowski J, Schauser I, Hupfer M. 2003. Long term effects of phosphorus precipitations with alum in hypereutrophic Lake Suesser See (Germany). *Water Research* 37(13):3194-3204.

Llewellyn DW, Shaffer GP. 1993. Marsh restoration in the presence of intense herbivory: the role of *Justicia lanceolata* (Chapm.) Small. *Wetlands* 13(3):176-184.

Loucks OL. 1990. Restoration of the pulse control function of wetlands and its relationship to water quality objectives. In: Kusler JA, Kentula ME, editors. *Wetland Creation and restoration: the status of the science*. Washington, DC: Island Press. p 467-477.

Lougheed VL, Crosbie B, Chow-Fraser P. 1998. Predictions on the effect of common carp (*Cyprinus carpio*) exclusion on water quality, zooplankton and submerged macrophytes in a Great Lakes wetland. *Canadian Journal of Fisheries and Aquatic Sciences* 55:1189-1197.

Lougheed VL, Chow-Fraser P. 2001. Spatial variability in the response of lower trophic levels after carp exclusion from a freshwater marsh. *Journal of Aquatic Ecosystem Stress and Recovery* 9:21-34.

Ludsin SA, Kershner MW, Blocksom KA, Knight RL, Stein RA. 2001. Life after death in Lake Erie: nutrient controls drive fish species richness, rehabilitation. *Ecological Applications* 11(3):731-746.

Macdonald RH, Lawrence GA, Murphy TP. 2004. Operation and evaluation of hypolimnetic withdrawal in a shallow eutrophic lake. *Lake and Reservoir Management* 20(1):39-53.

McKenna JE, Jr. 2003. Community metabolism during early development of a restored wetland. *Wetlands* 23(1):35-50.

McQueen DJ, Post JR, Mills EL. 1986. Trophic relationships in freshwater pelagic ecosystems. *Canadian Journal of Fisheries and Aquatic Sciences* 43:1571-1581.

Mehner T, Benndorf J, Kasprzak P, Koschel R. 2002. Biomanipulation of lake ecosystems: successful applications and expanding complexity in the underlying science. *Freshwater Biology* 47(12):2453-2465.

Middleton BA. 1999. *Wetland restoration: flood pulsing and disturbance dynamics.* New York: Wiley. 388 p.

Middleton BA. 2003. Soil seed banks and the potential restoration of forested wetlands after farming. *Journal of Applied Ecology* 40(6):1025-1034.

Miltner RJ, Rankin ET. 1998. Primary nutrients and the biotic integrity of rivers and streams. *Freshwater Biology* 40(1):145-158.

Mitsch WJ, Jorgensen SE. 2004. *Ecological engineering and ecosystem restoration.* Hoboken: Wiley. 411 p.

Moerke AH, Lamberti GA. 2003. Responses in fish community structure to restoration of two Indiana streams. *North American Journal of Fisheries Management* 23(3):748-759.

Moss B. 1983. The Norfolk Broadland: experiments in the restoration of a complex wetland. *Biological Reviews of the Cambridge Philosophical Society* 58:521-561.

Muhammetoglu A, Muhammetoglu H, Soyupak S. 2002. Evaluation of efficiencies of diffuse allochthonous and autochthonous nutrient input control in restoration of a highly eutrophic lake. *Water Science and Technology* 45(9):195-203.

Muotka T, Laasonen P. 2002. Ecosystem recovery in restored headwater streams: the role of enhanced leaf retention. *Journal of Applied Ecology* 39:145-156.

National Research Council. 1992. *Restoration of aquatic ecosystems: science, technology, and public policy.* Washington, DC: National Academy Press. 552 p.

Odum WE. 1987. Predicting ecosystem development following creation and restoration of wetlands. In: Zelazney J, Feierbend JS, editors. *Proceedings of the conference on wetlands: increasing our wetland resources.* Washington, DC: Corporate Conservation Council, national Wildlife Federation. p 67-70.

Ormerod SJ. 2003. Current issues with fish and fisheries: editor's overview and introduction. *Journal of Applied Ecology* 40:204-213.

Ortega-Mayagoita E, Rojo C, Rodrigo MA. 2002. Factors masking the trophic cascade in shallow eutrophic wetlands: evidence from a microcosm study. *Archiv fuer Hyrobiologie* 155(1):43-63.

Osmundson DB, Ryel RJ, Lamarra VL, Pitlick J. 2002. Flow-sediment-biota relations: implications for river regulation effects on native fish abundance. *Ecological Applications* 12(6):1719-1739.

Perrow MR, Jowitt AJD, Leigh SAC, Hindes AM, Rhodes JD. 1999a. The stability of fish communities in shallow lakes undergoing restoration: expectations and experiences from the Norfolk Broads (UK). *Hydrobiologia* 408-409:85-100.

Perrow MR, Jowitt AJD, Stansfield JH, Phillips GL. 1999b. The practical importance of the interactions between fish, zooplankton and macrophytes in shallow lake restoration. *Hydrobiologia* 395-396:199-210.

Perrow MR, Tomlinson ML, Zambrano L. 2002. Fish. In: Perrow MR, Davy AJ, editors. *Handbook of ecological restoration, Volume 1: Principles of restoration.* Cambridge: Cambridge University Press. p 324-354.

Pizzuto J. 2002. Effects of dam removal on river form and process. *Bioscience* 52(8):683-691.

Poff NL, Allan JD, Bain MB, Karr JR, Prestegaard, KL, Richter BD, Sparks RE, Stromberg JC. 1997. The natural flow regime: a paradigm for river conservation and restoration. *Bioscience* 47(11):769-784.

Pretty JL, Harrison SSC, Shepherd DJ, Smith C, Hildrew AG, Hey RD. 2003. River rehabilitation and fish populations: assessing the benefit of instream structures. *Journal of Applied Ecology* 40:251-265.

Primack RB. 1996. Lessons from ecological theory: dispersal, establishment, and population structure. In: Falk DA, Millar CI, Olwell M, editors. *Restoring diversity: strategies for reintroduction of endangered plants.* Washington, DC: Island Press. p 209-233.

Reinartz R, Bloesch J, Ring T, Stein H. 2003. Sturgeons are more than caviar: a plea for the revival of sturgeons in the Danube River. *Archiv fuer Hydrobiologie Supplement* 147:387-403.

Reitzel K, Hansen J, Jensen HS, Andersen FO, Hansen KS. 2003. Testing aluminum addition as a tool for lake restoration in shallow, eutrophic Lake Sonderby, Denmark. *Hydrobiologia* 506-509:781-787.

Robertson DM, Goddard GL, Helsel DR, MacKinnon KL. 2000. Rehabilitation of Delavan Lake, Wisconsin. *Lake and Reservoir Management* 16(3):155-176.

Royer T, Tank JL, David M. 2001. Is denitrification a major sink for nitrate in agricultural streams? *Bulletin of the North American Benthological Society* 18:208.

Scharf W. 1999. Restoration of the highly eutrophic Lingese Reservoir. *Hydrobiologia* 416:85-96.

Schauser I, Lewandowski J, Hupfer M. 2003. Decision support for the selection of an appropriate in-lake measure to influence the phosphorus retention in sediments. *Water Research* 37(4):801-812.

Scheffer M, Redelijkheid DE, Noppert F. 1992. Distribution and dynamics of submerged vegetation in a chain of shallow eutrophic lakes. *Aquatic Botany* 42(3):199-216.

Schrage LJ, Downing JA. 2004. Pathways of increased water clarity after fish removal from Ventura Marsh, a shallow, eutrophic wetland. *Hydrobiologia* 511(1):215-231.

Seabloom EW, van der Valk AG. 2003. Plant diversity, composition, and invasion of restored and natural prairie pothole wetlands: implications for restoration. *Wetlands* 23(1):1-12.

Shapiro J, Lamarra V, Lynch M. 1975. Biomanipulation—an ecosystem approach to lake restoration. In: Brezonic PL, Fox JL, editors. *Water quality management through biological control.* Gainsville, FL: University of Florida. p 85-96.

Shepard BB, Spoon R, Nelson L. 2002. A native westslope cutthroat trout population responds positively after brook trout removal and habitat restoration. *Intermountain Journal of Sciences* 8(3):191-211.

Sherfy MH, Kirkpatrick RL. 2003. Invertebrate response to snow goose herbivory on moist soil vegetation. *Wetlands* 23(2):236-249.

Skov C, Berg S. 1999. Utilization of natural and artificial habitats by YOY pike in a biomanipulated lake. *Hydrobiologia* 408/409:115-122.

Sondergaard M, Jeppesen E, Jensen JP. 2000. Hypolimnetic nitrate treatment to reduce internal phosphorus loading in a stratified lake. *Lake and Reservoir Management* 16(3):195-204.

Spieles D J, Mitsch WJ. 2000. Macroinvertebrate community structure in high and low nutrient constructed wetlands. *Wetlands* 20(4):716-729.

Spieles D J, Mitsch WJ. 2003. A model of secondary production and trophic structure in constructed wetlands. *Ecological Modelling* 161:183-194

Stanley EH, Doyle MW. 2002. A geomorphic perspective on nutrient retention following dam removal. *BioScience* 52(8):693-701.

Stanley EH, Luebke MA, Doyle MW, Marshall DW. 2002. Short-term changes in channel form and macroinvertebrate communities following low-head dam removal. *Journal of the North American Benthological Society* 21:172-187.

Strand JA, Weisner SEB. 2001. Dynamics of submerged macrophyte populations in response to biomanipulation. *Freshwater Biology* 46(10):1397-1408.

Throop W, editor. 2000. *Environmental restoration: ethics, theory, and practice*. Amherst, NY: Humanity Books. 240 p.

van der Valk AG, Pederson RL. 1989. Seed banks and the management and restoration of natural vegetation. In: Leck MA, Parker VT, Simpson RL, editors. *Ecology of soil seed banks*. San Diego: Academic Press. p 329-346.

Varjo E, Liikanen A, Salonen VP, Martikainen PJ. 2003. A new gypsum-based technique to reduce methane and phosphorus release from sediments of eutrophied lakes: gypsum treatment to reduce internal loadings. *Water Research* 37(1):1-10.

Villena MJ, Romo S. 2003. Phytoplankton changes in a shallow Mediterranean lake (Albufera of Valencia, Spain) after sewage diversion. *Hydrobiologia* 506-509:281-287.

Walker RE, Applegate RL. 1976. Growth, food, and possible ecological effects of young of the year walleyes in a South Dakota prairie pothole. *The Progressive Fish Culturist* 38:217-220.

Weiher E, Keddy PA. 1995. The assembly of experimental wetland plant communities. *Oikos* 73:323-35.

Welch EB, Cooke GD. 1999. Effectiveness and longevity of phosphorus inactivation with alum. *Lake and Reservoir Management* 15(1):5-27.

Welch EB, Weiher ER. 2004. Improvement in Moses Lake quality from dilution and sewage diversion. *Lake and Reservoir Management* 20(1):76-84.

Wen YH. 1992. Contribution of bacterioplankton, phytoplankton, zooplankton and detritus to organic seston carbon load in a Changjiang floodplain lake (China). *Archives Hydrobiology* 126(2):213-238.

Whillans TH. 1996. Historic and comparative perspectives on rehabilitation of marshes as habitat for fish in the lower Great Lakes basin. *Canadian Journal of Fisheries and Aquatic Sciences* 53:58-66.

Wilson MV, Ingersoll CA, Wilson MG, Clark DL. 2004. Why pest plant control and native plant establishment failed: a restoration autopsy. *Natural Areas Journal* 24(1):23-31.

Wind-Mulder HL, Rochefort L, Vitt DH. 1996. Water and peat chemistry comparisons of natural and post-harvested peatlands across Canada and their relevance to peatland restoration. *Ecological Engineering* 7:161-181.

Zalewski M, Bis B, Lapinska M, Frankiewicz P, Puchalski W. 1998. The importance of the riparian ecotone and river hydraulics for sustainable basin-scale restoration scenarios. *Aquatic Conservation* 8(2):287-307.

Zimmer KD, Hanson MA, Butler MG. 2002. Effects of fathead minnows and restoration on prairie wetland ecosystems. *Freshwater Biology* 47:2071-2086.

In: Progress in Aquatic Ecosystems Research
Editor: A. R. Burk, pp. 83-97

ISBN 1-59454-383-6

Chapter 4

Nitrogen Transformations in Riparian Wetland Microcosms Determined with Stable Isotopes

F. E. Matheson, M. M. Gibbs, M. L. Nguyen and A. B. Cooper
National Institute of Water & Atmospheric Research, Hamilton, New Zealand
T. P. Burt
Department of Geography, University of Durham, Durham, United Kingdom

Abstract

In developed catchments riparian wetlands can intercept runoff and potentially act as buffer zones protecting downstream waterways from excess N enrichment. However, our understanding of the processes involved in N transformation and the sustainability of the N removal function are far from complete. Due to its low natural abundance the stable isotope of nitrogen (^{15}N) can be used as a tracer to investigate N transformation processes in ecological systems. In separate laboratory experiments we investigated the fate of $^{15}NH_4^+$ in lake-edge wetland microcosms inhabited by *Salix cinerea* (grey willow) and *Typha orientalis* (raupo) plants and the fate of $^{15}NO_3^-$ in stream-side wetland microcosms inhabited by *Glyceria declinata* (glaucous sweetgrass) relative to unplanted microcosms. We found that *Salix* and *Typha* plants readily assimilated NH_4^+ (9-11% of applied $^{15}NH_4^+$), enhanced the natural diffusion of $^{15}NH_4^+$ up through organic soil into the overlying water (by 7 to 18-fold) and, via litterfall, provided a flocculent substrate for sequential mineralisation-nitrification-dentrification processes in the overlying water layer. In the other microcosm experiment *Glyceria* plants readily assimilated NO_3^- (15% of applied $^{15}NO_3^-$), but also inhibited dissimilatory NO_3^- reduction to NH_4^+ (from 49 to <0.1%) and enhanced denitrification (from 30 to 61% of applied $^{15}NO_3^-$) by increasing the oxidation level of microcosm soils presumably through root oxygen release. The results of both experiments highlight the key role played by emergent plants in sustainable N processing in wetland environments. Further research is required to determine whether these or other species of emergent plant might be used to maximise N removal by denitrification in those areas where wetlands are critical for protecting downstream water quality. Some of the advantages and limitations of using ^{15}N tracer microcosm experiments for investigating N transformations in riparian wetland systems are discussed.

INTRODUCTION

Riparian wetlands, adjacent to streams and lakes, play an important role in developed catchments by intercepting runoff (groundwater and overland flow) from upland catchments. Effective removal of N by riparian wetlands has been well-documented particularly for nitrate (NO_3^-), since this is often the main form of N present in overland flow and groundwater. However, there are some instances when ammonium (NH_4^+) and dissolved organic N (DON) may be the dominant N fractions present in groundwater throughflow; for example when catchment soils have a high clay content which limits oxygen penetration and promotes anoxia (Lusby et al. 1998; Matheson 2001).

Recent studies investigating NO_3^- removal in riparian wetlands have demonstrated the important role played by catchment and riparian wetland hydrogeology in regulating N removal, particularly the degree of runoff contact with wetland surface soils and the plant root zone, where there is typically the greatest potential for NO_3^- transformation, uptake and removal (Hill 1996, Devito et al. 2000, Maitre et al. 2003, Rutherford and Nguyen 2004). However despite major advances in this area, our understanding of the biotic processes involved in N transformations and the factors that control partitioning between competing pathways is far from complete. The key N transformation processes in riparian wetlands are nitrification, denitrification, plant uptake, microbial immobilisation, mineralisation, nitrogen fixation and dissimilatory nitrate reduction to ammonium (DNRA). These processes are all illustrated in Figure 1, which depicts the N cycle. It should be noted however that while NH_4^+ is generally considered the preferred form of N taken up by both plants and microbes, NO_3^- can also be assimilated directly, with transformation to NH_4^+ occurring intracellularly (Lin and Stewart 1998, Puri and Ashman 1999, Matheson et al. 2002). More is known about some processes (e.g. nitrification and denitrification) than others. Many studies have investigated NH_4^+ and NO_3^- transformation by one or more processes, often as measurements of "potential rates" using separate experiments, but few have considered the relative importance of all possible pathways of N removal and attempted to measure them simultaneously.

Use of ^{15}N stable isotope as a tracer enables simultaneous measurement of a number of N transformation processes and for the 'fate' of tracer N to be determined. A dose of highly ^{15}N-enriched (e.g. 99 atom %) substrate of interest (e.g. NO_3^-, NH_4^+ etc) is added to the wetland system and subsequent increases in the ^{15}N enrichment of other N pools above natural abundance levels (generally around 0.37 atom %) indicates the fate of the tracer and the transformation process responsible. This technique has previously been used to study the fate of N in a range of freshwater wetland systems (Hemond 1983, Moraghan 1993, Cooke 1994) including a riparian fen (Ambus et al. 1992) and a lake-edge wetland (Nijburg and Laanbroek 1997).

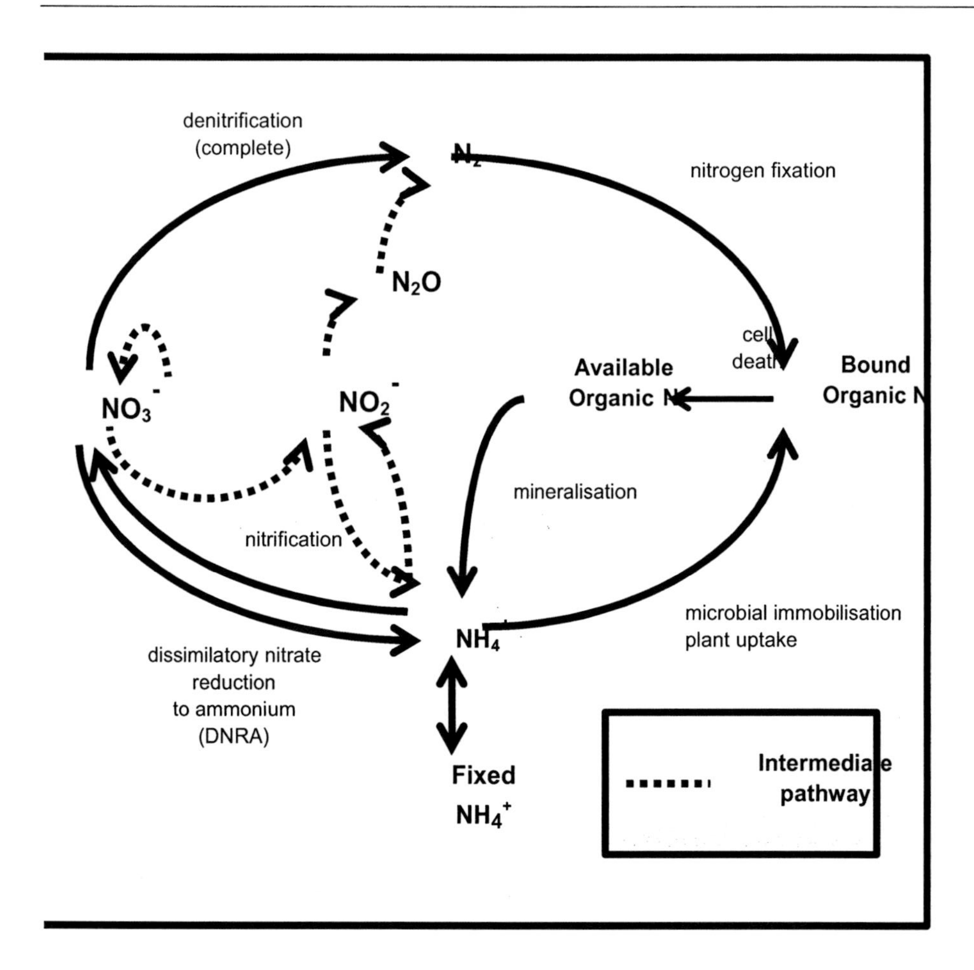

Figure 1. The nitrogen cycle.

In this chapter we report on the findings of two laboratory-based microcosm experiments that have investigated N removal and transformation processes in riparian wetlands (one lake-edge and one stream-side) using ^{15}N as a tracer. In the lake-edge wetland experiment the main component of catchment runoff and the substrate of interest was NH_4^+ while in the stream-side wetland experiment it was NO_3^-. In both experiments we based the microcosm experimental design as closely as possible on field conditions and sought to determine not only which N transformation processes were likely to predominate in these systems but also the influence of the wetland vegetation on N transformation processes and overall N removal. These experiments have previously been published separately elsewhere (Lusby et al. 1998, Matheson et al. 2002) but here we have attempted to synthesise the findings of these two experiments, add additional previously unpublished data from these studies where possible and interpret these findings in light of other more recent studies in order to improve our understanding of the N transformation processes operating in riparian wetland systems, the role played by wetland vegetation and how ^{15}N stable isotope techniques might be better used to study N cycling in these complex systems.

MATERIALS AND METHODS

Experiment 1: Fate of NH_4^+ in Lake-edge Wetland Microcosms Determined $^{15}NH_4Cl$

Full details of the method employed are detailed elsewhere (Lusby et al. 1998) but a summary is given here. Soil, water and plants were collected from a lake-edge wetland at Lake Okareka, New Zealand, which intercepted NH_4^+-contaminated groundwater from the surrounding urban catchment prior to discharging into the lake. The two uppermost soil layers (organic layer overlying pumice sand) were reconstituted in 12 clear, acrylic microcosms (65 cm length, 7 cm inner diameter, 10 cm depth organic layer, 40 cm pumice sand layer) with a 3 cm layer of overlying lake water. Moisture content was 67% for the organic layer and 29% for the sand layer, while bulk density was 0.45 g cm^{-3} for the organic layer (Lusby 1996) and a bulk density of 1 g cm^{-3} has been assumed for the sand layer. A single *Salix cinerea* (grey willow) plant was transplanted into three microcosms, a single *Typha orientalis* (raupo) plant into another three, and the remaining 6 were left unplanted.

Microcosms were incubated in a climate control room at 25°C under Gro-Lux lights (PAR 550 μmol m^{-2} s^{-1}; 16 h light to 8 h dark cycle) in a large water tank (40 cm depth) to maintain the high humidity of a natural wetland. Overlying water in each column was continuously recirculated using a peristaltic pump (1.5 ml min^{-1}). Plants were allowed to acclimatise in the microcosms for 16 d prior to the start of the experiment.

On Day 1 of the experiment 20 mg $^{15}NH_4Cl$ (2 ml of 10 mg N l^{-1}, 99 atom % ^{15}N) was injected into the six planted and three unplanted microcosms at 10 cm depth through a sampling septum in the side of each microcosm. Three leaf clip samples were taken from each plant on Day 1 then again on Days 3, 5, 7, 9 and 13. These samples were dried (65°C overnight) and analysed for ^{15}N content on a Finnigan MAT delta-C continuous flow isotope ratio mass spectrometer.

On Day 13 any remaining leaf (shoot) material on each plant was collected, dried overnight and finely shredded. All *Salix* leaf material was placed in the overlying water of one of the previously unused unplanted microcosms while all *Typha* leaf material was placed in another. The remaining unused unplanted microcosm was used as a control for this litter decomposition component of the study.

On Day 28 overlying water samples from each microcosm were collected and analysed for total dissolved nitrogen (TDN) and ^{15}N content. In the two litter decomposition columns remaining litter was collected and processed and analysed in the same manner as the leaf material (above). Soils in each microcosm were sectioned into 2-cm slices and plant root material (if any) removed. Soils were dried (65°C overnight) to determine moisture content, finely ground and analysed for total nitrogen (TN) and ^{15}N content. Root material was processed and analysed in the same manner as leaf and litter material.

Significant differences ($p<0.05$) in parameters measured between microcosm types were determined using the *F*-test of multiple μ's (Datadesk 6.0). Where data were not normally distributed, they were log-transformed prior to analysis.

Experiment 2: Fate of NO_3^- in Stream-side Wetland Microcosms Determined with $K^{15}NO_3$

Full details of this experiment are described in Matheson et al. (2002) but a summary is provided here. Soil (organic-rich surface layer to 20-30 cm depth), water and wetland plants (*Glyceria declinata;* glaucous sweetgrass) were collected from a stream-side wetland that intercepted NO_3^--rich groundwater and occasional surface runoff from a pasture prior to discharge into the adjacent stream. Soil was homogenised, passed through a 2 mm sieve, and placed in plastic, cylindrical microcosms (11.2 cm diameter x 18.2 cm depth) that were wrapped in aluminium foil to restrict light and oxygen penetration through the plastic. Soil depth in the microcosms was 15 cm.

Microcosms were incubated in a climate control room at 20 deg C with plant growth lighting (PAR 460 μmol m^{-2} s^{-1}) and a 12 h light to 12 h dark cycle. Three *Glyceria* plants (all of similar size) were placed in half of the microcosms while the remainder were left unplanted. After a 24 h stabilisation period, 10 ml of wetland water was added to each microcosm to create a 0.1 cm overlying water layer which was increased to 0.2 cm the following day and maintained for the duration of the experiment. Microcosm positions in the climate control room were marked out on a grid, were randomly chosen for each microcosm and were adjusted every 2 days. A 32-d stabilisation period was used to enable plants to establish prior to the start of the experiment.

On Day 0, 12 ml of $K^{15}NO_3^-$ (7.9 mg N l^{-1}, 99 atom % ^{15}N, 0.5 μg ^{15}N g^{-1}) was injected (in an even distribution) into each microcosm using a 20 cm long needle. The resulting input of NO_3^- was approximate to that normally received by the wetland (~0.7 μg N g^{-1}). These injections were repeated every 2 days (at the same time) for the duration of the experiment (32 days).

Four planted and four unplanted microcosms were destructively sampled on Days 0 and 32, and two of each type on Days 8, 16 and 24. Soil pH, oxygen saturation (by linear sweep voltammetry) and redox potential measurements were made in each microcosm prior to sectioning. Plant shoots were cropped then soils (and plant roots where present) were sectioned into three depth layers (0-5, 5-10 and 10-15 cm). A 10 g sub-sample of soil from each layer was collected and extracted with 2M KCl. Extracts were stored for analysis of NH_4^+, $^{15}NH_4^+$, NO_3^- and $^{15}NO_3^-$. Extracted soil was stored for total N and ^{15}N analysis. Plant roots were removed from remaining soil in the layer, and along with plant shoot samples were dried (60°C overnight) and stored for analysis of total N and ^{15}N.

Total N and ^{15}N analysis of samples was performed using a Finnigan MAT Delta-Plus mass spectrometer. ^{15}N content of NO_3^- and NH_4^+ in extracts was determined following sequential diffusion onto acidified filters.

Significant differences ($p<0.05$) in parameters measured between the two microcosm types were determined using two sample *t*-tests (Datadesk 6.0). Where data were not normally distributed, they were log-transformed prior to analysis or non-parametric Mann-Whitney tests were used.

RESULTS AND DISCUSSION

The fate of ^{15}N at the end of each experiment, and the transformation processes responsible, are summarised in Table 1.

Table 1. Percentages of Added ^{15}N in Various N Pools in Microcosms at the End of Experiments 1 and 2

N pool		Transformation process responsible Exp. 1/Exp. 2	Experiment 1 ($^{15}NH_4^+$)			Experiment 2 ($^{15}NO_3^-$)	
			Salix	*Typha*	No plants	Glyceria	No plants
Plant	All	Plant uptake	11 (± 4)^a	9 (± 6)^a	-	15 (± 4)	-
	Shoot only	Plant uptake	3 (± 2)^a	0.3 (± 0.2)^a	-	7 (± 2)	-
	Root only	Plant uptake	8 (± 2)^a	9 (± 6)^a	-	8 (± 1)	-
Soil	Total N	Various[1]	85 (± 6)^a	87 (± 7)^a	95 (± 3)^a	24 (± 2)^a	71 (± 4)^b
	Organic N	Immobilisation	-	-	-	24 (± 3)^a	22 (± 10)^a
	NH_4-N	No tr[2]. /DNRA	-	-	-	<0.1 (±<0.1)^a	49 (± 3)^b
	NO_3-N	Nit[3] /no tr[2].	-	-	-	<0.1 (±<0.1)^a	<0.1 (±<0.1)^a
Overlying water	Total N		0.4 (± 0.1)^b	1 (± 0.4)^b	<0.1 (±<0.1)^a	-	-
Unaccounted	Total N	Denitrification	6 (± 5)^a	3 (± 2)^a	5 (± 3)^a	61 (± 3)^b	30 (± 4)^a

[1] For Experiment 1 transformation processes responsible are NH_4^+ immobilisation and nitrification while for Experiment 2 they are NO_3^- immobilisation and dissimilatory NO_3^- reduction to NH_4+ (DNRA).

[2] No tr.; no transformation of added ^{15}N.

[3] Nit.; nitrification

Note: Values are means (± standard error). For each experiment different alphabetic superscripts indicate significant differences between treatments at $p<0.05$.

In Experiment 1 most $^{15}NH_4^+$ (85-95%) was retained in the soil N pool in both planted and unplanted microcosms but since the ^{15}N content of each individual soil N pool (organic N, NH_4^+ and NO_3^-) was not determined it is unclear how much of the $^{15}NH_4^+$ was immobilised or nitrified by soil microbes, or remained untransformed. Recognising this limitation, the ^{15}N content of each soil N pool was determined in Experiment 2. While considerable work is involved in yielding these separate determinations (soil is extracted and NH_4^+ and NO_3^- diffused in turn from extract solutions prior to ^{15}N analysis), much useful information on N transformation processes is gained as illustrated for Experiment 2. Had only the ^{15}N content of the total N pool been determined it would have been unclear whether the high retention of $^{15}NO_3^-$ in the soil in the unplanted microcosms (71%) was due to immobilisation, dissimilatory NO_3^- reduction to NH_4^+ (DNRA) or as a result of $^{15}NO_3^-$ remaining untransformed. As a result of the additional analysis it can be seen that a large proportion of $^{15}NO_3^-$ (49%) was transformed as a result of DNRA, while the remainder was due to microbial immobilisation (22% - a comparable figure to that in the *Glyceria* microcosms).

In general, DNRA and immobilisation processes have not often been measured in studies of N transformation processes in riparian wetlands (except e.g. Ambus et al. 1992, Nijburg and Laanbroek 1997) or other freshwater or terrestrial environments (except e.g. Buresh and Patrick 1978, D'Angelo and Reddy 1993, Bengtsson and Bergwall 2000, Silver et al. 2001). In contrast, the other two competing fates for NO_3^-, particularly denitrification, have usually been the main subjects of study. It is understandable that much attention is paid to denitrification since it is the only process that results in permanent loss of N from soil-water systems and therefore contributes directly to downstream water quality protection. However, laboratory based determinations of denitrification 'potential' (which optimise the physico-chemical conditions for the process) are unlikely to yield reliable estimates of actual *in situ* denitrification. Laboratory or *in situ* methods that attempt to mimic field conditions as closely as possible (e.g. Hanson et al. 1994, Lowrance et al. 1995, Addy et al. 2002, Clement et al. 2003, Mokkherji et al. 2003, Rutherford and Nguyen 2004) yield better estimates but simultaneous determination of other possible fates for NO_3^- enables a mass balance check to be made (ie. that observed NO_3^- removal equates to the NO_3^- transformed by the various processes).

The results of our experiments highlight the potentially important contributions made by DNRA and microbial immobilisation to the transformation of N in riparian wetland environments. Since these processes act to conserve N within the soil-water system the magnitude of these processes has important implications for the N removal capacity of riparian wetlands. The percentage of NO_3^- removal due to DNRA found in unplanted soil in Experiment 2 (49%) is larger than previously reported in riparian wetland and lake littoral sediments (3-9%) (Ambus et al. 1992; Nijburg and Laanbroek 1997) however the rates are lower (up to 0.005 μg N g^{-1} h^{-1} c.f. 0.005-0.184 μg N g^{-1} h^{-1} Nijburg and Laanbroek 1997 and 0.015-0.270 μg N g^{-1} h^{-1} Ambus et al. 1992). The higher percentages of NO_3^- removal attributable to DNRA but the lower DNRA rates in our study may be due to the low additions of NO_3^- used (0.5 μg N g^{-1} c.f. 2-3 μg N g^{-1} by Nijburg and Laanbroek 1997 and 24-228 μg N g^{-1} by Ambus et al. 1992). Higher, one-off NO_3^- additions are likely to stimulate rates of all N transformations but also oxidise the soil to some extent and favour denitrification over DNRA as a fate for NO_3^-. This is because denitrifiers are typically aerobic microbes with the capability of using NO_3^- as an alternative electron acceptor when oxygen becomes limiting while DNRA microbes are typically obligate anaerobes with a fermentative metabolism and therefore better adapted to more reducing conditions (Tiedje 1988).

In the present study denitrification accounted for much more ^{15}N removal in Experiment 2 (30-61%) than in Experiment 1 (3-6%). This is not surprising given that in Experiment 1 added $^{15}NH_4^+$ would first have to be nitrified, and this aerobic process was almost certainly limited by the highly reducing microcosm soils (*in situ* redox potentials of +50 at surface to – 400 at 30 cm depth, Lusby et al. 1998). Results indicate that a small amount of coupled nitrification-denitrification of $^{15}NH_4^+$ occurred at aerobic-anaerobic interfaces in the microcosms (ie. in the oxidised rhizosphere of wetland plant roots and close to the soil surface). In Experiment 2, denitrification as a fate for $^{15}NO_3^-$ was greatly enhanced by the presence of the wetland plant, *Glyceria declinata*, relative to unplanted conditions. Redox potential measurements (Figure 2) and oxygen saturation profiles (see Matheson et al. 2002) in the microcosms show that the planted soils were less reducing compared to the unplanted soils, presumably as a result of root oxygen release by the plants. The planted soils were still sufficiently anoxic to enable denitrification but presumably the competing DNRA process, which dominated in the more reducing unplanted soils, was hindered under the more oxidised conditions.

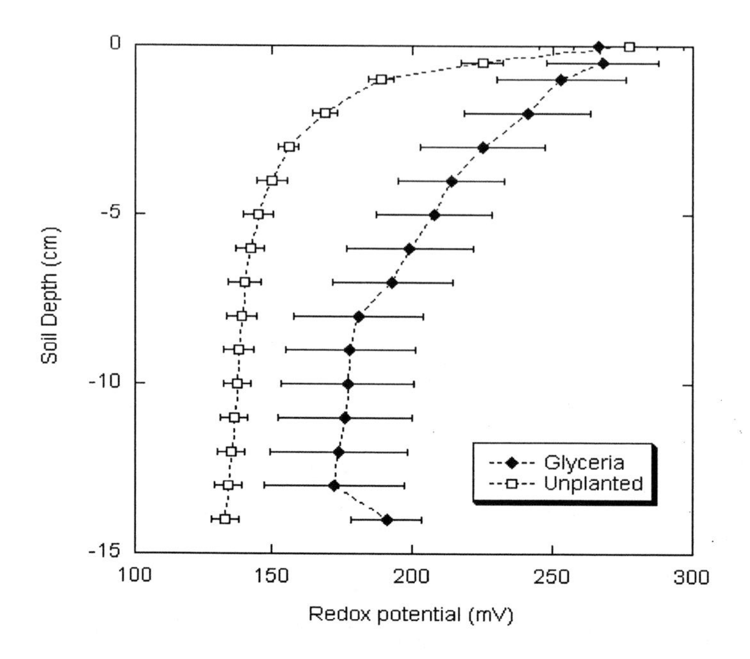

Figure 2. Redox potential of soil measured at 0.5-1 cm intervals in *Glyceria* and unplanted microcosms at the end of the Experiment 2 (Day 32). Values are means (± standard error, n = 4).

The denitrification rate estimated for all microcosms in Experiment 1 of 0.002 μg ^{15}N g^{-1} h^{-1} corresponds to an areal rate of 8-9 mg N m^{-2} d^{-1} (Lusby et al. 1998) while in Experiment 2 the denitrification rates in unplanted and planted microcosms of 0.003 μg ^{15}N g^{-1} h^{-1} and 0.006 μg ^{15}N g^{-1} h^{-1} correspond to areal rates of ~1.5 mg N m^{-2} d^{-1} and ~3 mg N m^{-2} d^{-1}, respectively. The rates found in both experiments are comparable to those reported for two riparian wetlands by Hanson et al. (1994) of 1.4-4.3 mg N m^{-2} d^{-1} without nitrate addition and 1.9-10.4 mg N m^{-2} d^{-1} with nitrate addition but lower than the average rate of 18.6 mg N m^{-2} d^{-1} reported by Lowrance et al. (1995) for a restored riparian forest wetland.

The amount of immobilisation of $^{15}NO_3^-$ in Experiment 2 (22-24% or 0.002 μg ^{15}N g^{-1} h^{1}) did not differ significantly between planted and unplanted microcosms. Higher immobilisation was expected in the planted microcosms due to the more oxidised soil conditions, which would enhance the activity of aerobic microbes that have greater N requirements than their anaerobic counterparts (Reddy and Patrick 1984). Moreover, it was thought that the significantly higher levels of NH_4^+ found in unplanted microcosms (Figure 3) would inhibit microbial immobilisation of NO_3^- since this process has been shown to be repressed by NH_4^+ concentrations as low as 0.1 μg N g^{-1} (Rice and Tiedje 1989). However a number of studies, including this one, have found considerable proportions of NO_3^- removal attributable to immobilisation (5-24%) in the presence the relatively high concentrations of NH_4^+ (>5 μg N g $soil^{-1}$) (Ambus et al. 1992, D'Angelo & Reddy 1993, Cooke 1994). A number of explanations can be put forward to explain these observations including: (1) NO_3^-

immobilisation occurring in NH_4^+-free soil microsites (Ambus et al. 1992); (2) the residual rate of NO_3^- immobilisation in the presence of NH_4^+ being sufficient to account for the accumulation of organic N (Rice and Tiedje 1989); and (3) NO_3^- being first dissimilated to NH_4^+ (extracellularly by DNRA), followed by immobilisation of NH_4^+, as opposed to NO_3^-, into the cell. However, a study by McCarty and Bremner (1992) suggests that apparent inhibition of assimilatory nitrate reductase activity in the presence of NH_4^+ is due to the products formed by microbial assimilation of NH_4^+ as opposed to NH_4^+ *per se*. In addition some of this apparent immobilisation may be due to abiotic fixation since biotic and abiotic components of the soil N pool were not separately analysed. However, abiotic fixation of NO_3^- is likely to be low as NO_3^- is generally considered to be highly water-soluble and mobile in soils, as illustrated by the problems with high losses of N due to NO_3^- leaching in many agricultural catchments (Davies 2000).

Plant uptake accounted for comparable proportions of ^{15}N removal in both experiments (9-15%). The upper estimates of ammonium uptake rates for *Salix cinerea* in Experiment 1 of 8.8-46 mg N m^{-2} d^{-1} (Lusby et al. 1998) are comparable with rates of 34-69 mg N m^{-2} d^{-1} reported for a range of *Salix* varieties by Riddell-Black et al. (1997). Uptake rates were not calculated for *Typha* in Experiment 1 since it was felt that the leaf clipping technique, which sampled shoot tips, underestimated ^{15}N uptake in this species (see Lusby et al. 1998). In Experiment 2 the NO_3^- uptake rates by *Glyceria* were ~1 mg N m^{-2} d^{-1}, which is considerably lower than the NH_4^+ uptake rates measured for *Salix*. Around one third less ^{15}N was added to microcosms in the *Glyceria* study (0.5 μg $^{15}NO_3^-$ g^{-1} $2d^{-1}$ or 8 μg $^{15}NO_3^-$ g^{-1} over 32d c.f. one dose of 12 μg $^{15}NH_4^+$ g^{-1}). The higher availability of inorganic N in Experiment 1 combined with the preference of many wetland plants for assimilating NH_4^+ over NO_3^- (Gunterspergen et al. 1991) may explain the much lower rates of ^{15}N uptake by *Glyceria* plants compared to *Salix* plants. Another possible explanation is the generally lower biomass of *Glyceria* attained during experimentation (~0.8 kg dry matter (dm) m^{-2}) relative to *Salix* (2.2-4.6 kg dm m^{-2}). However the result may also reflect inherent differences in N assimilation capacity by different wetland plant species as demonstrated by Kao et al. (2003) for five species of wetland plant (*Sparganium americanum*, *Scripus cyperinus*, *Juncus effusus*, *Phalaris arundinacea*, *Calamagrostis canadensis*) at the same site.

In Experiment 1 measurements made of N in the overlying water layer of microcosms indicated that the presence of both plant species, but especially *Typha*, may enhance the upward diffusion of reduced forms of N, particularly NH_4^+. The mean amounts (± SE) of ^{15}N detected in the overlying water of *Salix*, *Typha* and unplanted microcosms at the end of the experiment were 85 (± 15), 240 (± 65) and 13 μg (± 4), respectively. The different fractions of ^{15}N were not determined but measurements of the different fractions of total (^{14}N + ^{15}N) N in the overlying water (Figure 4) show significantly higher NH_4^+ concentrations in the overlying water of *Typha* microcosms, and higher (although not significantly) DON concentrations in the overlying water of *Salix* and *Typha* microcosms relative to unplanted microcosms. Reduced forms of N (ie. DON and NH_4^+ are likely to be stable under reducing soil conditions such as those found in all of the microcosms. Their enhanced upward movement in the planted microcosms may be due to root growth increasing the porosity of the soils.

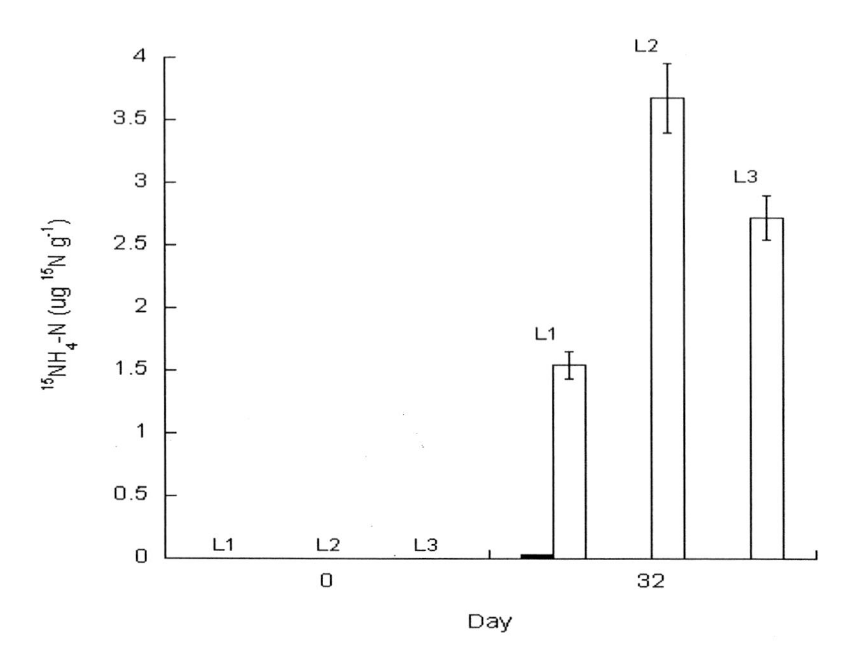

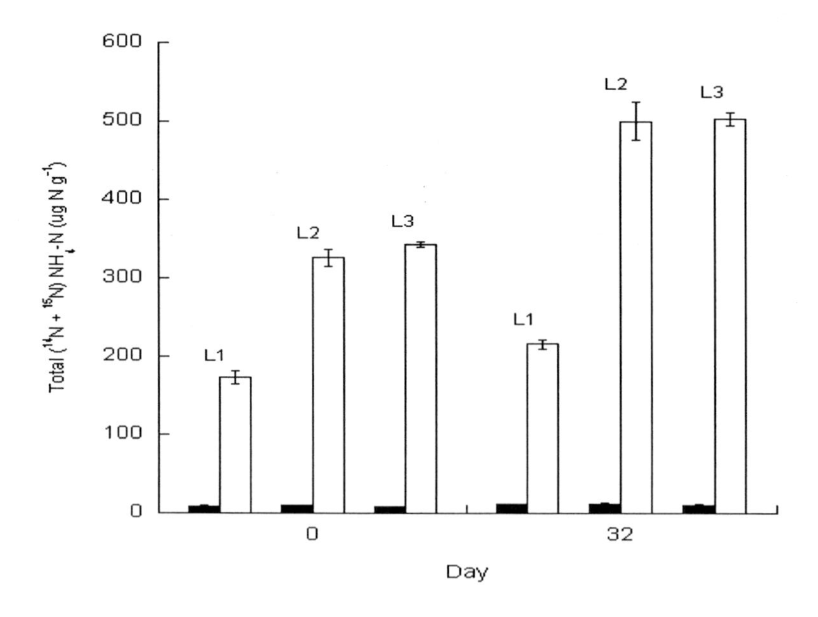

Figure 3. $^{15}NH_4^+$ and total (^{14}N+^{15}N) $^{15}NH_4^+$ in soil layers (L1 = 0-5 cm; L2 = 5-10 cm; L3 = 10-15 cm) of *Glyceria* (filled) and unplanted (no fill) riparian wetland microcosms at the start (Day 0) and finish (Day 32) of Experiment 2. Values are means (± standard error, n = 4).

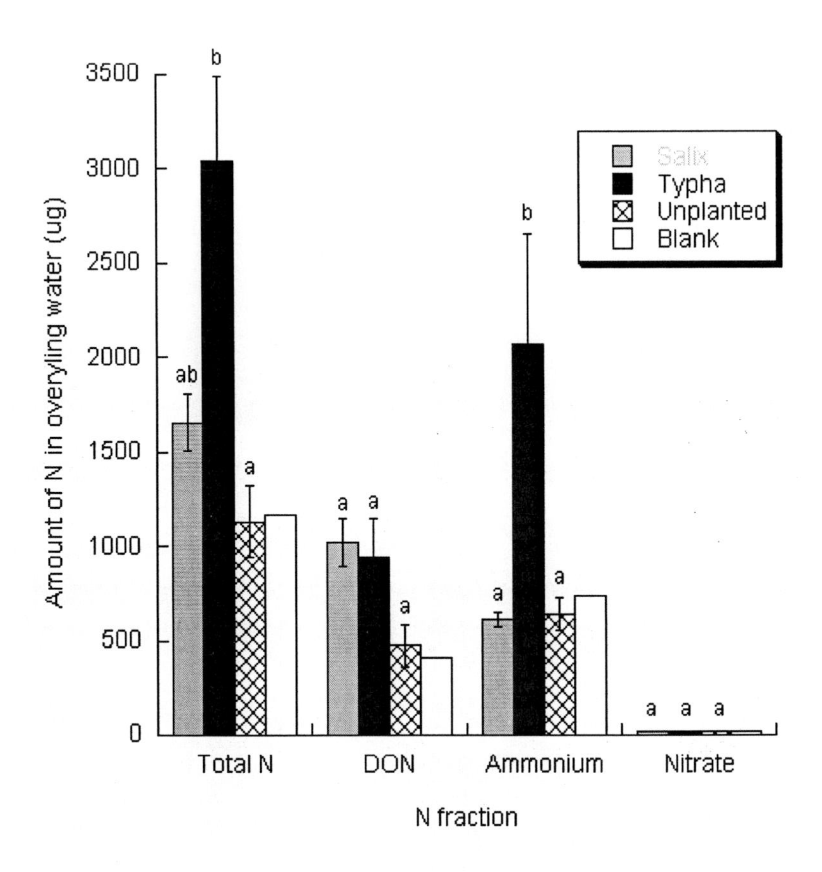

Figure 4. Amount of N ($^{14}N + ^{15}N$) detected in the overlying water layer of Experiment 1 microcosms at the end of the experiment. Values are means (± standard error, n=3) except for the blank microcosm (no ^{15}N added, n=1). For each N fraction, significant differences between microcosms at $p<0.05$ are indicated by different alphabetic superscripts.

As part of Experiment 1 we also investigated the release and transformation of ^{15}N from plant shoot material resulting from simulated litterfall and decomposition (Table 2). Mass balance analysis for the *Typha* litter microcosm found more ^{15}N than was estimated to be present in *Typha* shoots supporting the notion that the leaf clipping technique had underestimated ^{15}N uptake for this plant species. For the *Salix* litter microcosm results show that a large proportion (65%) of the $^{15}NH_4^+$ present in *Salix* shoots was unaccounted for after 14 d and was presumably transformed through sequential mineralisation, nitrification and denitrification processes. This corresponds to a removal rate of ~24 mg N m^{-2} d^{-1}. This result highlights a potentially important plant-mediated mechanism for permanent N removal in wetlands. N assimilated by plant roots in deeper soils is ultimately moved to the soil surface through the uptake and litterfall processes and thereafter may be rapidly transformed in the more microbially-active surface soils. Moreover, the greater N losses associated with denitrification in the *Salix* litter column compared to the litter-less (*Salix*, *Typha* and unplanted) microcosms suggests that the presence of a flocculant substrate (ie. the decaying leaf litter) may facilitate these sequential transformations. Where soils are highly reducing (as demonstrated for the lake edge wetland soils by the redox potential measurements), the aerobic-anaerobic interface may be located in the overlying water as opposed to in the soils. In these circumstances coupling of nitrifying and denitrifying activity is likely to be enhanced by the presence of a water-borne substrate for microbes involved in these transformations. The availability of a water-borne substrate (resuspended sedimented phytoplankton) was also considered important in the regulating of denitrification N losses in a shallow Danish lake (Jensen et al. 1992).

Table 2. Mass Balance of ^{15}N (mg) in the Two Plant Litter Microcosms in Experiment 1 after 14 d

Litter microcosm	Litter	Overlying water	Soil	Total	Amount added[1]	Unaccounted
Salix	0.03	0.34	0.27	0.64	1.85	1.21
Typha	0.02	0.13	0.12	0.27	0.15	+(0.12)[2]

[1] Estimated from ^{15}N analysis of leaf clips collected from planted microcosms at the end of the experiment.

[2] More ^{15}N detected in the microcosm than amount added estimate. ^{15}N in *Typha* leaves probably underestimated due to shoot leaf clipping technique (assimilated N likely to have accumulated in the intercalary meristem as opposed to the shoots).

Earlier in the chapter the usefulness of separate measurements of ^{15}N content for the different soil and water N pools was highlighted. A number of other useful modifications were applied to the design of Experiment 2 based on lessons learnt from Experiment 1. In Experiment 2 we used destructive sampling of replicates, as opposed to repetitive subsampling of a single set of microcosms (ie. leaf-clipping) to enable us to monitor temporal changes. While this meant that many more microcosms had to be set up and analysed, it enabled us to more accurately determine the ^{15}N (and total N) content of the various N pools and avoids any disturbance of the microcosms associated with repetitive sub-sampling. In Experiment 2 we also tried to minimise the disruption and priming of microcosms associated with additions of N substrate. For example, in the case of NO_3^- additions these can potentially oxidise the soil thus favouring the activity of those microbes preferring less reducing conditions (ie. denitrifiers over DNRA microbes). To do this we added ^{15}N in multiple, low-level doses (every 2 days) that only increased ambient levels from 1.1 to 1.6 μg N g^{-1}.

Despite our best efforts, we acknowledge that experimental design is these types of experiments could still be improved. In both of our experiments we attempted to measure ^{15}N gas fluxes from the microcosms by periodically covering with closed chambers but were unsuccessful in doing so due to the high background N concentrations in the atmosphere (~ 80% N). Instead we had to rely upon mass balance analysis to estimate gaseous ^{15}N losses (ie. as a result of denitrification). It is possible to detect gaseous ^{15}N losses if ^{15}N additions and emissions are sufficient to raise the enrichment of atmospheric N above natural abundance levels (e.g. Delaune et al. 1998) or if the background N in the closed chamber atmosphere into which the gas is diffusing can be reduced by flushing with an inert gas such as helium (e.g. Matheson et al. 2003). However, there are issues with both techniques since higher ^{15}N additions may artificially stimulate N transformation rates and altering atmospheric pressure and composition using closed chambers is likely to alter diffusion rates of gases from the soil. In addition there also remains the problem that some gaseous N is likely to remain dissolved in soil porewater and this must also be taken account.

ACKNOWLEDGEMENTS

Support for the research was provided by Environment Bay of Plenty (Exp. 1), the New Zealand Foundation for Research, Science and Technology (Exps. 1 & 2) and the European Commission DG XII-funded NICOLAS project (Nitrogen Control by Landscape Structures in Agricultural Environments, Scientific Advisor H. Barth, Grant no. ENV4-CT97-0395) (Exp. 2).

REFERENCES

Addy, K., Kellogg, D.Q., Gold, A.J., Groffman, P.M., Ferendo, G., Sawyer, C. (2002). In situ push pull method to determine ground water denitrification in riparian zones. *Journal of Environmental Quality 31*, 1017-1024.

Ambus, P., Mosier, A., Christensen, S. (1992). Nitrogen turnover rates in a riparian fen determined by ^{15}N dilution. *Biology and Fertility of Soils 14*, 230-236.

Bengtsson, G., Bergwall, C. (2000). Fate of ^{15}N labelled nitrate and ammonium in a fertilized forest soil. *Soil Biology and Biochemistry 32*, 545-557.

Buresh, R.J., Patrick, W.H. Jr., (1978). Nitrate reduction to NH_4^+ in anaerobic soils. *Soil Science Society of America Journal 42*, 913-918.

Clement, J.C., Holmes, R.M., Peterson, B.J., Pinay, G., (2003). Isotopic investigation of denitrification in a riparian ecosystem in western France. *Journal of Applied Ecology 40*, 1035-1048.

Cooke, J.G. (1994). Nutrient transformations in a natural wetland receiving sewage effluent and the implications for waste treatment. *Water Science and Technology 29*, 209-217.

D'Angelo, E.M., Reddy, K.R. (1993). Ammonium oxidation and nitrate reduction in sediments of a hypereutrophic lake. *Soil Science Society of America Journal 57*, 1156-1163.

Davies, D.B. (2000). The nitrate issue in England and Wales. *Soil Use and Management* 16, 142-144.

Delaune, R.D., Lindau, C.W., Sulaeman, E., Jugsujinda, A. (1998). Nitrification and denitrification estimates in a Louisiana swamp forest soil as assessed by ^{15}N isotope dilution and direct gaseous measurements. *Water Air and Soil Pollution 106*, 149-161.

Devito, K.J., Fitzgerald, D., Hill, A.R., Aravena, R. (2000). Nitrate dynamics in relation to lithology and hydrologic flow path in a river riparian zone. *Journal of Environmental Quality 29*, 1075-1084.

Gunterspergen, G.R., Stearns, F., Kadlec, J.A. (1991). Wetland vegetation. In D.A. Hammer (Ed.), *Constructed wetlands for wastewater treatment* (3rd Edition, pp 73-88).

Hanson, G.C., Groffman, P.M., Gold, A.J. (1994). Denitrification in riparian wetlands receiving high and low groundwater nitrate inputs. *Journal of Environmental Quality 23*, 917-922.

Hemond, H. (1983). The nitrogen budget of Thoreau's bog. *Ecology 64*, 99-109.

Hill, A.R. (1996). Nitrate removal in stream riparian zones. *Journal of Environmental Quality 25*, 743-755.

Jensen, J.P., Jeppesen, E., Kristensen, P., Christensen, P.B., Sondergaard, M. (1992). Nitrogen loss and denitrification as studied in relation to reductions in nitrogen loading in a shallow hypertrophic lake (Lake Sobygard, Denmark). *Internationale Revue der Gesamten Hydrobiologie 77*, 29-42.

Kao, J.T., Titus, J.E., Zhu, W. (2003). Differential nitrogen and phosphorus retention by five wetland plant species. *Wetlands 23*, 979-987.

Lin, J.T., Stewart, V. (1998). Nitrate assimilation by bacteria. *Advances in Microbial Physiology 39*, 2-31.

Lowrance, R., Vellidis, G., Hubbard, R.K. (1995). Denitrification in a restored riparian forest wetland. *Journal of Environmental Quality 24*, 808-815.

Lusby, F.E. 1996. *Nitrogen transformations in a lake-edge wetland.* Unpublished MSc thesis, University of Waikato, New Zealand.

Lusby, F.E., Gibbs, M.M., Cooper, A.B., & Thompson, K. (1998). The fate of groundwater ammonium in a lake-edge wetland. *Journal of Environmental Quality 27*, 459-466.

Maitre, V., Cosandey, A., Desagher, E., Parriaux, A. (2003). Effectiveness of groundwater nitrate removal in a river riparian area: the importance of hydrogeological conditions. *Journal of Hydrology 278*, 76-93.

Matheson, F.E. (2001). *Nitrogen removal and the fate of nitrate in riparian buffer zones.* Unpublished PhD thesis, University of Durham, United Kingdom.

Matheson, F.E., Nguyen, M.L., Cooper, A.B., Burt, T.P., Bull, D.C. (2002). Fate of ^{15}N-nitrate in unplanted, planted and harvested riparian wetland soil microcosms. *Ecological Engineering 19*, 249-264.

Matheson, F.E., Nguyen, M.L., Cooper, A.B., Burt, T.P. (2003). Short-term nitrogen transformation rates in riparian wetland soil determined with nitrogen-15. *Biology and Fertility of Soils 38*, 129-136.

McCarty, G.W., Bremner, J.M. (1992). Regulation of assimilatory nitrate reductase activity in soil by microbial assimilation of ammonium. *Proceedings of the National Academy of Sciences 89*, 453-456.

Mookherji, S., McCarty, G.W., Angier, J.T. 2003. Dissolved gas analysis for assessing the fate of nitrate in wetlands. *Journal of the American Water Resources Association 39*, 381-387.

Moraghan, J.T. (1993). Loss and assimilation of ^{15}N-nitrate added to a North Dakota cattail marsh. *Aquatic Botany 46*, 225-234.

Nijburg, J.W., Laanbroek, H.J. (1997). The fate of ^{15}N-nitrate in healthy and declining Phragmites australis stands. *Microbial Ecology 34*, 254-262.

Puri, G., Ashman, M.R. (1999). Microbial immobilization of ^{15}N-labelled ammonium and nitrate in a temperate woodland soil. *Soil Biology and Biochemistry 31*, 929-931.

Reddy, K.R., Patrick, W.H. Jr. (1984). Nitrogen transformations and loss in flooded soils and sediments. *Critical Reviews in Environmental Control 13*, 273-307.

Rice, C.W., Tiedje, J.M. (1989). Regulation of nitrate assimilation by ammonium in soils and in isolated soil microorganisms. *Soil Biology and Biochemistry 21*, 597-602.

Riddell-Black, D., Alker, G., Mainstone, C.P., Smith, S.R., Butler, D. (1997). Economically viable buffer zones – the case for short rotation forest plantations. In N.E. Haycock, T.P.

Burt, K.W.T. Goulding, G. Pinay (Eds.) *Buffer zones: their processes and potential in water protection* (pp 228-235). Quest Environmental.

Rutherford, J.C., Nguyen, M.L. (2004). Nitrate removal in riparian wetlands: interactions between surface flow and soils. *Journal of Environmental Quality 33*, 1133-1143.

Silver, W.L., Herman, D.J., Firestone, M.K. (2001). Dissimilatory nitrate reduction to ammonium in upland tropical forest soils. *Ecology 82*, 2410-2416.

Tiedje, J.M. (1988). Ecology of denitrification and dissimilatory nitrate reduction to ammonium. In A.J.B. Zehnder (Ed.). *The Biology of Anaerobic Microorganisms* (pp 179-245). Wiley.

In: Progress in Aquatic Ecosystems Research
Editor: A. R. Burk, pp. 99-120
ISBN 1-59454-383-6

Chapter 5

LONG-TERM CHANGES IN NET ECOSYSTEM METABOLISM AND NET DENITRIFICATION IN THE OHTA RIVER ESTUARY OF NORTHERN HIROSHIMA BAY – AN ANALYSIS BASED ON THE PHOSPHORUS AND NITROGEN BUDGETS

Tamiji Yamamoto,* Asako Kubo and Toshiya Hashimoto
Graduate School of Biosphere Sciences, Hiroshima University, Higashi-Hiroshima, Japan
Yoshinori Nishii
Hiroshima Fisheries Experimental Station, Ondo-cho, Aki-gun, Hiroshima, Japan

ABSTRACT

Long-term changes in net ecosystem metabolism (NEM) and net denitrification (ND) in the Ohta River estuary of northern Hiroshima Bay were analyzed based on the phosphorus (P) and nitrogen (N) budgets, using the method recommended by the Land-Ocean Interaction in the Coastal Zone (LOICZ) Working Group. From 1987 to 1997, data were collected as part of several monitoring programs conducted by governmental organizations. The NEM of the estuary was calculated to be 0.52 g C m^{-2} day^{-1} on average, with a positive average value in the upper layer (0.71 g C m^{-2} day^{-1}) and a negative average value in the lower layer (-0.20 g C m^{-2} day^{-1}), thus indicating production in the upper layer and active decomposition in the lower layer. The ND was calculated to be -90 and 72 mg N m^{-2} day^{-1} for the upper layer and lower layer, respectively, suggesting that nitrogen fixation was occurring in the upper layer while denitrification dominated the lower layer, perhaps in the surficial sediments. From the long-term analyses of ND, we detected a shift from the first nitrogen fixation phase (1987-1991) to the denitrification phase (1992-1995), followed by a transition to a second nitrogen fixation phase (1996-1997). These shifts in ND coincided with the trend in NEM, which

* corresponding author: Tel: +81-824-24-7945; Fax: +81-824-24-7945; Email: tamyama@hiroshima-u.ac.jp

showed a significant decrease since 1992, followed by a slight recovery since 1996. We discuss these long-term trends in the context of mitigation efforts undertaken by the Japanese government to reduce the anthropogenic release of P and N into the Ohta River estuary.

Keywords: denitrification, net ecosystem metabolism, nitrogen, phosphorus
Regional Index Terms: Japan; Hiroshima: Hiroshima Bay

INTRODUCTION

The environmental conditions in estuaries are highly variable in space and time, and must be considered when estimating average values for primary production based on in situ measurements. Discrete measurements of primary production are inherently restricted to the time and place at which they were collected. These data cannot reflect environmental variations with timescales of weeks to months, such as changes in weather conditions and the succession of phytoplankton species. For example, nutrient inputs from rivers following rainfall events should significantly affect primary production in estuaries. Discrete measurements are not representative of a given month or season unless the measurements are carried out frequently and according to a sampling scheme that reflects the temporal and spatial variations of the estuarine environment (Yamamoto et al., 1994).

An alternative method that could provide conservative average values of primary production in highly variable aquatic environments is the budgeting approach. The Land-Ocean Interaction in the Coastal Zone (LOICZ) Working Group recommends this approach based on a box model for estuaries where fluctuations in environmental conditions are large in time and space (Gordon et al., 1996). In the LOICZ method, net ecosystem metabolism (NEM) and net denitrification (ND) are estimated. The NEM is the net amount of primary production minus respiration, and the ND is the net amount of denitrification minus nitrogen fixation. To determine whether or not a given estuary is productive, and how it will respond to eutrophication processes or reduction measures for nutrients is a fundamental problem for the management of the ecosystem.

The Ohta River estuary of northern Hiroshima Bay is one of the most enclosed estuaries in the Seto Inland Sea, Japan. Eutrophication of the estuary could be attributed to phosphorus and nitrogen transported from Hiroshima City, which has a population of about 1.1 million. To alleviate the eutrophication, a directive regarding phosphorus reduction was officially put in place since 1980 along with the regulation of chemical oxygen demand (COD) by the establishment of a special law - the Law Concerning Special Measures for Conservation of the Environment of the Seto Inland Sea. Later in 1995, nitrogen was also targeted as another element to be reduced. The parameters required for the box model analyses (i.e. temperature, salinity and nutrient concentrations) have been collected by both the central government and local governments following legislation. These databases have been used to a limited degree by governmental organizations, however, recent disclosure of the information has now made them available for general scientific inquiry.

Yamamoto et al. (2002c) demonstrated that the phosphorus concentration in the Ohta River has been decreased by the governmental phosphorus reduction measure since 1980, which resulted in an increased N/P ratio in the river water. It has been reported that the change in the C: N: P ratio of the phytoplankton community in Hiroshima Bay reflected the reduction of only phosphorus (Yamamoto et al., 2002b). The decline in oyster production in Hiroshima Bay and the decline in fishery production in the entire Seto Inland Sea are clearly shown in statistics. Several reasons have been proposed to explain the causative mechanisms. One factor is the direct effect of reduction measures of phosphorus on primary production and another is the change in ecosystem structure through the alteration in phytoplankton species composition due to the change in the N/P ratio in seawater (Yamamoto, 2002, 2003). The toxic/harmful phytoplankton blooms really have damaged the oyster cultivation in Hiroshima Bay. Given this background, the authors' concern is to know how the Ohta River estuary has responded to the phosphorus and/or nitrogen reduction measures in terms of changes in the NEM and the ND in relation to phytoplankton species succession.

In the present study, the method recommended by the LOICZ Working Group is applied to the Ohta River estuary. We discuss long-term changes in the ecosystem of the Ohta River estuary within the context of governmental mitigation efforts to alleviate eutrophication.

Materials and Methods

Budget analyses of phosphorus (P) and nitrogen (N) for the Ohta River estuary of northern Hiroshima Bay were calculated using data acquired over an approximately 10 year period (from May 1987 to November 1997). Data for both total dissolved forms (TDP and TDN) and particulate forms (PP and PN) are required for estimating net ecosystem metabolism (NEM) and net denitrification (ND), because the transformation of these materials between dissolved forms and particulate forms occurs via oceanic organisms through uptake and decomposition. According to Gordon et al. (1996), utilizing only DIP and DIN is a second best alternative, because the main organisms in coastal areas which uptake dissolved forms in the water column are believed to be phytoplankton, which mainly uptake DIP and DIN. Databases from the government monitoring programs usually have dissolved inorganic forms for the same reason; these substances are utilized by phytoplankton and are closely related to eutrophication problems such as harmful algal blooms. Based on our knowledge that the major paths of transformation of P and N in coastal waters are likely from DIP and DIN to PP and PN, respectively, by phytoplankton (the reverse path applies to decomposition of organic matter via bacteria), and that the remaining portions comprising dissolved organic forms (DOP and DON) are relatively stable in time and space, provides the rational basis for using the DIP and DIN data.

Determination of Diffusion Coefficients

Box model analyses were carried out using the monthly salinity data collected by the Sixth Regional Coast Guard Headquarters (1987-1997). From the vertical profiles of density, we assumed that on average the pycnocline was at a depth of 5 m during the stratified period (cf. Yamamoto et al., 2000a, 2002a, 2002e). A two-layer box model was applied to the Ohta River estuary for an entire year, dividing the water column into an upper layer (extending from the surface to 5 m; BoxU) and a lower layer (ranging from 5 m to the bottom; BoxL). The dimensions of the boxes are summarized in Table 1. Southern Hiroshima Bay and Kure

Bay were treated as boundaries for the model calculations. Average salinity values in BoxU and BoxL were calculated for each month as the arithmetic averages of the upper and the lower layers of five stations, respectively (Figure 1). At each station, the average values were calculated as the geometric means for the upper (0, 3 and 5 m) and lower layers (5, 10, 20 and 30 m; if 30 m depth data is available). We assumed that Etauchi Bay, which is surrounded by Eta Sima and Nishi-Nomi Sima, was closed, due to the narrow width of its mouth (500 m).

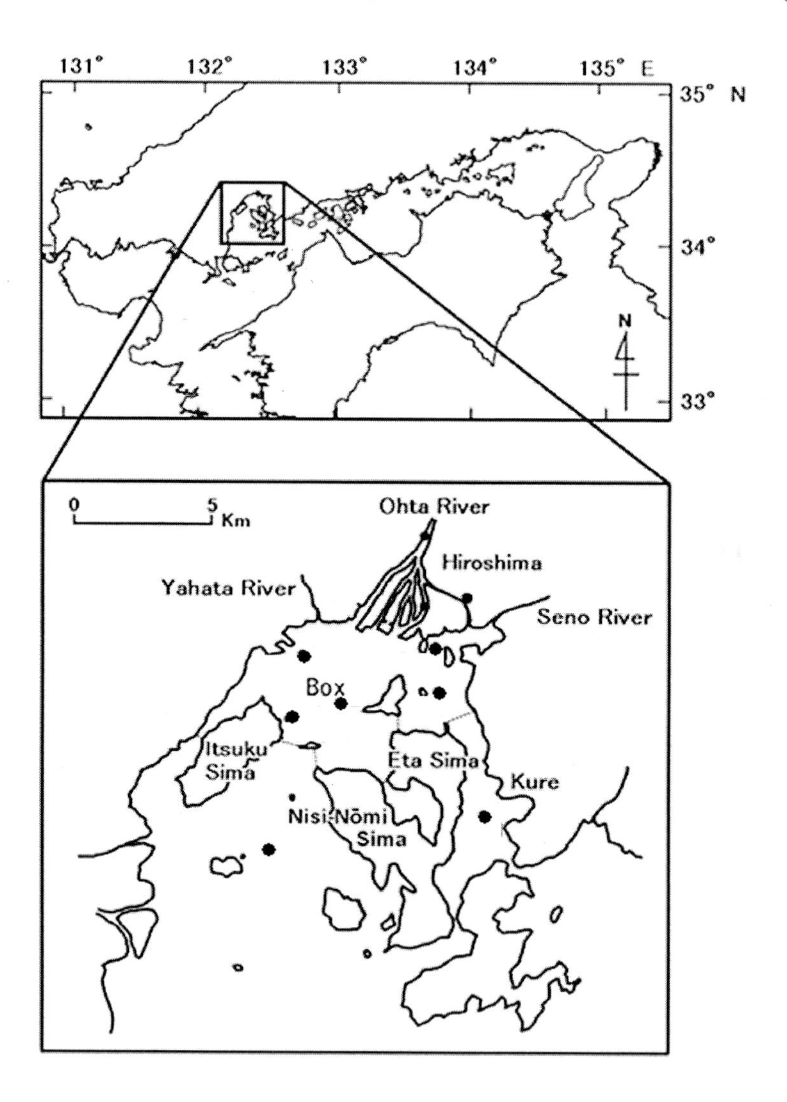

Figure 1. Map showing the location of Hiroshima Bay, Japan. Enlarged map shows the location of sampling stations with major rivers.

In the model, water exchange between the boxes and between the boxes and the boundaries was considered to occur through advection and diffusion (Figure 2). The temporal changes of salinity in BoxU and BoxL are expressed in the following equations.

$$V_u \frac{S_{1u,i+1}-S_{1u,i}}{dt} = K_{h12u,i}\frac{S_{2u,i}-S_{1u,i}}{dx_{12}}A_{h12u} + K_{h13u,i}\frac{S_{3u,i}-S_{1u,i}}{dx_{13}}A_{h13u} - \left(Q_{R,i}+Q_{P,i}+Q_{EC,i}-Q_{E,i}\right)S_{1u,i} + K_{v,i}\frac{S_{1l,i}-S_{1u,i}}{dz}A_v + Q_{EC,i}\times S_{1l,i} \quad (1a)$$

$$V_l \frac{S_{1l,i+1}-S_{1l,i}}{dt} = K_{hl2l,i}\frac{S_{2l,i}-S_{1l,i}}{dx_{12}}A_{hl2l} + K_{hl3l,i}\frac{S_{3l,i}-S_{1l,i}}{dx_{13}}A_{hl3l} + Q_{EC,i}S_{2l,i} + K_{v,i}\frac{S_{1u,i}-S_{1l,i}}{dz}A_v - Q_{EC,i}S_{1l,i} \quad (1b)$$

where, V_u and V_l are the volumes of BoxU and BoxL (m^3), respectively, A_{hu} is the cross sectional area between BoxU and the boundary, and A_{hl} is the cross sectional area between BoxL and the boundary (m^2). Subscripts indicate upper layer (u), lower layer (l), box (1), the southern boundary (2), and the Kure Bay boundary (3). A_v is the area between BoxU and BoxL (m^2), and S_i, S_{i+1} are the observed salinity at month 'i' and the following month 'i+1'. K_{hu}, K_{hl} and K_v are the horizontal diffusion coefficients and vertical diffusion coefficient (m^2 sec^{-1}), respectively. dx is the horizontal distance between the boxes and the corresponding boundaries, and dz is the vertical distance between BoxU and BoxL (m). Q_R, Q_P, Q_{EC} and Q_E are the water volumes supplied per unit time by the rivers, precipitation, estuarine circulation and evaporation (m^3 sec^{-1}), respectively. The K_{hu}, K_{hl} and K_v were determined simultaneously, changing serially from $1.0\text{x}10^0$ to $1.0\text{x}10^3$ m^2 sec^{-1} for K_h and from $1.0\text{x}10^{-5}$ to $1.0\text{x}10^{-2}$ m^2 sec^{-1} for K_v, with a time step of 600 sec, so that the difference between the estimated and observed salinities in the following month were minimized.

The Ohta River is the major river emptying into the estuary, and data on freshwater discharge were cited from reports of the Japan River Association (1989-1999). The station where the flow rate of river water is monitored is located at ca. 15 km upstream from the river mouth. The flow rate was multiplied by 1.11 according to Yamamoto et al. (1996), in order to include the catchment area downstream of the station. Freshwater discharge from two other small rivers, the Seno River and the Yahata River (Figure 1), were also included in the calculation of freshwater discharge. This was done by multiplying the amount of freshwater discharge from the Ohta River by the areal ratio in the tributary areas of these rivers to that of the Ohta River (Yamamoto et al., 2000a). Data on precipitation measured at Hiroshima City were cited from the annual reports issued by the Japan Meteorological Agency (1988-1998). Evaporation was estimated from wind velocity, air temperature, and humidity using the formula provided by the Japanese Society of Construction (1985).

The estuarine circulation driven by discharge from the Ohta River was considered an important advective term as suggested by Yamamoto et al. (2000a) as follows.

$$U_h = 4.852\log\left(Q_{OR}\right) - 7.788 \quad (2)$$

where, U_h is the observed current velocity (m sec^{-1}) at the center of the lower layer, Q_{OR} is the volume inflow from the Ohta River (m^3 sec^{-1}), indicating the non-linear relationship between the river inflow and the current driven velocity. In the case of negative values for U_h, this term was assumed to be zero. Since the cross sectional area between BoxL and the southern boundary is $73.5\text{x}10^3$ m^2, the volume entrained by the estuarine circulation (Q_{EC}) was estimated as,

$$Q_{EC} = U_h \times 73.5 \times 10^3 . \tag{3}$$

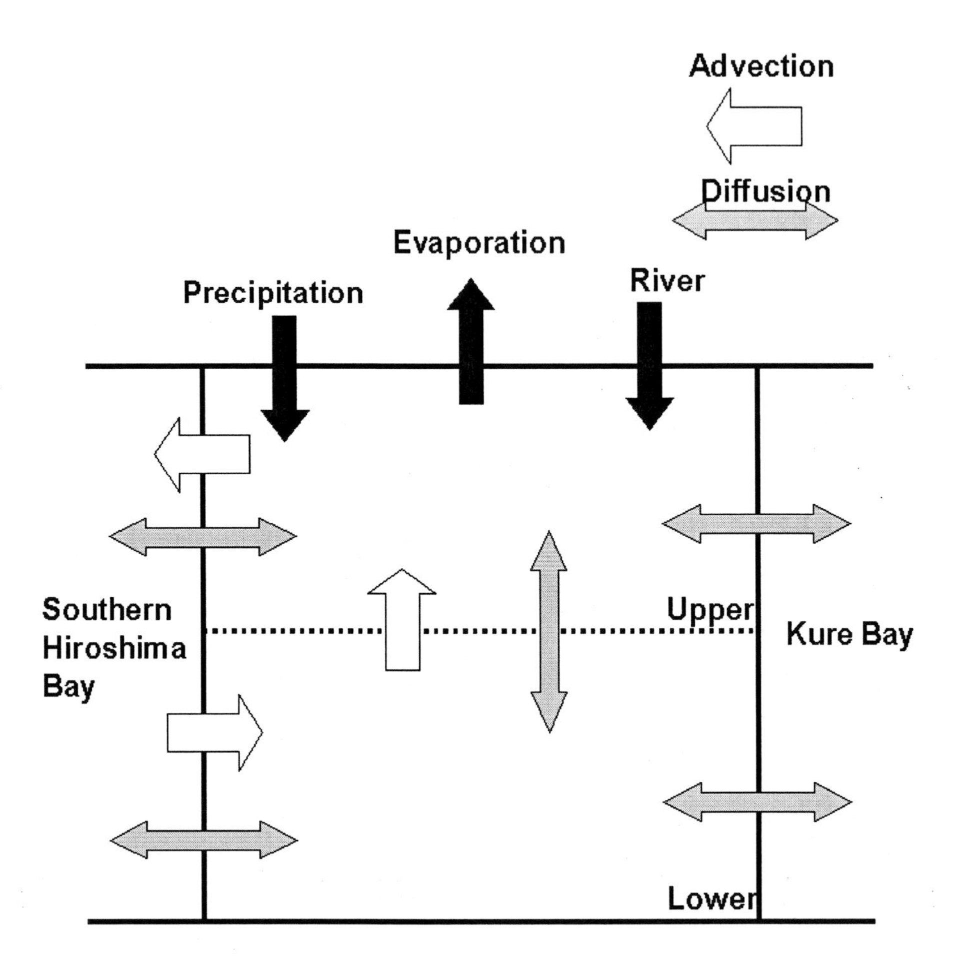

Figure 2. Schematic diagram of the box model used in the present study.

We did not consider the estuarine circulation for the Kure Bay area, because the freshwater discharged from the Ohta River flows to the right (northwest part of the northern Hiroshima Bay) in a density current with the Coriolis force (Yamamoto et al., 2000a). The upwelling velocity from BoxL to BoxU (U_v) was also calculated as,

$$Uv = Q_{EC} / Av . \tag{4}$$

The residence time of water (T_R; days) in the estuary was calculated as,

$$T_R = \left(\frac{V_u + V_l}{Q_R + Q_P + Q_{EC} - Q_E} \right) \tag{5}$$

Estimation of Net Ecosystem Metabolism (NEM)

Using the diffusion coefficients ($K_{h12u,i}$, $K_{h13u,i}$, $K_{h12l,i}$, $K_{h13l,l}$, and $K_{v,i}$) obtained above, DIP and DIN budgets in the system were calculated. The time rate of change of DIP or DIN in BoxU and BoxL is expressed by the following equations (Eqs. 6a and 6b). From these equations, the amount of particulate phosphorus or particulate nitrogen transformed from DIP or DIN per unit time (P in mg at sec^{-1}) was estimated.

$$V_u \frac{C_{1u,i+1} - C_{1u,i}}{dt} = K_{h12u,i} \frac{C_{2u,i} - C_{1u,i}}{dx_{12}} A_{h12u} + K_{h13u,i} \frac{C_{3u,i} - C_{1u,i}}{dx_{13}} A_{h13u} + Q_{R,i} C_{R,i} + Q_{P,i} C_{P,i} + Q_{EC,i} C_{1l,i}$$

$$+ Q_{E,i} C_{1u,i} + K_{v,i} \frac{C_{1l,i} - C_{1u,i}}{dz} A_v - (Q_{R,i} + Q_{P,i} + Q_{EC,i} - Q_{E,i}) C_{1u,i} - P_{1u,i} \quad (6a)$$

$$V_l \frac{C_{1l,i+1} - C_{1l,i}}{dt} = K_{h12l,i} \frac{C_{2l,i} - C_{1l,i}}{dx_{12}} A_{h12l} + K_{h13l,i} \frac{C_{3l,i} - C_{1l,i}}{dx_{13}} A_{h13l} + K_{v,i} \frac{C_{1u,i} - C_{1l,i}}{dz} A_v + Q_{EC,i} C_{2l,i}$$

$$- Q_{EC,i} C_{1l,i} + B_i A_v - P_{1l,i} \quad (6b)$$

where C is the observed concentration of DIP or DIN (mg at m^{-3}) in the boxes and in the boundaries. The concentrations in the rivers and precipitation have subscripts R and P, respectively. B is the benthic flux of DIP or DIN from a unit area of the sediment surface (mg at m^{-2} s^{-1}).

The values of C in BoxU and BoxL were estimated by averaging the values from five stations in the upper 5 m layer and the lower layer at which the water samples were collected at 0, 5, 10, 20 and 30 m (Hiroshima Fisheries Experimental Station, 1988-1998) by the trapezoidal procedure, as was done for salinity. Concentrations of DIP or DIN in the Ohta River water were cited from Yamamoto et al. (2002c). Concentrations of DIP or DIN from two small rivers were cited from Hiroshima Prefecture (1988-1998). The DIP loads from the industrial factories situated along the coastline of the estuary are taken into consideration, by multiplying the values for the rivers by 2.0 for DIP and 1.3 for DIN, according to Nakanishi (1977). Since these coefficients are those estimated in 1972 when the Japanese economy was booming, applying the same coefficient values to the entire period of our analyses may cause an overestimate of the loads in recent years of economic decline. Concentrations of DIP and DIN in rainwater were cited from Yuasa (1994). For the benthic flux of DIP and DIN, the values that were measured seasonally in 1996 and 1997 (Yamamoto et al. 2000b) were applied to the entire period of the analyses, although it might be a source of error for the other years. The contribution of the benthic flux to the total budget was substantially smaller than other sources as shown later.

P in equations 6a and 6b is the net rate that accounts for processes mainly due to utilization by phytoplankton and release by decomposition of organic matter by bacteria. Although each process is not identifiable in the present approach, it can be said that particulate production is prevailing when P has a positive value, while particulate decomposition is prevailing when P is negative. Assuming that the major primary producers in the water column of coastal areas are phytoplankton, we can apply the Redfield ratio $(C/P)_{phyto}$ conversion factor (=106; Redfield, 1934) to the DIP decrease/increase (ΔDIP) to estimate the carbon-based production, referred to as "net ecosystem metabolism (NEM)" (mg C m^{-2} d^{-1}).

$$NEM = -\Delta DIP \times (C/P)_{phyto} \quad (7)$$

NEM is calculated from the DIP budget, but not from DIN budget because of its inclusion of gas phases as explained below.

Estimation of Net Denitrification (ND)

Nitrogen is very different from phosphorus in terms of the utilization of gaseous states such as N_2 and N_2O in biogeochemical cycles. In the biogeochemical cycling of nitrogen, the production and consumption of such gasses are generally the result of microbial activities. For example, under reduced conditions some groups of *Pseudomonas* use NO_3 as an oxygen source instead of O_2 and produce N_2 as a byproduct; this process is called "denitrification" (Seitzinger, 1988). There are also groups of blue-green algae which utilize N_2 gas as a nitrogen source for their growth. It is well known that *Trichodesmium* is capable of a process called "nitrogen fixation" in open oceans where nitrogenous nutrients are usually deficient (Capone et al., 1997). *Microcystis*, a freshwater microalgae that has the ability to fix N_2 gas, forms noxious blooms in lakes and ponds, causing serious contamination of drinking water (Sakamoto, 1973). Furthermore, there are abundant small-sized planktonic blue-green algae in seawater called *Synechococcus* that have been shown to perform nitrogen fixation (e.g. Xiuren et al., 2000).

Since there is no gaseous exchange of phosphorous between seawater and the atmosphere, the difference between the amounts of nitrogen and phosphorus transformed from DIN and DIP in the seawater could be attributed to "net denitrification" (ND; denitrification minus nitrogen fixation) (Gordon et al., 1996).

$$ND = -\Delta DIN - (-\Delta DIP) \times (N/P)_{part} \quad (8)$$

Nitrogen fixation is said to be prevailing if the DIN decrease is smaller than the DIP decrease, while denitrification is said to be prevailing for the opposite scenario.

RESULTS

Advection and Diffusion Fields of the Ohta River Estuary

The amount of precipitation is usually high in early summer and low in winter reflecting the monsoon weather patterns in this area (Figure 3, top panel). In the summers of 1993 and 1997, record high precipitation was observed; 646 and 458 mm/month, respectively. There was a drought in the summer of 1994. The evaporation showed a seasonal variation of increasing from autumn to winter and decreasing from spring to summer. The average amount of freshwater supply by precipitation during the entire period of analysis was 2.0 m^3 sec^{-1}, and is comparable to the average value for evaporation which was 1.7 m^3 sec^{-1}. The river water discharge was highly variable between 9.1-380 m^3 sec^{-1} (avg. 80 m^3 sec^{-1}), and peaked in the summer of 1989, 1993, 1995 and 1997 (Figure 3, middle panel). Record low precipitation occurred from July to December of 1994. The contribution by rivers to total freshwater input into the estuary was estimated to be 93%, and that by precipitation was only 7%.

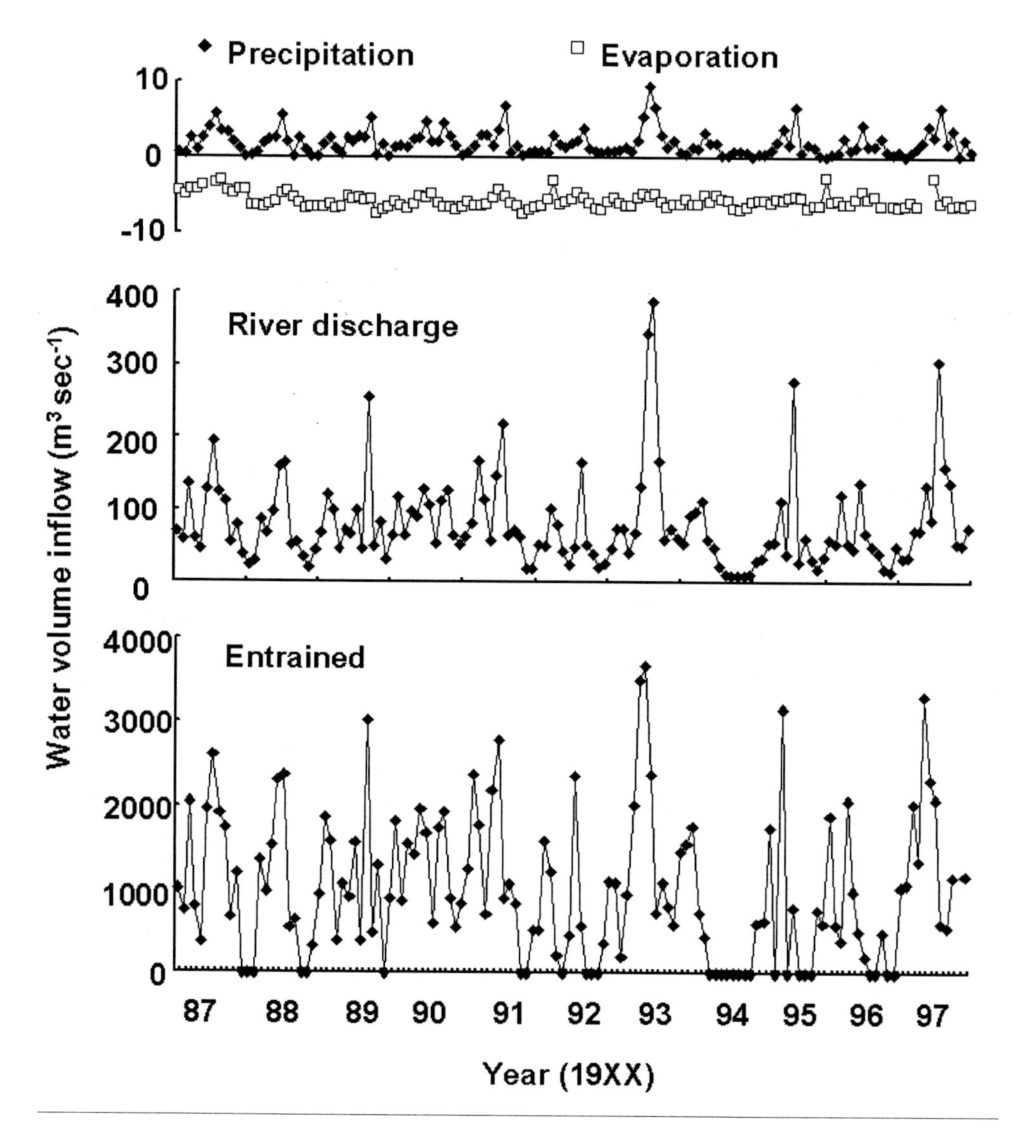

Figure 3. Temporal variations in observed precipitation, and estimated evaporation (top), and observed river inflow (middle) from 1987 to 1997. Entrained water volume due to estuarine circulation is also shown (bottom).

The amount of seawater entrained by the estuarine circulation driven by the river discharge varied from 0 to 3700 m^3 sec^{-1}, and the average was 1000 m^3 sec^{-1} (Figure 3, bottom panel); this amounts to about 12 times the river water discharge. Yamamoto et al. (2000a) have preliminarily reported that the volume entrained by the estuarine circulation was about 7 times the river water discharge for 1993-1995. Their underestimation might be a result of the drought in 1994. The estimated velocity of horizontal advection (*Uh*) ranged from 0 to 29×10^{-2} m sec^{-1} in the upper layer and the average was 4.3×10^{-2} m sec^{-1}. In the lower layer, there were 0-5.0×10^{-2} m sec^{-1} and 0.7×10^{-2} m sec^{-1}, respectively. The estimated upward velocity, *Uv*, was 0.071×10^{-2} m sec^{-1} on average (range: 0-0.26×10^{-2} m sec^{-1}).

The salinity values in the upper and the lower layers calculated from the salinities of the previous month using equations (1a) and (1b) could be fitted to the observed values of the next month, except for several months that showed extremely low salinities due to heavy rains (e.g. August 1987, July 1990, July 1991 and August and September 1993) (Figure 4). The accuracy of our calculations is limited to a minimum resolution of one month because the sampling interval was monthly. Actually, the estimated average residence time of seawater in the estuary is 24 days, using equation (5). This could be a source of inconsistency between the calculated values and the observed values when the system received large amounts of freshwater input in a short period of time.

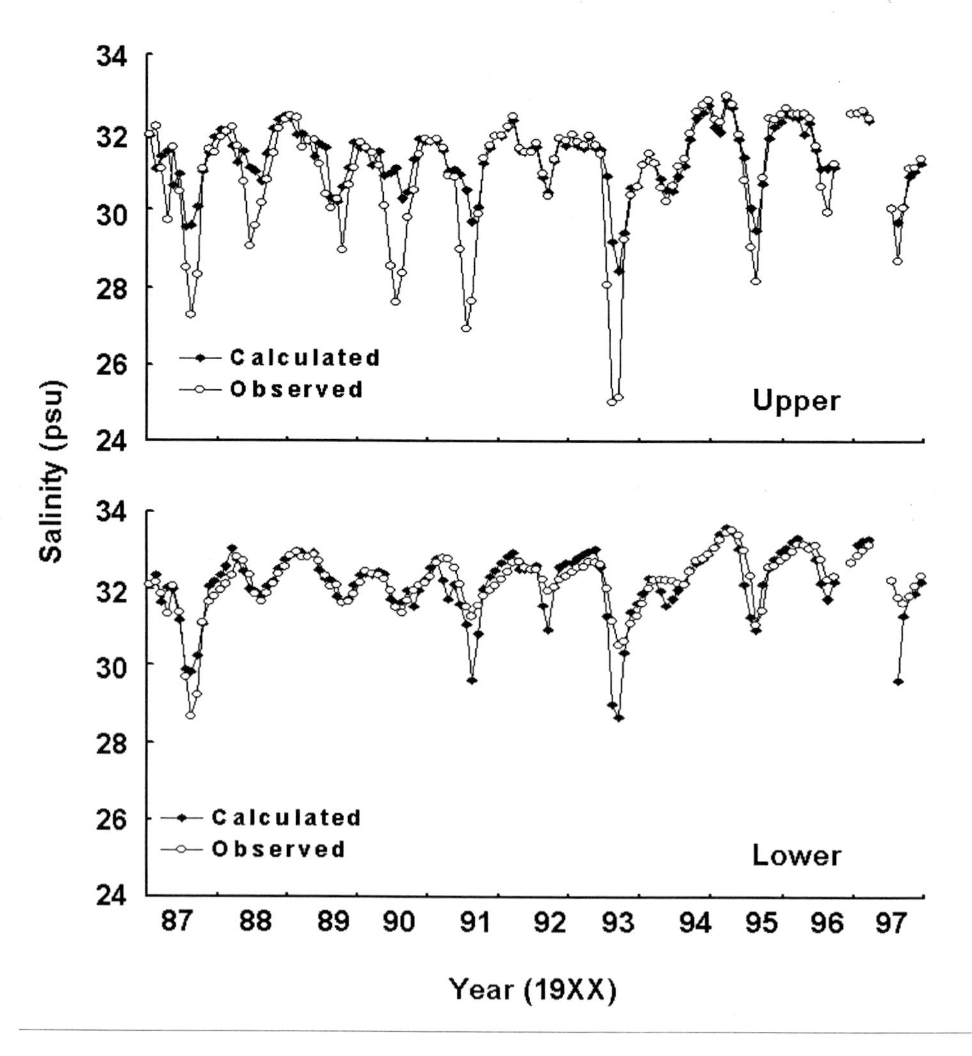

Figure 4. Comparison between calculated salinity and observed salinity in the upper and lower boxes from 1987 to 1997.

The estimated values for the vertical diffusion coefficients (*Kv*) are in the range of 0.1-25.0×10^{-4} m^2 sec^{-1}, and showed an obvious trend that increased in winter and decreased in summer, which reflects mixing of the water column in winter and stratification of the water column in summer. However, specific temporal trends were not clear in the horizontal diffusion coefficients (*Kh*) with the exception of those between BoxL and the lower layer of the southern boundary; it showed higher values in summer and lower values in winter. *Kh* was estimated to be 1.0-11×10^2 m^2 sec^{-1}.

DIP and DIN Budgets

The concentration of DIP in the estuary ranged between 0.01-1.02 mg at P m^{-3} with an average of 0.32 mg at P m^{-3} (Figure 5, top panel). The concentration of DIN was in the range of 0.3-12.3 mg at N m^{-3} with an average of 3.7 mg at N m^{-3} (Figure 5, bottom panel). During the period of the analyses, an increasing trend was clearly observed for both DIP and DIN (Kendall's rank correlation coefficients for a 6-month running average were $\tau=0.644$, $z=10.341$, $p<0.0001$ for DIP and $\tau=0.466$, $z=7.485$, $p<0.0001$ for DIN, respectively).

DIP loads from different sources into the estuary are shown in Figure 6. The DIP load through the rivers ranged from 0.25-8.0 ton P day^{-1}, and the average for the entire period of the analyses was 1.2 ton P day^{-1}. The DIP supplied through the rivers has significantly decreased over the 10 years of the study period (Kendall's rank correlation for a 6-month running average; $\tau=0.285$, $z=4.559$, $p<0.0001$). This is attributed to the phosphorus reduction measures that started in 1980 following the governmental directive (Yamamoto et al., 2002c). The DIP load by precipitation was 1.2x10^{-3} ton P day^{-1} on average (range: 0.09x10^{-3} to 5.8x10^{-3} ton P day^{-1}), which was three orders of magnitude lower than that of riverine input, indicating that the DIP input from the atmosphere is negligible in the total phosphorus cycle for the system. The amount of DIP entrained by the process of estuarine circulation varied from 0 to 6.2 ton P day^{-1} with an average of 0.87 ton P day^{-1}, which is up to 73 % of the DIP input from the river. The transportation of DIP by horizontal diffusion showed a negative value, -0.25 ton P day^{-1}, on average, indicating net outflow of DIP from the estuary to the southern area by this process. The inter-annual variation shows a trend whereby net inflow was prevailing before June of 1992 and turned to net outflow afterwards with a range of -6.6-4.9 ton P day^{-1}. This is consistent with the phenomenon that the concentration of DIP has become higher since 1992 as shown in Figure 5, and suggests that a net loss of DIP was occurring due to outflow especially in winter when utilization by phytoplankton does not occur. DIP was likely transported from Kure Bay with an average 0.18 ton P day^{-1}, suggesting that Kure Bay is a source of DIP for northern Hiroshima Bay.

The DIN input from the atmosphere varied from 0.01 to 0.59 ton N day^{-1} with an average of 0.12 ton N day^{-1} (Figure 7), which was exactly two orders higher than that of the phosphorus input from the same source. Nitrate and ammonium gasses can be dissolved in rainwater, thus the atmospheric input of nitrogen was much higher compared to phosphorus. The benthic flux of DIN was small compared to the total budgets as was also found for DIP. The DIN input from the rivers was estimated to be 8.3 ton N day^{-1} on average with a range of 0.97-35.7 ton N day^{-1}. It also showed a significant decreasing trend (Kendall's rank correlation for a 6-month running average; $\tau=0.452$, $z=6.902$, $p<0.0001$) as found for DIP. Governmental regulation of nitrogen started in 1995. This could be the major cause of decline in the DIN load from the rivers. The DIN input by the estuarine circulation was estimated to be 4.0 ton N day^{-1} on average (range: 0-27.7 ton N day^{-1}), and this accounts for 48% of the riverine DIN input. The DIN was transported out to the southern area since 1992 (Figure 7), which was also the case for the DIP budgets.

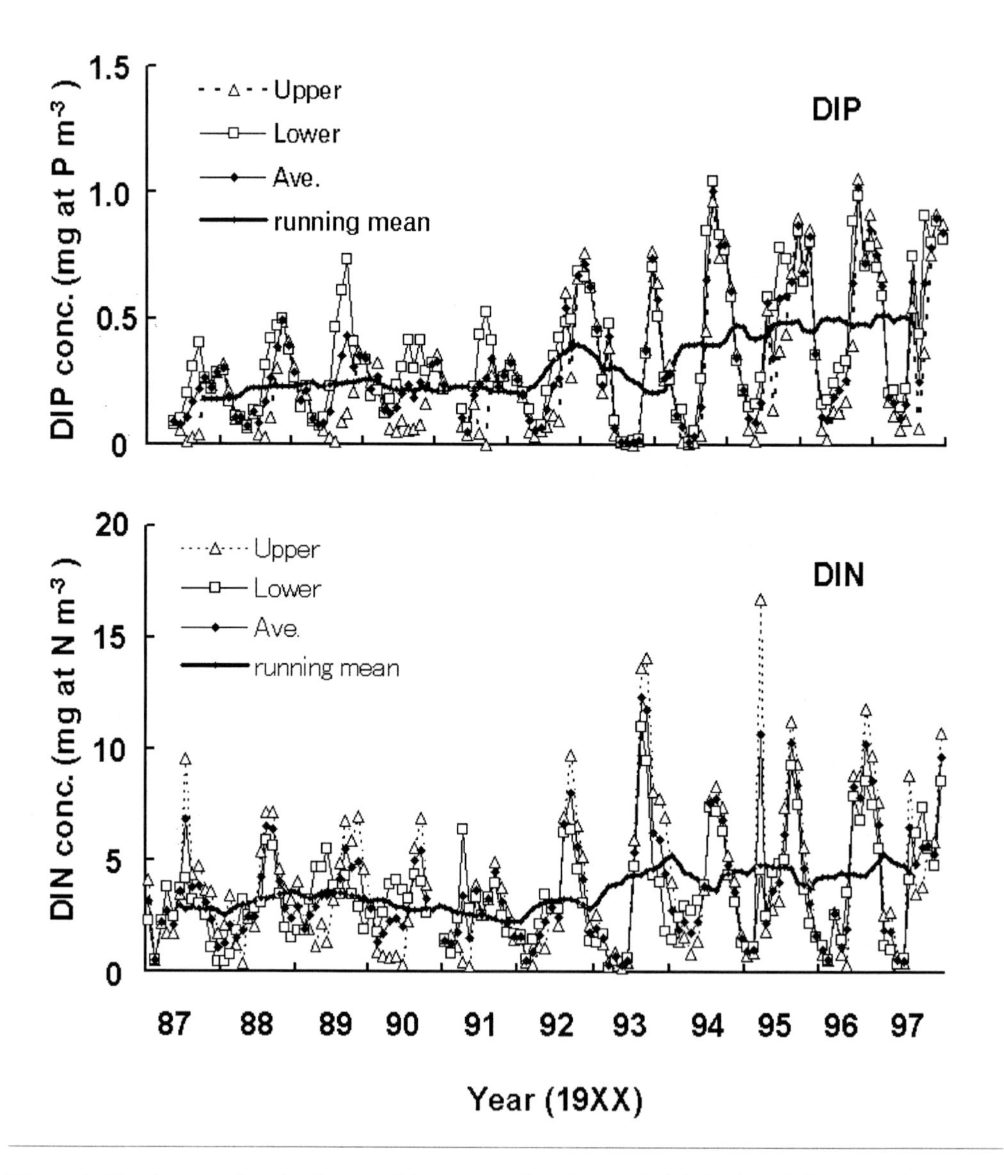

Figure 5. Yearly variations in dissolved inorganic phosphorus (DIP) (top) and dissolved inorganic nitrogen (DIN) (bottom) concentrations in northern Hiroshima Bay water during 1987-1997. The thick line shows 6-month running mean.

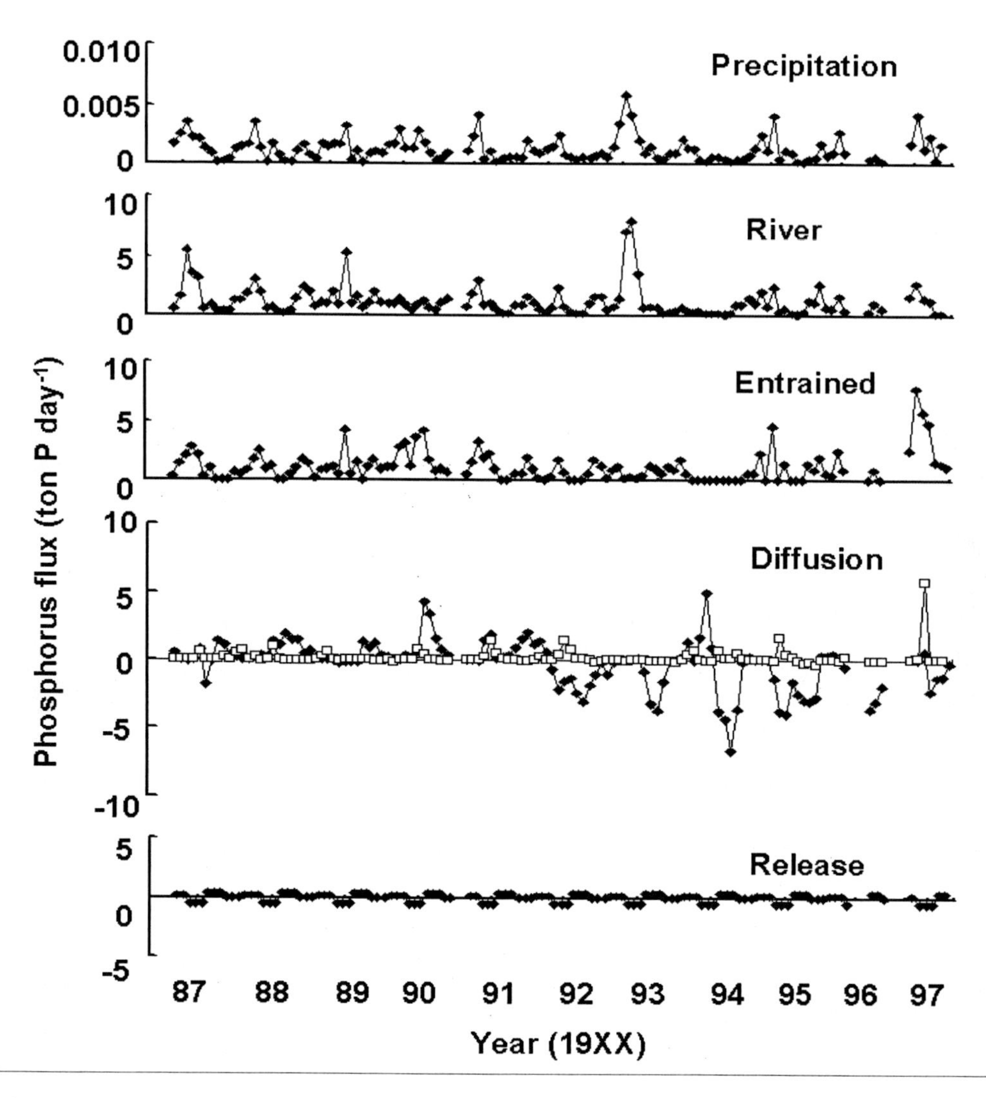

Figure 6. Yearly variations of phosphorus inputs from various sources to northern Hiroshima Bay during 1987-1997. In diffusion, the filled lozenges and the open squares denote the exchange to the southern boundary and to the Kure Bay boundary, respectively. Negative is outflow to the boundaries.

Net Ecosystem Metabolism

The NEM calculated for BoxU ranged from -1.1 to 5.5 g C m^{-2} day^{-1}, and the average was 0.71 g C m^{-2} day^{-1} (Figure 8). In BoxL, the NEM ranged from -3.9 to 1.6 g C m^{-2} day^{-1} with an average of -0.20 g C m^{-2} day^{-1}. This means that production is prevailing in the upper layer while decomposition is prevailing in the lower layer, and the total is 0.52 g C m^{-2} day^{-1} on average (range: -1.6-3.0 g C m^{-2} day^{-1}), indicating that the Ohta River estuary has been productive as a whole.

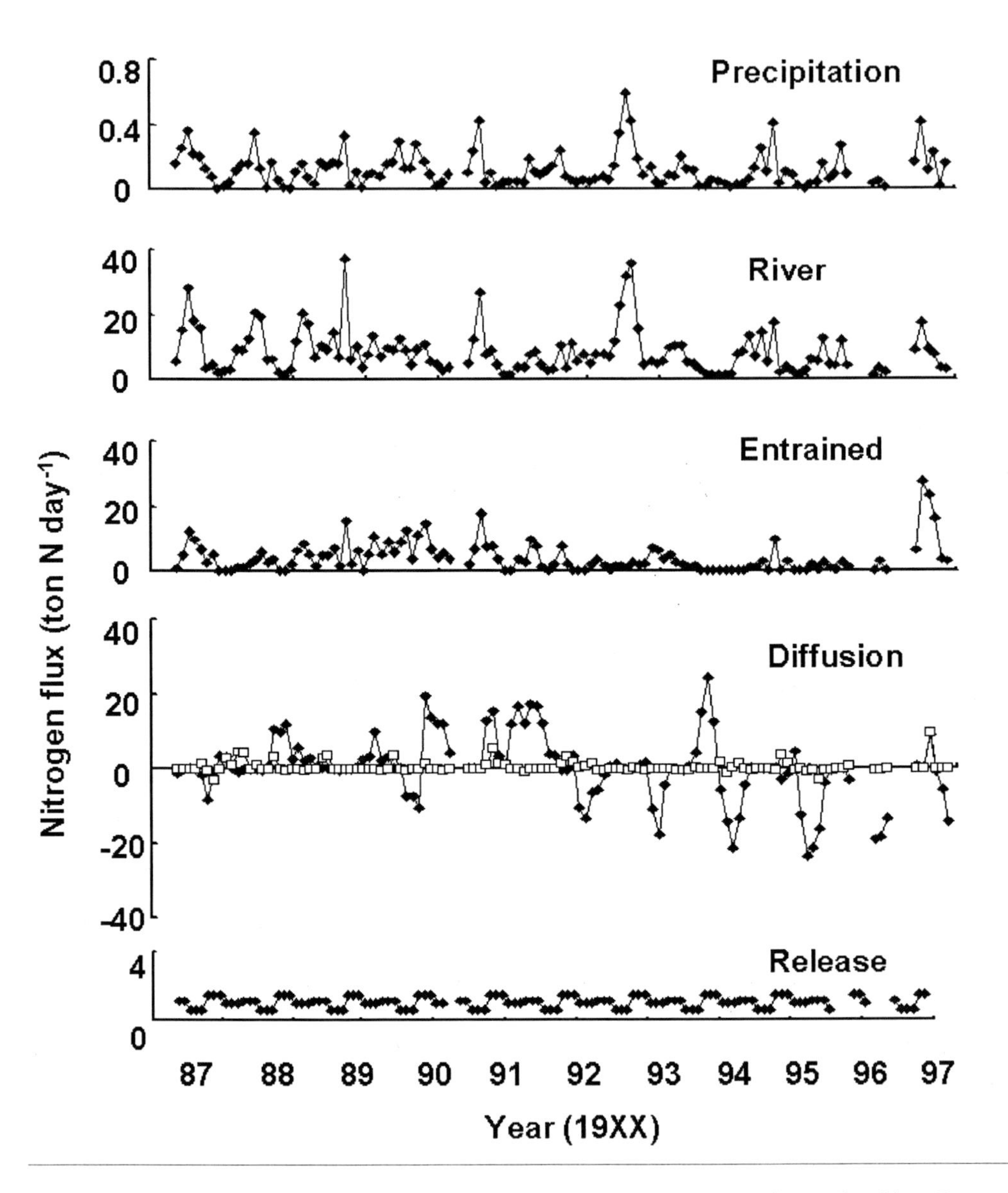

Figure 7. Yearly variations of nitrogen inputs from various sources to northern Hiroshima Bay during 1987-1997. In diffusion, the filled lozenges and the open squares denote the exchange to the southern boundary and to the Kure Bay boundary, respectively. Negative is outflow to the boundaries.

The NEM was high from spring to summer and low in winter (Figure 8). This could be a result of the seasonal variation of DIP inflow that generally increases in spring to summer due to the monsoon rains. Our calculations also show reasonable results for annual climatic changes. For example, in 1993 and 1997, we had large amounts of rain in the summer. On these occasions, the NEM clearly have higher total values that were not restricted to the upper layer (Figure 8). In the summer of 1994 during the drought, the NEM showed negative values in total accompanied by values approaching zero in the upper layer.

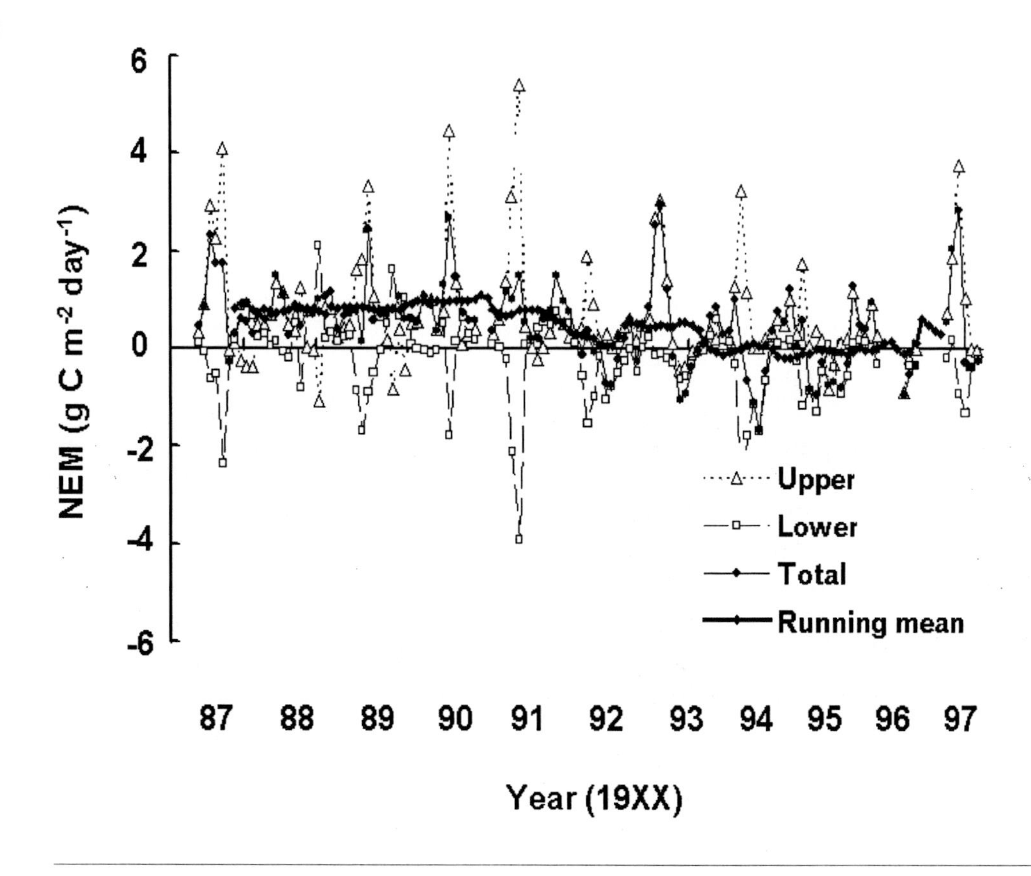

Figure 8. Yearly variations of estimated net ecosystem metabolism (NEM) in northern Hiroshima Bay during 1987-1997. The thick line shows 6-month running mean.

The NEM also showed a significant decrease since 1992 (Kendall's rank correlation for a 6-month running average for the entire period; $\tau=0.540$, $z=8.558$, $p<0.0001$) with values approaching zero from 1994 to 1996, and then a small increase from 1996 (Figure 8).

Net Denitrification (ND)

The calculated net denitrification (ND) varied from -660 to 46 mg N m^{-2} day^{-1} in BoxU and from -80 to 590 mg N m^{-2} day^{-1} in BoxL, with average values during the entire analytical period of -90 and 72 mg N m^{-2} day^{-1}, respectively (Figure 9). This indicates that nitrogen fixation was prevailing in the upper layer while denitrification was prevailing in the lower layer. In the whole water column, it varied between -210 and 170 mg N m^{-2} day^{-1}, and the grand average was -14 mg N m^{-2} day^{-1}, showing net nitrogen fixation on the whole.

The variation in the 6-month running means shows that there were roughly three phases; Phase I: the first nitrogen fixation phase from 1987 to 1991, Phase II: the denitrification phase from 1992 to 1995, and Phase III: the second nitrogen fixation phase from 1996 to 1997 (Figure 8). The increasing trend in ND from Phase I to II and the decreasing trend from Phase

II to III were statistically significant; Kendall's rank correlation coefficients were $\tau=0.480$, $z=6.546$, $p<0.0001$ for the former and $\tau=0.771$, $z=5.871$, $p<0.0001$, respectively.

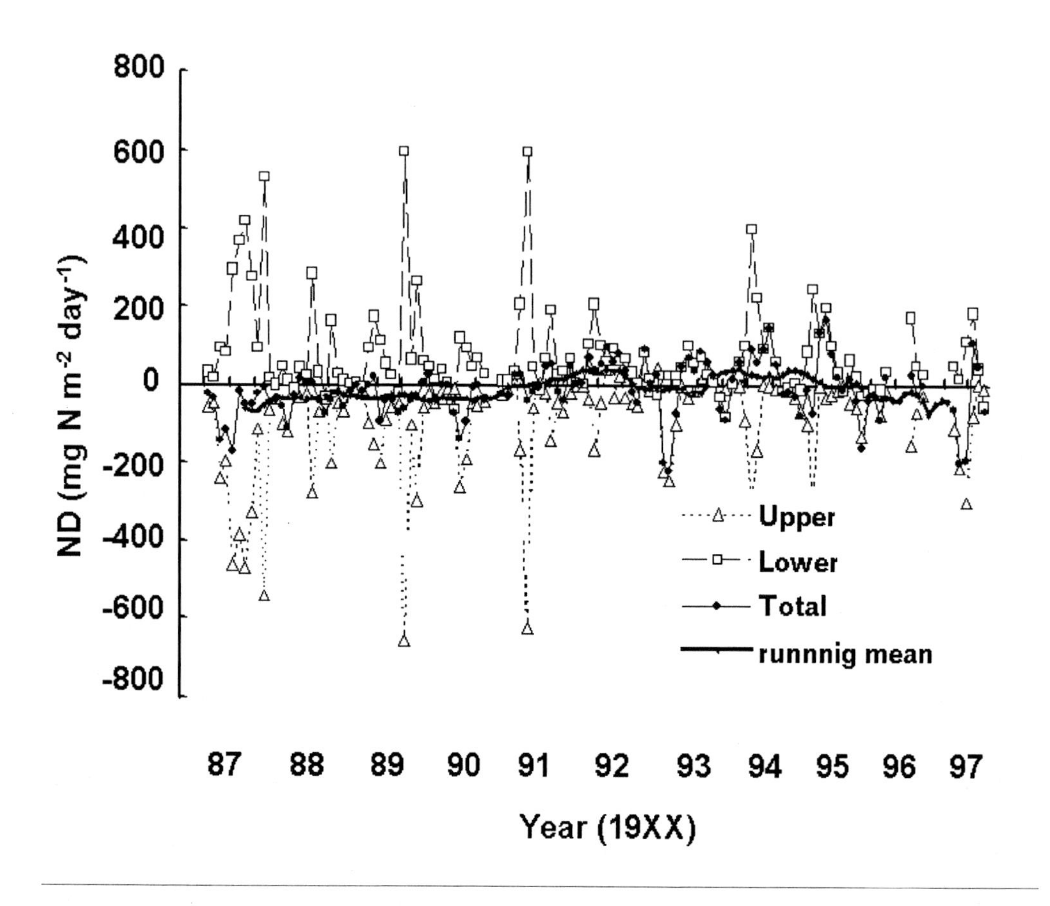

Figure 9. Yearly variations in estimated net denitrification (ND) in northern Hiroshima Bay during 1987-1997. The thick line shows 6-month running mean.

Discussion

An average ND rate of -14 mg N m^{-2} day^{-1} was obtained for the Ohta River estuary, which indicates net nitrogen fixation. Nitrogen fixation in estuaries and coastal areas is reported to range from 0.2 to 7.5 mg N m^{-2} day^{-1} where the higher extremes are the cases where there are microbial mats and symbiotic microbes that are able to fix nitrogen (Howarth, 1988). However, recent research carried out in San Fransisco Bay showed that 95% of the picophytoplankton consisted of cyanobacteria which may be able to fix nitrogen (Xiuren et al., 2000). It is also known that some bacteria that are parasitic on vertebrates and invertebrates possess nitrogenase activity (Jonathan et al., 2000). In the Baltic Sea, a high nitrogen fixation rate (3-36 mg N m^{-2} day^{-1}) has been reported recently (Rahm et al., 2000). The average value (14 mg N m^{-2} day^{-1}) estimated here seems to be reasonable in the light of this recent research. A possible explanation for the stronger nitrogen fixation compared to denitrification in this estuary is the limited supply of nitrate (Kim et al., 1997) to act as a major electron donor in denitrification. The supply rate of nitrate is expected to be suppressed particularly in the summer because of the low oxygen concentration in the bottom layer in summer (often <2 mg l^{-1}; Hashimoto et al., 1994), which is known to be oxygen-deficient water, and would inhibit the nitrification process that produces nitrate from ammonia (Seitzinger, 1988).

There are several positive peaks (denitrification dominated) and negative peaks (nitrogen fixation dominated) found in 1987, 1989 and 1991 which might be errors due to the longer sampling interval (one month) compared to the residence time of water (calculated as 24 days in this study). If these values are omitted, we still obtain a negative value (-7.8 mg N m^{-2} day^{-1}) on average. This suggests that nitrogen fixation is slightly stronger than denitrification in the water column in the Ohta River estuary. Using this value and the surface area of the estuary, the net nitrogen fixation is estimated to be 1.1 ton N day^{-1}, which corresponds to ca. 10% of the total DIN input of 12.4 ton N day^{-1}. Only one research paper is available on nitrogen fixation in the Ohta River estuary so far (Lee et al., 1996). According to this study, the nitrogen fixation was zero in their measurements from 1993 to 1994. If we calculate the ND for the same period, we can get 6.6 mg N m^{-2} day^{-1}, suggesting that denitrification could actually have taken place at this rate during this period. Only one report is available on the denitrification rate in the estuary (Kim et al., 1997). They reported that the denitrification varied between 0-23.2 mg N m^{-2} day^{-1} in their seasonal measurements conducted from the summer of 1994 to the spring of 1995. Since the ND was estimated for the lower layer at 29.9 mg N m^{-2} day^{-1} for the same period in the present study, our calculation on ND does not significantly deviate from the measured range.

The average NEM value of 0.52 g C m^{-2} day^{-1} is a conservative value. The average value of primary production was measured by the ^{13}C method 4 times from autumn 1993 to summer 1994, and was 1.013 g C m^{-2} day^{-1} (Hashimoto et al., 1997). In our present analysis, the NEM for the same period was 0.17 g C m^{-2} day^{-1}. According to Gordon et al. (1996), the seasonal average NEM amounts to about 10% of primary production, because NEM is the net value of primary production minus respiration and decomposition. Given that Hiroshima Bay is more productive than other areas, our calculation of the NEM is considered to have provided reasonable values.

The temporal variation of the NEM (Figure 8) and that of the ND (Figure 9) seem to be coincident with each other. The change in the ND from Phase I (nitrogen fixation) to Phase II (denitrification) corresponds with the decrease in the NEM which began in 1992. Phase III (nitrogen fixation) of the ND corresponds to the recovery period of the NEM from 1996. This could be explained by the phosphorus reduction measures enacted since 1980 and the additional reduction measures for nitrogen since the end of 1995. The reduction of phosphorus is expected to have decreased NEM and might have accelerated the denitrification process with excess nitrogen. Nitrogen reduction from the end of 1995 could have shifted the system to the prevalence of nitrogen fixation again. These relationships suggest that the microbial community in the estuarine system is very sensitive to the artificial regulation of these elements and responds quickly to changes in nutrient supply.

The underlying mechanisms responsible for the steady increase in the concentrations of DIP and DIN in northern Hiroshima Bay seem to be quite complicated (Figure 5). The direct causes are not major sources accompanied by climatic or physical processes, i.e. the riverine input and associated estuarine circulation and diffusion. As shown in Figs. 6 and 7, the riverine input has in fact been decreasing in both DIP and DIN, and the entrained amounts of DIP and DIN have also been decreasing. Diffusion can be a cause of increases in DIP and DIN when the concentrations in the southern boundary are higher than those in the estuary. However, the pattern of transportation of these materials by diffusive processes is simply the result of the increase in their concentrations in the northern area, and the resulting outflow from the estuary to the boundary.

In addition to the change of the system in terms of NEM and ND, the long-term reduction of phosphorus since 1980 also appears to have been effective in reducing the number of algal blooms in the estuary (Yamamoto et al., 2002c). Furthermore, diatom-dominance of the algal community before 1992 was overtaken by dinoflagellates thereafter with a decrease in the total number of occurrences. In addition to the direct effect of the DIP reduction on the NEM, the change in species composition would have led to the decrease of NEM due to the generally low growth rates of dinoflagellates.

According to the scientific literature, several dinoflagellate species can utilize DOP for their growth because of the simultaneous requirement of carbon as an energy source to compensate for their high energy consumption due to swimming, while diatoms generally utilize only DIP. The dominant dinoflagellate species forming blooms in Hiroshima Bay were *Gymnodinium mikimotoi* and *Alexandrium tamarense* with *Heterocapsa circularisquama* having also appeared since 1992 and 1995. These dinoflagellates possess an enzyme, alkaline phosphatase, that can decompose organic phosphorus (DOP) to DIP. For example, *A. tamarense* prefers to utilize DOP rather than DIP (Oh et al., 2002). In addition, the maximum uptake rate, ρ_{max}, is much higher for ATP (5.0 pmol cell^{-1} hr^{-1}), which is a DOP commonly found in seawaters, than for DIP (1.4 pmol cell^{-1} hr^{-1}), although the half-saturation constant is higher for ATP (5.63 μM) than that for DIP (2.60 μM) (Yamamoto and Tarutani, 1999; Oh et al., submitted). The half-saturation constant for DIP of *H. circularisquama* is 1.90 μM (Yamamoto, unpublished), and this species is known to utilize a variety of DOP (Yamaguchi et al., 2001), although the kinetic parameters for DOP have not been investigated yet. The predominance of dinoflagellates and resultant inefficient utilization of DIP is considered to be a plausible cause for the increase in DIP in the estuarine water as already suggested by Yamamoto et al. (2002b). Although their affinity for DIN and DON has not been investigated yet, a similar explanation can be expected for nitrogen.

Compared to the large contribution to the phosphorus budget by the estuarine circulation (73% of the riverine input), the contribution to DIN circulation is low (48%). This is due to the difference in the DIN/DIP ratios in the river water and the seawater in the southern boundary area. Summarizing all the data used in this study, the average DIN/DIP ratio in the river water is calculated to be 20, which is obviously higher than the value of 12 for the seawater in the southern boundary area. This means that an increase in the river water discharge would lead to an excess of DIN, while an increase in estuarine circulation or diffusion would cause an excess in DIP in the Ohta River estuary. Since the intensity of the estuarine circulation is accelerated with the freshwater discharge in a curvilinear fashion (Yamamoto et al., 2000a), it is not easy to assess the outcome without a computer model. Future directions for this study include the development of a numerical model that can account for the mechanisms by which the Ohta River estuary responds to the river input and the estuarine circulation.

A change in the DIN/DIP ratio in the seawater could change the species composition in the estuary, and this would further alter the total ecosystem structure. Changes in ecosystem structure due to a change in the elemental ratio is exemplified by the whole lake experiments (Schindler, 1977). Similarly, we suggest that this phenomenon may also apply to estuaries. For example, the biomass of jellyfish is reported to have been increasing in the western part of the Seto Inland Sea which is located next to Hiroshima Bay (Uye, 2002). In the Ohta River estuary, the ratio of nitrogen to phosphorus would be the essential factor determining the ecosystem structure because there are sufficient quantities of silicate for the growth of phytoplankton in the seawater of the estuary (Yamamoto et al., 2002b). From the viewpoint of conserving the ecosystem of the Ohta River estuary, regulating the N/P loading ratio will be of paramount importance to control the ecosystem structure and to maintain sustainable production.

Several shortcomings of the present analyses should also be pointed out. The data used in this study were collected by several public organizations. Most of these monitoring programs do not take the residence time of water into consideration in the design of their sampling scheme. Theoretically, the sampling interval should be determined in accordance with the residence time of water because it is the most important physical parameter that governs the total aquatic ecosystem; however, most of the data were collected on a monthly interval. As a result, these analyses cannot account for short-term fluctuations such as flushing events of river water which usually occur with at least a one week timescale. The most important contribution from these kinds of analyses is to quantify the long-term variations and the average values of the NEM and ND of the system, which are quite valuable considering that field measurements of primary production, nitrogen fixation and denitrification are discrete measurements as well as labor intensive.

ACKNOWLEDGEMENTS

We thank Emeritus Professor Osamu Matsuda of the Graduate School of Biosphere Sciences, Hiroshima University, Japan for his valuable contributions to discussions concerning this work. We also thank Miss Catherine Brown of the Department of Oceanography, Dalhousie University, Canada for her critical reading and correction of our English. Financial support was provided by a fellowship program for Japanese scholars and researchers to study abroad from the Ministry of Education, Culture, Sports, Science and Technology of Japan.

REFERENCES

Capone, D.G.; Jonathan, P.Z., Hans, W.P., Birgitta, B., & Edward, J.C. (1997). *Trichodesmium*, globally significant marine cyanobacterium. *Science 276*, 1221-1229.

Gordon, D.C., Boudreau, P. R., Mann, K. H., Ong, J. E., Silvert, W. L., Smith, S.V., Wattayakorn, G. Wulff, F., & Yanagi, T. (1996). *LOICZ Biogeochemical Modelling Guidelines, LOICZ Report and Studies, No.5*, 96 pp.

Hashimoto, T., Matsuda, O., Yamamoto, T., & Yonei, Y., (1994). Oceanographic characteristics of Hiroshima Bay. - Seasonal and spatial variation from 1989 to 1993.

Journal of Faculty of Applied Biological Science, Hiroshima University 33, 9-20. (in Japanese with English abstract).

Hashimoto, T., Yamamoto, T., Tada, K., Matsuda, O., & Nagasue, T. (1997). Primary production and physical structure of the Seto Inland Sea. *Bulletin on Coastal Oceanography 35*, 109-114. (in Japanese with English abstract).

Hiroshima Fisheries Experimental Station (1988-1998). *Oceanographic surveys in the region of Hiroshima Prefecture. Annual reports of Hiroshima Fisheries Experimental Station, FY1987-1997.* (in Japanese).

Hiroshima Prefecture (1988-1998). *Report on the Water Quality of the Areas for Public Use, 1987-1997.* (in Japanese).

Howarth, W. H., Marino, R., & Lane, J. (1988). Nitrogen fixation in freshwater, estuarine, and marine ecosystems. 1. Rates and importance. *Limnology and Oceanography 33*, 669-687.

Japanese Society of Construction (1985). *Hydrographic Formulae.* Japanese Society of Construction, Tokyo, 625 pp. (in Japanese).

Japan River Association (1989-1999). *Annual Report on the Major River in Japan During 1987-1997.* (ed.) The River Bureau, Ministry of Construction, Japan. (in Japanese).

Japan Meteorological Agency (1988-1998). *Annual Report on the Weather of Japan in 1987-1997.* (in Japanese).

Jonathan, P. Z., Carpenter, E. J., & Villareal, T. A. (2000). New perspective on nitrogen-fixing microorganisms in tropical and subtropical oceans. *Trends in Microbiology 8*, 68-73.

Kim, D.-H., Matsuda, O., & Yamamoto, T. (1997). Nitrification, denitrification and nitrate reduction rates in the sediment of Hiroshima bay, Japan. *Journal of Oceanography 53*, 317-324.

Lee, Y. S., Seiki, T., Mukai, T., Takimoto, K., & Okada, M. (1996). Limiting nutrient of phytoplankton community in Hiroshima Bay, Japan. *Water Research 30*, 1490-1494.

Nakanishi, H. (1977). General report on water pollution in Seto Inland Sea. *Yamaguchi Sangyo Igaku Nenpo 22*, 16-33. (in Japanese).

Oh, S.J., Yamamoto, T., Kataoka, Y., Matsuda, O., Matsuyama, Y., & Kotani, Y. (2002). Utilization of dissolved organic phosphorus by the two dinoflagellates: *Alexandrium tamarense* and *Gymnodinium catenatum* (Dinophyceae). *Fisheries Science 68*, 416-424.

Rahm, L., Jonsson, A., & Wulff, F. (2000). Nitrogen fixation in the Baltic Proper: empirical study, *Journal of Marine Systems 25*, 239-248.

Redfield, A. C. (1934). On the proportions of organic derivatives in sea water and their relation to the composition of plankton. *James Johnstone Mem. Vol.* (pp. 177-192). Liverpool.

Sakamoto, M. (1973). Eutrophication in fresh water. In Japanese Society of Fisheries Science (Ed.), *Eutrophication in Aquatic Environments and Aquaculture* (pp. 9-28). Tokyo, Japan: Kouseisha-Kouseikaku. (in Japanese).

Schindler, DW. (1977). Evolution of phosphorus limitation in lakes. *Science* 195, 260-262.

Seitzinger, S. P. (1988). Denitrification in freshwater and coastal marine ecosystems: Ecological and geochemical significance. *Limnology and Oceanography 33*, 702-724.

Sixth Regional Coast Guard Headquarters (1987-1997). *Kaiyo Gaiho* (Report on sea conditions of Hiroshima Bay). *No. 1-12.* (in Japanese).

Uye, S., Fujii, N., & Takeoka, N. (2002). Unusual aggregations of the scyphomedusa *Aurelia aurita* in coastal waters along western Shikoku, Japan. *Plankton Biology and Ecology 50*, 17-21.

Xiuren N., Cloern, J. E., & Cole, B. E. (2000). Spatial and temporal variability of picocyanobacteria *Synechococcus* sp. in San Francisco Bay. *Limnology and Oceanography 45*, 695-702.

Yamaguchi, M., Itakura, S., & Uchida, T. (2001) Nutrition and growth kinetics in nitrogen- or phosphorus-limited cultures of the 'novel red tide' dinoflagellate *Heterocapsa circularisquama* (Dinophyceae). *Phycologia 40*, 313-318.

Yamamoto, T. (2002) Proposal of mesotrophication through nutrient discharge control for sustainable fisheries. *Fisheries Science 68*, 538-541.

Yamamoto, T. (2003) The Seto Inland Sea-Eutrophic or oligotrophic? *Marine Pollution Bulletin 47*, 37-42.

Yamamoto, T., Hashimoto, T., & Tada, K. (1994) Primary production supporting the fisheries of the Seto Inland Sea, Japan. *Scientific Forum of the Seto Inland Sea 6*, 150-153. (in Japanese).

Yamamoto, T., Hashimoto, T., Tarutani, K., & Kotani, Y. (2002a) Effects of winds, tides and river water runoff on the formation and disappearance of the *Alexandrium tamarense* bloom in Hiroshima Bay, Japan. *Harmful Algae 1*, 301-312.

Yamamoto, T., Hashimoto, T., Tsuji, K., Matsuda, O., & Tarutani, K. (2002b) Spatial and temporal variations of biophilic elements in Hiroshima Bay, Japan, during 1991-2000, with special reference to the deviation of phytoplanktonic C: N: P ratio from the Redfield ratio. *Bulletin of Coastal Oceanography 39*, 163-169. (in Japanese with English abstract).

Yamamoto, T. Ikeda, H. Hara, T., & Takeoka, H., (2000b) Applying heat and mass balance theory to determine the flow rate for the measurement of benthic material flux in a flow-through system. *Hydrobiologia 435*, 135-142.

Yamamoto, T., Ishida, M., & Seiki, T. (2002c) Long-term variation in phosphorus and nitrogen concentrations in the Ohta River water, Hiroshima, Japan as a major factor causing the change in phytoplankton species composition. *Bulletin of Japanese Society of Fisheries Oceanography 66*, 102-109. (in Japanese with English abstract).

Yamamoto, T., Kitamura, T., & Matsuda, O. (1996) Riverine inputs of fresh water, total nitrogen and total phosphorus into the Seto Inland Sea. *Journal of Faculty of Applied Biological Science, Hiroshima University 35*, 81-104. (in Japanese with English abstract).

Yamamoto, T., Seike, T., Hashmoto, T., & Tarutani, K. (2002e) Modelling the population dynamics of the toxic dinoflagellate *Alexandrium tamarense* in Hiroshima Bay, Japan. *Journal of Plankton Research 24*, 33-47.

Yamamoto, T., & Tarutani, K. (1999) Growth and phosphate uptake kinetics of the toxic dinoflagellate *Alexandrium tamarense* from Hiroshima Bay. *Phycological Research 47*, 27-32.

Yamamoto, T., Tarutani, K., Kawahara, M., & Oh, S.J. (1999) Utilization and excretion of dissolved organic phosphorus by *Aexandrium tamarense* (Hiroshima Bay Strain). *Journal of Faculty of Applied Biological Science, Hiroshima University 38*, 151-159. (in Japanese with English abstract).

Yamamoto, T., Yoshikawa, S., Hashimoto, T., Takasugi, Y., & Matsuda, O., (2000a) Estuarine circulation processes in the northern Hiroshima Bay, Japan. *Bulletin on Coastal Oceanography 37*, 111-118. (in Japanese with English abstract).

Yuasa, I., (1994). Influence of rain for the eutrophication of estuary and coastal embayment. *Man and Environment 16*, 161-164. (in Japanese).

In: Progress in Aquatic Ecosystems Research
Editor: A. R. Burk, pp. 121-137
ISBN 1-59454-383-6

Chapter 6

CONTAMINANT REMOVAL PROCESSES AND DESIGN OF HORIZONTAL SUBSURFACE FLOW CONSTRUCTED WETLANDS

Joan García,* Paula Aguirre,
Eduardo Álvarez and Rafael Mujeriego
Environmental Engineering Division. Department of Hydraulics,
Coastal and Environmental Engineering. School of Civil Engineering.
Technical University of Catalonia, Barcelona. Spain
Josep M. Bayona and Laura Ortiz
Department of Environmental Chemistry. Instituto de Investigaciones Químicas y Ambientales.
Consejo Superior de Investigaciones Científicas, Barcelona. Spain

ABSTRACT

The objective of this paper is to review current knowledge of the design of horizontal subsurface flow constructed wetlands planted with common reed when constructed to reach secondary treatment. Observations included in that paper are based on our own previous studies and those to be found in the literature. TSS removal in these systems is usually high and effluents have concentrations lower than 35 mg/L. To avoid clogging and ensure effluent quality the surface TSS loading rate should be lower than 20 g $TSS/m^2.d$. BOD_5 removal is highly dependent on the surface organic loading rate, which should be lower than 4-6 g $BOD_5/m^2.d$ to reliably obtain effluents with a concentration of less than 25-30 mg/L. Nutrient removal is not significant (10 to 40%) when the systems are designed for secondary treatment. Microbial removal ranges from 2 to 4 log-units for faecal bacteria and 1 to 3 for faecal viruses. Further improvements to both nutrient and microbial removal require additional steps beyond those provided by an unique horizontal subsurface flow constructed wetlands. The use of small sized granular medium does not appear to have significant advantages in pollutant removal. For this reason the

* Author for correspondence: joan.garcia@upc.es

use of media ranging from 3 to 8 mm is recommended, to avoid clogging and allow plant growth. With regard to bed sizing, the apparent hydraulic conductivity should be reduced by a factor of 10 to 100 to guarantee long-term performance. The aspect ratio is not critical (assuming it is above 1:1), and long beds do not show advantages in pollutant removal. The impact of water depth on water treatment is still not well known. Our results indicate that shallow beds (27 cm) yield better results than deep beds (50 cm).

Key words: wastewater, constructed wetlands, subsurface flow, reed beds, organic matter, design.

INTRODUCTION

Subsurface flow (SSF) constructed wetlands (CWs) are beds that are usually dug in the ground, lined, filled with a granular medium (gravel with a diameter usually between 3 and 20 mm) and planted with macrophytes. In horizontal SSF CWs the wastewater flows from the inlet zone through the bed in contact with the granular medium and the roots and rhizomes of the macrophytes. Treated water exits the system by means of a perforated drainage pipe located on the bottom of the bed at the end opposite the water inlet. The outlet pipe controls the level of the water in the bed, that is usually maintained approximately 5 cm below the surface. The water level is usually set at a value that ranges between 0.4 and 0.6 m. The inlet and outlet zones are filled with stones or coarse gravel (50 to 100 mm in diameter) to ensure water distribution and collection.

In SSF CWs water treatment is performed mainly by the biofilm that grows on the granular medium and on the roots and rhizomes of the macrophytes (Polprasert *et al.*, 1998). This biofilm is responsible for the decomposition of the organic matter. It is also involved in the removal of nutrients, especially nitrogen, although nutrient removal is not significant in systems designed only for TSS and organic matter removal (Tanner, 2001). Surface phenomena seem to be the most important processes for phosphorus, faecal bacteria and virus removal in this type of systems (Kadlec and Knight, 1996). During the 1980s and 1990s there was much controversy on the role of the macrophytes in the treatment, but it is now widely accepted by the scientific community that they have a direct albeit minor influence on the removal processes (Tanner, 2001; US EPA, 2000). For example, it is commonly accepted that macrophytes only remove 10-20% of the nitrogen load.

The main application of SSF CWs is the treatment of urban wastewater from small rural communities to a secondary level. These systems commonly involve pretreatment, primary treatment in a septic tank or a similar unit process, and secondary treatment in the beds (with one or more steps). Simplicity of operation, low costs of operation and maintenance, and the lack of sludge production (at least in the secondary step) are characteristics that make SSF CWs especially suitable for small communities. SSF CWs are well known in some countries in central and northern Europe, and are becoming very popular in other European countries, especially in places where the sanitation programmes for small communities are still incomplete, as is the case in Spain (García *et al*, 2001; Haberl *et al.*, 1995).

In the mid-1990s the Environmental Engineering Division of the Technical University of Catalonia initiated a series of research projects on horizontal SSF CWs in order to obtain design criteria specifically for local conditions, due to the lack of experience in this field in Spain. This paper is a review of current knowledge on the capacity of SSF CWs for contaminant removal when they are designed to reach secondary treatment. The final objective is to offer design criteria derived from our experiences, and contrast them with the commonly accepted standards in the literature.

METHODS

The data used in this paper comes from three small treatment plants. One system treats the domestic wastewater of Can Massaguer youth hostel (Sant Feliu de Buixalleu, Barcelona). This system was put into operation in 1997 to treat 6 m^3/d and consists of a septic tank and a SSF CW with two beds in parallel (total surface area of 172 m^2) containing gravel (D_{60}=4 mm, C_u=2). The second system treats the domestic wastewater from a hotel called Hostal del Carme (Vilagrassa, Lleida). This system was started up in 1999 to treat 11 m^3/d, and is formed by two septic tanks in series and a SSF CW with two parallel beds (total surface area of 375 m^2) containing gravel (D_{60}=4 mm, C_u=1.8). The third system was constructed in 2001 to treat a design flow of 20 m^3/d produced in the Can Suquet housing development (Les Franqueses del Vallès, Barcelona). This system is a pilot plant and was specially designed to check the influence of hydraulic load, aspect ratio (length to width), depth and the size of the granular medium on treatment. Wastewater is screened and flows to an Imhoff tank for primary treatment (Figure 1). This tank is connected with another tank in series with a pump, and from there wastewater is channelled to a distribution chamber and split into eight by means of a weir. Between the pump and the distribution chamber there is a valve and flowmeter. The wastewater is then conveyed to eight beds of equal surface area (total surface area of 430 m^2) but with different aspect ratio and depth. The two beds of the type known as A have an aspect ratio of 1:1, B of 1.5:1, C of 2:1 and D of 2.5:1. Type A, B and C beds have an average water depth of 50 cm and type D of 27 cm. Furthermore, the size of the granitic gravel in each pair also varies. Thus, type 1 beds contain coarser gravel (D_{60}=10 mm, C_u=1.6) while type 2 beds contain finer gravel (D_{60}=3.5 mm, C_u=1.7). The flow pumped to the beds (and consequently the surface hydraulic load) can be adjusted by combining the use of a timer to control the operation time of the pump and the aperture of the valve. The three treatment systems considered in this paper were constructed with piezometers uniformly distributed throughout the beds for sampling.

All these systems were planted with the common reed (*Phragmites australis*). Detailed descriptions of the systems, sampling procedures and analyses can be found in Arias *et al.* (1998) and García *et al.* (1997; 2003a, b; 2004).

RESULTS AND DISCUSSION

TSS Removal

TSS removal in SSF CWs occurs by means of a combination of physical mechanisms including flocculation, sedimentation and solids entrapment (Kadlec and Knight, 1996). It is not clear which is the main mechanism; nevertheless, it is known that well designed SSF CWs are very effective for TSS (Cooper *et al.*, 1996). For example, as can be seen in Figure 2, all thc systems studied systematically produced effluents with a TSS concentration lower than 35 mg/L. Even when the system at the youth hostel received, for a short time period, influents with a very high TSS concentration (because the septic tank was full of slude) the effluents had a low TSS concentration. This is related to the fact that effluent TSS concentration is independent of the TSS load, at least up to a maximum load value. This can be explained by the primary physical nature of the mechanisms involved in TSS removal (as is also the case in water filters). The US EPA (2000) report indicates that the maximum TSS load for a good performance that will avoid clogging phenomena is 20 g TSS/m^2.d.

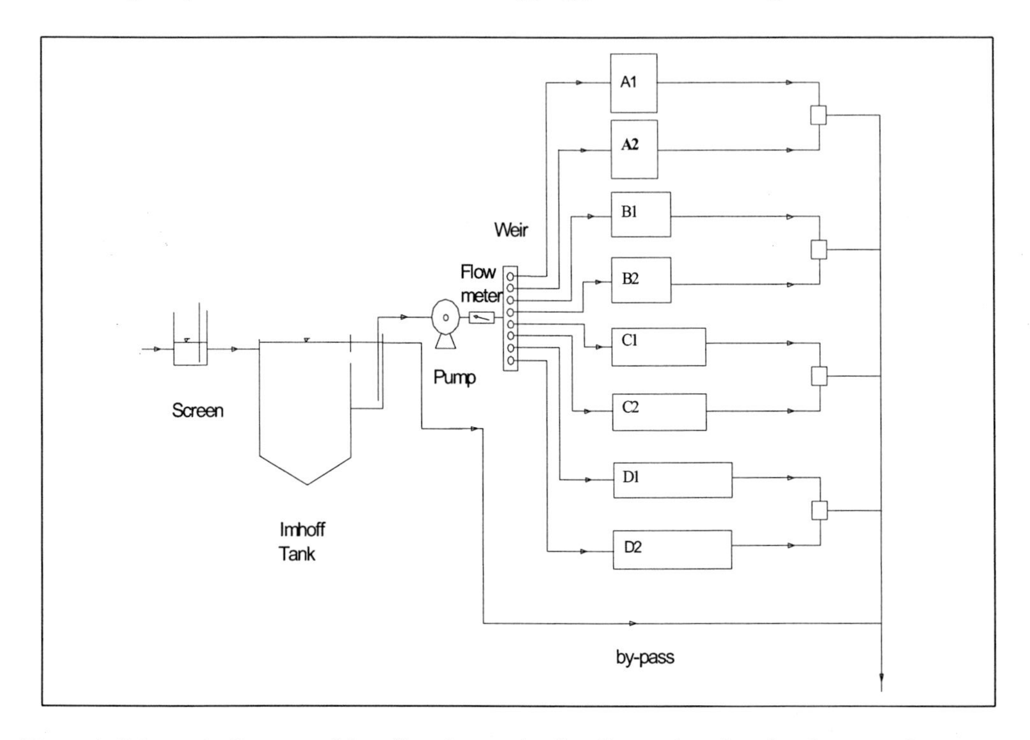

Figure 1. Schematic diagram of the pilot plant at the Can Suquet housing development, Les Franqueses del Vallès, Barcelona. Type A beds have an aspect ratio of 1:1, B of 1.5:1, C of 2:1 and D 2.5:1. Type 1 beds contain a coarser gravel (D_{60}=10 mm, C_u=1.6) while type 2 contain a finer gravel (D_{60}=3.5 mm, C_u=1.7).

After the solids have been retained in the beds, they begin to decompose and thus represent an internal source of organic matter (Tanner *et al.*, 1998). A recent hypothesis described in the US EPA (2000) report indicates that solids retention could be related to the fact that SSF CWs do not show systematic improvements in BOD_5 removal during warmer periods (i.e., there is no relationship between BOD_5 removal efficiency and temperature). It is assumed that during cold periods solids accumulate and do not decompose quickly as happens during warmer periods. Thus the greater internal generation of dissolved organic matter during warmer periods serves to mask the effect of the temperature. This hypothesis should be verified.

Numerous studies have demonstrated that TSS removal occurs near to the inlet (Kadlec and Knight, 1996). Figure 3 illustrates this well-known characteristic for one of the beds of the hotel. As a result of this principle, TSS accumulate near to the inlet, and unless the system has enough surface area perpendicular to the flow, clogging may appear in a short period of time. Surface area perpendicular to the flow is commonly calculated using Darcy's Law, which takes into account the flow, the slope of the beds and the hydraulic conductivity of the medium (Cooper *et al.*, 1996). Hydraulic conductivity declines over time and therefore it is strongly recommended that security factors be used during design. Security factors reduce the apparent (or initial) hydraulic conductivity; values for the security factors of 10 to 100 are strongly recommended (US EPA, 2000). During construction, it is necessary to ensure that clean granitic gravel of the grain size stated in the project is used.

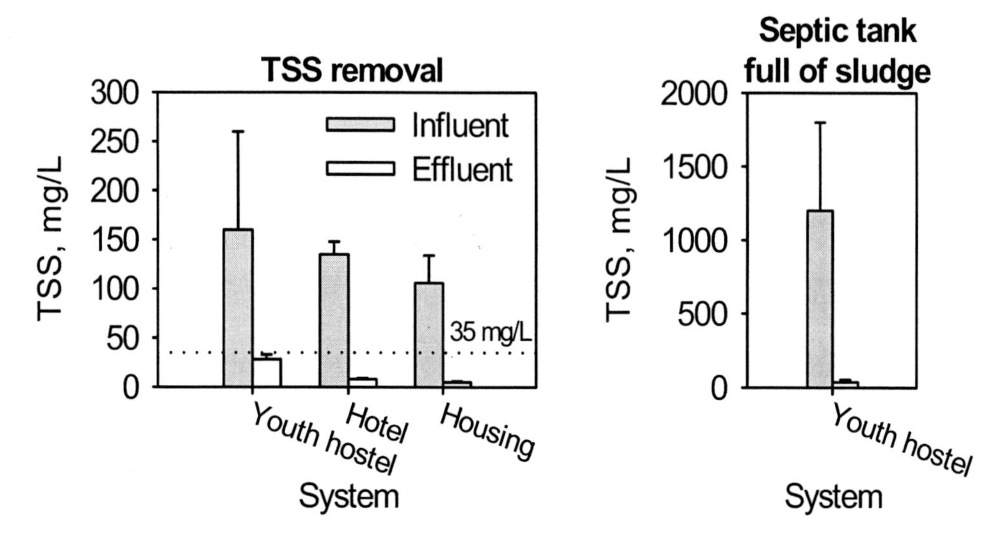

Figure 2. Average and standard deviation of influent and effluent TSS concentrations in the three systems studied. The dashed line in the graph on the left corresponds to a TSS concentration of 35 mg/L. The graph on the right shows the results obtained over a short period of time when the youth hostel system received sludge from the septic tank, which was full. Drawn using the data of Arias *et al.* (1998), García *et al.* (2004) and Píriz (2000).

In short, SSF CWs are good systems for TSS removal when they are designed to reach secondary treatment. Solids removed decompose in part; several studies have demonstrated that there is a net long-term accumulation of recalcitrant solids at a low rate (Nguyen, 2001, Tanner *et al.*, 1998). This means that even well-designed SSF CWs have an unknown expected life-span because of clogging; this drawback is not serious enough to make the system unfeasible. Primary treatment is essential for this type of facilities.

BOD_5 Removal

Although research on organic matter removal in SFF CWs has declined in recent years, the mechanisms for removal are still not well understood. For design purposes the "black box" concept is the most commonly used (Kadlec and Knight, 1996). Mechanisms for organic matter removal include aerobic and anaerobic pathways (Burgoon *et al.*, 1991). The predominance of one or other biochemical reaction depends on environmental factors such as wastewater composition, the depth of the bed and temperature. In fact, the mechanisms can vary in time and space in the same bed.

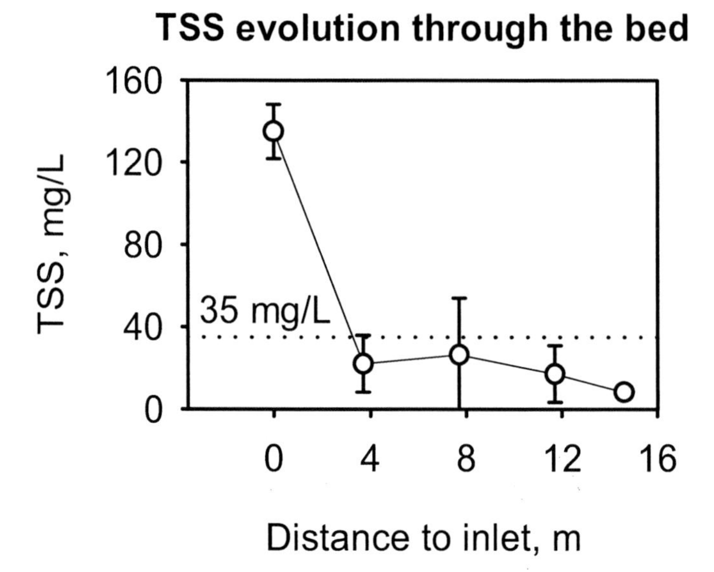

Figure 3. Evolution of TSS through one of the beds of the system at the hotel Hostal del Carme. Average and standard deviations are shown. The dashed line corresponds to a TSS concentration of 35 mg/L. Intermediate data were obtained from samples taken from piezometers. Drawn from Píriz (2000).

From our data, it is clear that strongly reducing conditions predominate in the beds. As can be seen in the profiles in Figure 4, DO concentrations are very low in the beds and the redox potential is highly negative. In fact, the DO profiles shown in Figure 5 are only indicative, as the concentration was so low that in all cases it was below the oxymeter limit of detection. The redox potential increases through the beds as a result of the oxidation of organic matter; however, the effluents still are in very reducing conditions. Lack of DO and the presence of intermediate and final products of sulphate reduction (dimethylsulphide (DMS), dimethyldisulphide (DMDS) and sulphide) may cause odours in the effluents of SSF CWs. For example, according to our data DMS and DMDS are in a concentration of 1.09±0.68 mg/L and 0.21±0.4 mg/L respectively. Odours can be minimized through the reduction of the organic load and if after the beds the water flows to a maturation pond with microalgae growing in it. Microalgae will increase the TSS of the final effluent, but in most cases a well oxidised effluent with TSS is environmentally preferable to a strongly reduced effluent without TSS. In each particular project all these considerations should be taken into account.

Figure 4 shows that the redox potential decreases in depth over the whole bed length. This characteristic could be related to two phenomena. First, aerobic respiration occurs mainly in the few first centimetres of the water depth because the DO diffusion rate in the bulk water is lower than the DO consumption (taking into account that stagnant conditions prevail in the beds and mixing is negligible). Aerobic respiration is a more efficient biochemical reaction than anaerobic pathways, and for this reason the organic matter content in the water is lower at the surface levels of the bed. Second, the location of the water outlet drainage pipe seems to encourage preferential circulation at the bottom of the beds (Breen and Chick, 1995). This could occur especially in warmer periods, when water temperature can differ by more than 1 °C between the surface and the bottom. Differences in temperature may lead to strong water density gradients which, in conjunction with the location of the drainage pipe, could promote circulation at the bottom of the beds. This short-circuiting will reduce the mean hydraulic retention time and water could exit the system with a higher organic matter concentration than expected. It seems that macrophytes may help to avoid thermal gradients in the water and, therefore, to reduce short-circuiting (Brix, 1994). Nevertheless, to our knowledge no study has proven this hypothesis.

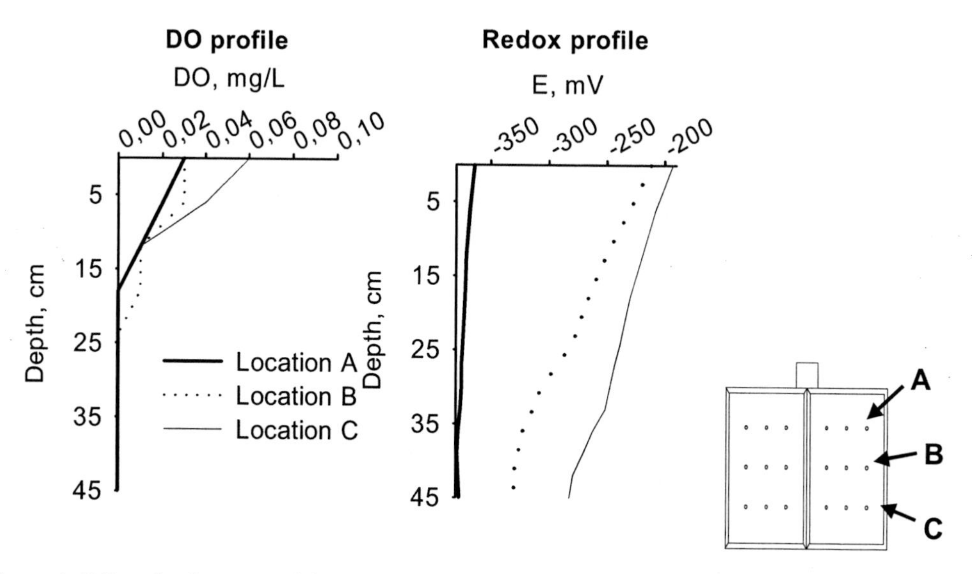

Figure 4. DO and redox potential profiles through the length of one of the beds of the hotel Hostal del Carme. The profile for each level is the average of the data obtained from three piezometers. Note that location A is nearer to the inlet and C is nearer to the outlet. Drawn from Píriz (2000).

Table 1 shows the relative importance of the biochemical reactions involved in the degradation of organic matter in the beds at the pilot plant of the Can Suquet housing development. The assumptions made for the calculations were as follows: 1) the amount of total organic matter removed was estimated by the difference between dissolved influent and effluent COD, 2) the amount of organic matter removal by sulphate reduction was estimated by the difference between influent and effluent sulphate and taking into account the stoichiometric relationship between glucose and sulphate, 3) organic matter removal by denitrification was quantified by the difference between influent and effluent ammonia, assuming that 10% of the ammonia was removed by biological assimilation and the rest was oxidized to nitrate; finally, nitrate was stoichiometrically related to glucose (note that nitrate concentration in the influent and the effluent was always undetectable), 4) the amount of organic matter removed by aerobic respiration was evaluated using the Tenessee surface reaeration models (Tchobanoglous and Schroeder, 1987) enabling the estimation of the DO concentration in the bulk water, DO concentration being stoichiometrically related to glucose, and 5) organic matter removed by methanogenesis was calculated by the difference between the total organic matter removal and the amount removed by the biochemical reactions mentioned above. Note that all theoretical glucose concentrations derived from the calculations were transformed to COD to allow summation.

Table 1. Relative Importance of the Biochemical Reactions Involved in the Degradation of Organic Matter in Shallower and Deeper Beds of the Pilot Plant at the Can Suquet Housing Development

Type of bed	Aerobic respiration	Denitrification	Sulphate reduction	Methanogenesis
Shallow (27 cm)	9.9	56.9	33.2	0
Deep (50 cm)	5.7	0	89.4	4.9

Note: The values represent the average percentage of organic matter removed by each reaction. The values were estimated with the assumptions indicated in the text, and are only indicative.

As can be observed in Table 1 the most important reaction for organic matter removal was sulphate reduction in deep beds, while it was denitrification in shallow beds. This result had major consequences on the water treatment, because shallow beds were more effective than deeper ones in all the situations tested. The reason for this is that denitrification is a faster and more energetically favourable biochemical reaction than sulphate reduction. As can be seen, aerobic respiration is a minor reaction in both beds. This is consistent with findings indicating that the role of the macrophytes in oxygen diffusion through the rhizosphere is not so active as was assumed in the 1980s and early 1990s (Brix, 1993). In fact, most studies have demonstrated that oxygen transport through stems and rhizomes is mainly used for root respiration (Tanner, 2001). Our results for the biochemical reactions involved in organic matter removal are not generalizable because the prevalence of one or another reaction depends on environmental conditions and even vary in the same bed in time and space. The work of Burgoon *et al.* (1991) illustrates this statement.

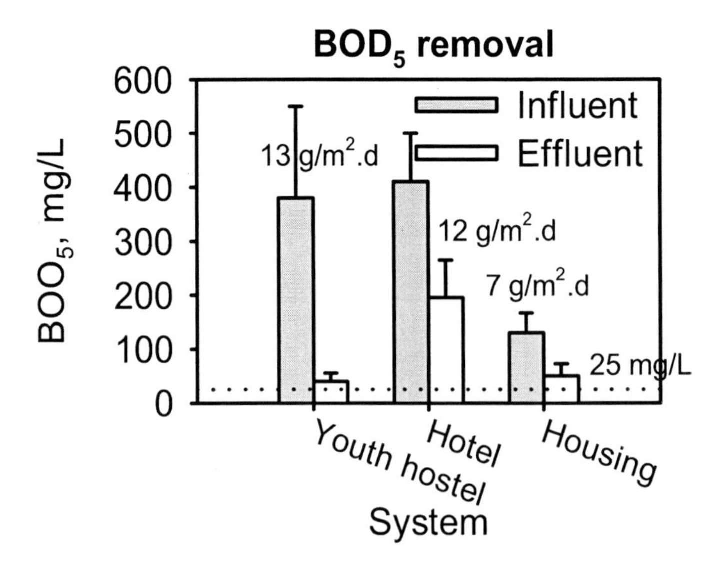

Figure 5. Average and standard deviation of influent and effluent BOD_5 concentrations in the three systems studied. The dashed line corresponds to a BOD_5 concentration of 25 mg/L. The estimated surface organic loading rate is shown (in terms of BOD_5). Note the lack of relationship between the organic loading rate and effluent concentration. Drawn using the data of Arias *et al.* (1998), García *et al.* (2004) and Píriz (2000).

As can be seen in Figure 5, all the systems studied had average effluent BOD_5 concentrations above 25 mg/L. In the case of the youth hostel and the hotel, the systems were underdesigned. The particularly poor results obtained in the hotel were due to the extensive use of disinfectants for cleaning and washing, which was not taken into account during the design process. The pilot plant of the housing development had an average BOD_5 higher than 25 mg/L because the system is tested with extreme loads for research purposes.

An important issue arising from Figure 5 is that when the results of the systems are considered as a whole, there is no relationship between the surface organic load and the effluent BOD_5 concentration. This trend occurs in many studies and can be attributed to the lack of good quality data. Often the loads are not well estimated because the flows and the concentrations are extremely variable. However, when the loads are well known there is a very good correlation with the effluent BOD_5 concentration. Figure 6 shows the accumulated frequency distributions of the influent and effluent BOD_5 concentrations grouped by different ranges of surface hydraulic loads from the data obtained in the pilot plant of the housing development. We prefer to use hydraulic instead of organic loads because they have a good controlled adjustment through the pump. Organic loads depend on influent concentrations that are extremely variable. As can be seen, the effluent BOD_5 concentration is directly related to the surface hydraulic loading rate. To obtain an effluent BOD_5 concentration of 25-30 mg/L reliably (i.e., most of the time) it is necessary to use a surface hydraulic load of 18-22 mm/d.

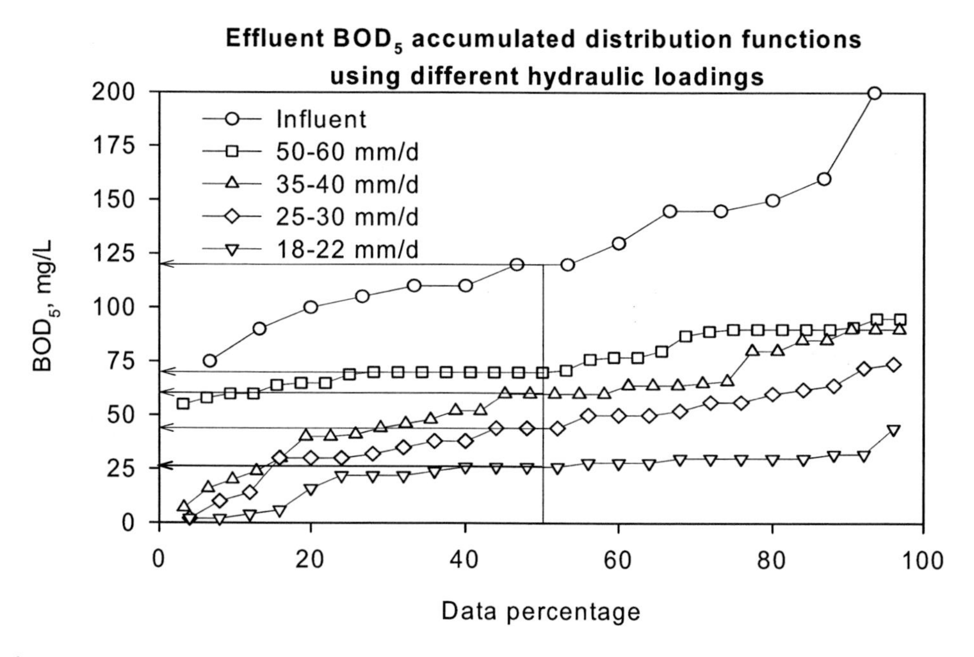

Figure 6. Accumulated frequency distributions of the influent and effluent BOD_5 concentrations grouped by different ranges of surface hydraulic loads from the data obtained in the pilot plant of the housing development. Note that arrows indicate the BOD_5 values corresponding to the percentile 50, which is the median (not the average). Drawn using the data of García *et al.* (2004).

Considering the results shown in Figure 6, assuming that one person-equivalent is 40 g BOD_5/d (according to Barrera (1999) this is a typical value for small communities in Spain), and taking into account that the primary treatment removes 30% of the BDO_5, the surface needed to obtain an effluent with a BOD_5 with 50% of the samples lower than 25 mg/L and 90% of the samples lower than 30 mg/L is 5 to 8 m^2/person-equivalent. In other words, to comply with secondary treatment effluent quality standards, SSF CWs should be designed with a surface organic load not exceeding 4-6 g BOD_5/m^2.d. This result is consistent with the recommendations given by the US EPA report (2000). Note that the surface organic load that can be obtained from other standard references is greater: 9 g BOD_5/m^2.d (Cooper, 1990), 35 g BOD_5/m^2.d (Kadlec and Knight, 1995) and 8 to 14 (Reed *et al.*, 1995).

In short, the effluents of a well-designed horizontal SSF CW can reliably be lower than 25-30 mg/L. To obtain these results the surface of the system should be carefully sized to avoid excessive loads. An SSF CW designed for BOD_5 removal also produces effluents with suitable TSS concentrations.

Nutrient Removal

Nutrient removal is not compulsory when SSF CWs are designed to reach secondary treatment; however, a certain degree of removal should be considered as desirable with regard to the environment. Most studies have demonstrated that systems designed for secondary treatment only remove nutrients in part (10-40%) (Tanner, 2001). For example, Figure 7 shows the average influent and effluent concentration in the youth hostel and housing development systems. As can be seen, the removal of nitrogen and phosphorus is small. The mechanisms involved in nitrogen removal are very complex, and include biological (microbial and plant assimilation, nitrification and denitrification) and physico-chemical (ammonia adsorption and stripping) pathways (Kadlec and Knight, 1996). Phosphorus is removed mainly by adsorption and precipitation. Most systems attain good removal of phosphorus early in their working life; nevertheless, as the adsorption capacity of the granular medium declines, the removal efficiency reduces to a minimum that remains steady over time (Kadlec and Knight, 1996). Much work is being done in nutrient removal using SSF CWs.

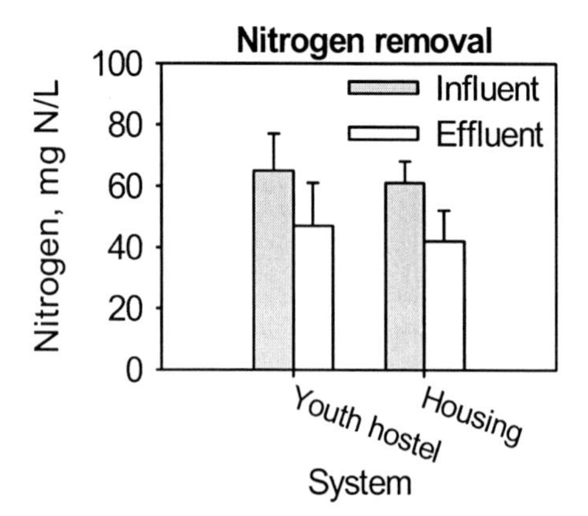

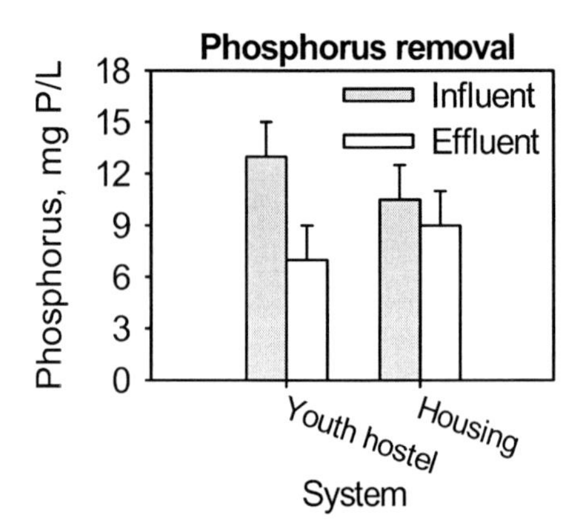

Figure 7. Average and standard deviation of influent and effluent BOD_5 concentrations in two of the systems studied. Drawn using the data of Arias *et al.* (1998) and García *et al.* (2002).

In short, with SSF CWs designed for secondary treatment it is not possible to remove nutrients to the extent required to meet current standards (for example, those of the European

Directive 91/271 for water bodies sensitive to eutrophication). Nutrient removal can be reached by a combination of different unit processes (for example, vertical SSF CWs followed by horizontal SSF CWs, or the use of specific granular media with high affinity for phosphorus).

Microbial Removal

As with nutrients, microbial removal is not compulsory to comply with secondary treatment standards, but it is recommendable, especially if the effluents are planned to be reused. Research with experimental, pilot and full scale SSF CWs has shown that faecal coliform bacteria inactivation usually ranges between 2 and 4 log-units (Rivera *et al.*, 1995) while coliphage removal ranges between 1 and 3 log-units (Gerba *et al.*, 1999). Our results agree with these well-known orders of magnitude (Figure 8).

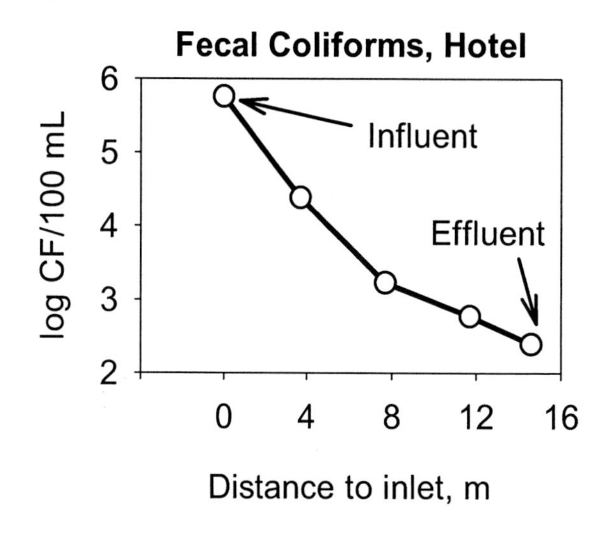

Figure 8. Evolution of faecal coliforms through one of the beds of the system of the hotel Hostal del Carme. The influent is the effluent of the septic tank. Intermediate data were obtained from samples taken from piezometers. Note that faecal coliform inactivation was greater than 3.5. Drawn using the data of Graus (2000).

The mechanisms for microbial inactivation in SSF CWs include entrapment, sedimentation, flocculation, oxidation, adsorption, toxicity, predation and natural death (Kadlec and Knight, 1996). However, to date it is unknown which of these are more significant. Faecal bacteria and virus removal has been related to environmental factors such as hydraulic retention time, granular medium and type of plant. However, several researchers have obtained contradictory results, and links between microbial removal and environmental factors have still not been definitively proved. This may be due to the inordinate variation in influent faecal microorganism concentration during the experimentation process.

In a study on a pilot plant (different from that of the Can Suquet housing development) treating secondary effluent with four beds with different sized granular medium, in which we took special care in operation and sampling procedure, there was a very good correlation between microbial inactivation, hydraulic retention time and granular medium (García *et al.*, 2003a) (Figure 9). In that study, to reduce variations in microbial influent concentrations and to obtain representative influent and effluent samples we took the following precautions: 1) samples were not taken until the beds reached 3 times the stated hydraulic retention time (which was controlled with a pump), 2) influent samples were always taken on weekdays, except on Monday, to avoid the interference of the weekend flow and mass loading reduction, 3) influent samples were not taken if there had been rainfall episodes during the previous three days, 4) effluent samples were taken after a number of days equivalent to the hydraulic retention time, and 5) effluent samples were only taken if no rain occurred during this period and if the flow pumped to each bed had an error of less than 10% of the necessary flow to obtain the predicted theoretical hydraulic retention time.

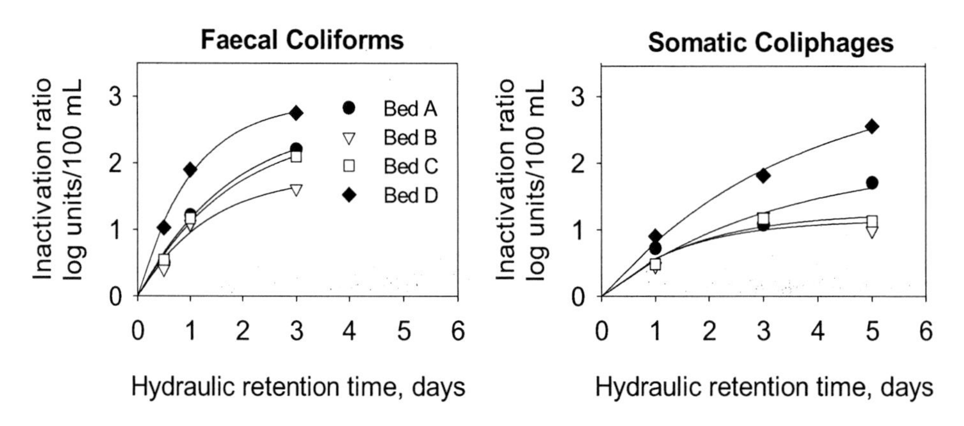

Figure 9. Faecal coliform and somatic coliphage inactivation ratio as a function of the HRT and the type of gravel media with an exponential linear adjustment. Data were obtained from the pilot plant located inside the wastewater treatment plant in Montcada i Reixac, Barcelona. Bed A contained granitic gravel (D_{60}=9.5 mm, C_u=2), B also granitic gravel (D_{60}=17 mm, C_u=2), C rolling stones (D_{60}=12.7, C_u=2.7), and D granitic gravel with 20% sand (D_{60}=9.52, C_u=8.1). Note that bed D has the highest microbial inactivation. From García *et al.* (2003a).

The results showed in Figure 9 indicate that there is a positive relationship between microbial inactivation ratio and hydraulic retention time, but in general when the hydraulic retention time is approximately 3 days, the microbial inactivation ratio reaches saturation values (both in the case of faecal coliforms and somatic coliphages). This means that any increase in the hydraulic retention time over 3 days will not result in a significantly higher inactivation ratio and a consequent lower microbial concentration in the effluents. As can be seen, the value of the microbial inactivation ratio at the saturation level depends on the granular medium contained in the bed. Beds containing smaller sized particles (D and A, note that D has 20% sand) are more efficient in both coliform and coliphage removal. In contrast, the bed containing the largest granular medium (D) showed the worst results in microbial removal.

In short, SSF CWs designed for secondary treatment have some capacity for microbial removal. The degree of microbial inactivation depends on hydraulic retention time (or surface hydraulic loading rate) and the granular medium used. The effect of the type of plant is still not well understood.

Design Considerations

The results shown in the above sections of this paper demonstrate that to provide useful information for the design of SSF CWs it is necessary to have a good amount of good quality data. This was the main objective of the research done in the pilot plant of the Can Suquet housing development (García *et al.*, 2004). In that study we tested the influence of surface area (varying the hydraulic loading rate), shape, depth, and granular medium on water treatment. Figure 10 shows the average effluent concentrations of BOD_5, ammonia and orthophosphate grouped according to the different characteristics tested.

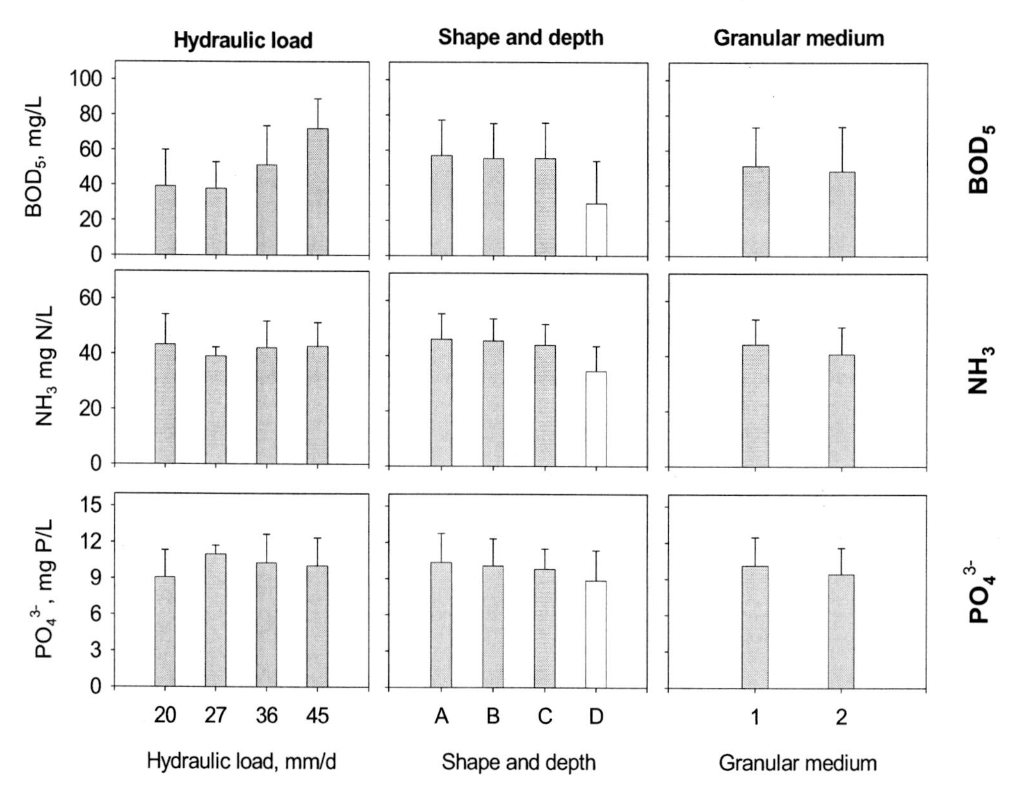

Figure 10. Average and standard deviation of the effluent BOD_5, ammonia and orthophosphate concentrations grouped according to the different characteristics tested in the pilot plant of the Can Suquet housing development. The values of the hydraulic load correspond to mean values. Note the shape (aspect ratio) of type A beds is 1:1, type B 1.5:1, type C 2:1 and type D 2.5:1. The bars in type D beds are drawn in white to indicate that they are shallower (average depth of 27 cm) than the others (average depth of 50). The granular medium in type 1 beds is d_{60}=10 mm and that of type 2 d_{60}=3.5 mm. Drawn using the data of García *et al.* (2004).

As can be seen in Figure 10, hydraulic load has a significant influence on effluent BOD_5, but none on ammonia and orthophosphate. This is because the hydraulic loads used were not low enough to significantly reduce influent nutrient concentrations, and as a result the removal efficiency was in all cases low. The Can Suquet study confirms that to obtain effluents with low BOD_5 concentrations it is necessary to operate the SSF CWs with a suitable range of hydraulic loads (according to our data 18-22 mm/d for domestic wastewater), in other words, to design the systems with enough surface area for a given flow.

Water depth is an important factor for water treatment because, as can be seen in Figure 10, shallower beds (type D) produced effluents with clearly lower concentrations of BOD_5 and ammonia. Orthophosphate concentration was also lower, although not significantly. It is paradoxical that shallower beds achieve the best results in water treatment, because they have less volume and therefore operate with a lower hydraulic retention time (note that all these beds receive the same flow). As has been described in the BOD_5 removal section, according to our data a shallow water depth promotes surface reaeration and as a result there is greater denitrification (note that nitrate is formed by nitrification in the presence of oxygen) and aerobic respiration than in deep beds (see Table 1). These biochemical reactions are faster and more energetically favourable than sulphate reduction and methanogenesis. The final result is that shallower beds are more efficient for organic matter and ammonia removal. In a study reported by the US EPA (2000) removal was found to be slightly better with greater media depth, when comparing 45 cm with 30 cm systems operated at the same surface organic loading rate. This difference was attributed to the higher hydraulic retention time attained in the deeper beds. We do not know the reasons for the different results obtained in our research and the US EPA (2000) study; however, the depth of the beds and their influence on water treatment efficiency deserves more attention than they have been given to date. Kadlec and Knight (1996) make an observation to the same effect.

The shape of the beds do not significantly affect the effluent contaminant concentrations (Figure 10). Note that there were no significant differences in effluent concentrations between beds with different aspect ratio but with the same water level. The same results were obtained in a study in which very different aspect ratios (from 4:1 to 30:1) were evaluated (Bounds *et al.*, 1998). The assumption that a high aspect ratio improves plug flow conditions is true, but not to the extent that it would cause significant differences with beds with a lower aspect ratio. This is a well-known property and is due to the internal mixing undergone by wastewater in the voids between the granular medium, the roots and the rhizomes (Kadlec and Knight, 1995). Thus, it is not necessary to design beds much longer than they are wide. To prevent short-circuiting, the beds should be at most as wide as they are long (squared), and preferably longer. Our results indicate that the use of a gravel with a smaller grain size (3.5 mm instead of 10 mm) does not increase significantly the efficiency of the systems. This is in agreement with the findings reported in other studies (US EPA, 2000). In fact, that report recommends the use of large gravels to avoid clogging and increase the expected life-span of the system, as the size of the granular medium is not a determinant factor for wastewater treatment.

Summary and Recommendations

Horizontal SSF CWs can be designed to reliably meet secondary effluent standards. For this to be achieved, however, some guidelines should be taken into account (in a mediterranean climate):

1. TTS surface loading rate should not be greater than 20 g $TSS/m^2.d$ to avoid clogging and ensure the system's life-span.
2. BOD_5 surface loading rate should not be greater than 6 g $BOD_5/m^2.d$ to obtain effluents with a concentration lower than 25-30 mg BOD_5/L.
3. The granular medium should be granitic gravels or rocks (limestone should be avoided because of the potential for dissolution under the strongly reducing environment in the beds). The medium should be coarse enough to avoid clogging, but not so coarse as to hinder plant growth. Moreover, there is no clear advantage in pollutant removal with gravels of different grain size. Therefore, we recommend using diameters ranging from 3 to 8 mm.
4. For bed sizing, the apparent hydraulic conductivity should be reduced by a factor ranging from 10 to 100. During construction the gravels should be washed to eliminate fine material.
5. The aspect ratio is not critical. The minimum length-to-width ratio should be 1:1 to avoid short-circuiting. The number of beds in a particular system depends on this aspect.
6. The impact of water depth is still not clear. Our results indicate that shallow beds (27 cm on average) obtain better results than deeper ones (50 cm on average). More experiments are needed to expand the knowledge on the effect of water depth.

If a horizontal SSF CWs is designed and operated with all these recommendations, the TSS and BOD_5 of the effluents will be lower than 25-30 mg/L, the effluent nutrient concentration will be 10 to 40% lower than that of the influent, and the microbial inactivation will range from 2 to 4 log-units for faecal bacteria and 1 to 3 for faecal viruses. Higher nutrient removal and microbial inactivation requires more intensive treatment than that provided by an unique horizontal SSF CW.

Acknowledgements

This paper has been made possible by the work of many people involved in the projects of the Environmental Engineering Division, in particular undergraduate and graduate students, and technical staff. In addition, the contribution of Hostal del Carme, Casa de Colònies Can Massaguer, Consorci per a la Defensa dels Rius de la Conca del Besòs, EMSSA, Ecomoll, and Les Franqueses del Vallès Town Council has been very helpful. We are grateful to all of them. The results of this study were obtained in part through a grants awarded by the Spanish Department of Science and Technology, Research Project REN2002-04113.

References

Arias, C., García, J. and Mujeriego, R. (1998). Saneamiento autónomo con humedales construidos de flujo subsuperficial: evaluación de su capacidad de tratamiento. *Equipamiento y Servicios Municipales*, 25, 25-31.

Barrera, A. (1999). *Análisis y Caracterización de los Parámetros de las Aguas Residuales Necesarios para el Dimensionamiento de Estaciones Depuradoras de Menos de 2000 hab-eq*, ETSECCPB dissertation, Technical University of Catalonia. 110 pp.

Breen, P.F. and Chick, A.J. (1995). Rootzone dynamics in constructed wetlands receiving wastewater: a comparison of vertical and horizontal flow systems. *Wat. Sci. Tech.*, 32(3), 281-290.

Brix, H. (1993). Macrophyte-mediated oxygen transfer in wetlands: transport mechanisms and rates. In: *Constructed Wetlands for Water Quality Improvement*, Moshiri, G.A. (ed.), Lewis Publishers, Boca Raton, Florida. pp. 391-398.

Brix, H. (1994). Use of constructed wetlands in water pollution control: historical development, present status, and future perspectives. *Wat. Sci. Tech.*, 30(8), 209-224.

Bounds, H.C., Collins, J., Liu, Z., Qin, Z. and Sasek, T.A. (1998). Effects of length-width ratio and stress on rock-plant filter operation. *Small Flow Journal*, 4(1), 4-14.

Burgoon, P.S., DeBusk, T.A., Reddy, K.R. and Koopman, B. (1991). Vegetated submerged beds with artificial substrates. I: BOD removal. *J. Env. Eng. ASCE*, 117, 394-407.

Cooper, P.F. (1990). *European Design and Operations Guidelines for Reed Bed Treatment Systems*, Report UI17, Water Research centre, Swindon, UK.

Cooper, P.F., Job, G.D., Green, M.B. and Shutes, R.B.E. (1996). *Reed Beds and Constructed Wetlands for Wastewater Treatment*, WRc Publications. 184 pp.

García, J., Ruiz, A. and Junqueras, X. (1997). Depuración de aguas residuales urbanas mediante humedales construidos. *Tecnología del Agua* 165, 58-65.

García, J., Mujeriego, R., Obis, J.M and Bou, J. (2001). Wastewater treatment for small communities in Catalonia (Mediterranean Region). *Water Policy*, 3(4), 341-350.

García, J., Vivar, J., Aromir, M. and Mujeriego, R. (2003a). Role of hydraulic retention time and granular medium in microbial removal in tertiary treatment reed beds. *Wat. Res.* 37, 2645-2653.

García, J., Ojeda, E., Sales, E., Chico, F., Píriz, T., Aguirre, P. and Mujeriego, R. (2003b). Spatial variations of temperature, redox potential, and contaminants in horizontal flow reed beds. *Ecol. Eng.* 21, 129-142.

García, J., Aguirre, P., Mujeriego, R., Huang, Y., Ortiz, L. and Bayona, J.M. (2004). Initial contaminant removal performance factors in horizontal flow reed beds used for treating urban wastewater. *Wat. Res.* 38, 1669-1678.

Gerba, C. P., Thurston, J.A., Falabi, J.A., Watt, P.M. and Karpiscak, M.M., (1999). Optimization of artificial wetland design for removal of indicator microorganisms and pathogenic protozoa. *Wat. Sci. Tech.*, 40(4-5), 363-368.

Graus, S. (2000). *Desinfecció en aiguamolls de flux subsuperficial*, ETSECCPB dissertation, Technical University of Catalonia. 77 pp.

Haberl, R., Perfler, R. and Mayer, H. (1995). Constructed wetlands in Europe. *Wat. Sci. Tech.*, 32(3), 305-316.

Kadlec, R.H. and Knight, R.L. (1996). *Treatment Wetlands*, Lewis Publishers, New York. 893 pp.

Nguyen, L. (2001). Accumulation of organic matter fractions in a gravel-bed constructed wetland. *Wat. Sci. Tech.*, 44(11-12), 281-287.

Píriz, A.J. (2000). *Condiciones de Óxido-Reducción en Humedales Construidos de Flujo Subsuperficial*, ETSECCPB dissertation, Technical University of Catalonia. 104 pp.

Polprasert, C., Khatiwada, N.R. and J. Bhurtel. (1998). Design model for COD removal in constructed wetlands based on biofilm activity. *J. Env. Eng. ASCE*, 124, 838-843.

Reed, S.C., Crites, R.W. and Middlebrooks, E.J. (1995). *Natural Systems for Waste management and Treatment*, McGraw Hill, New York.

Rivera, F., Warren, A., Ramírez, E., Decamp, O., Bonilla, P., Gallegos, E., Calderón, A. and Sánchez, J.T. (1995). Removal of pathogens from wastewaters by the root zone method (RZM). *Wat. Sci. Tech.*, 32(3), 211-218.

Tanner, C.C. (2001). Plant as ecosystem engineers in subsurface-flow treatment wetlands. *Wat. Sci. Tech.*, 44(11-12), 9-18.

Tanner, C.C., Sukias, J.P.S. and Upsdell, M.P. (1998). Organic matter accumulation during maturation of gravel-bed constructed wetlands treating farm dairy wastewaters. *Wat. Res.*, 32, 3046-3054.

Tchobanoglous, G. and Schroeder, E.D. (1987). *Water Quality*. Addison-Wesley Publishing, California, 768 pp.

US EPA. (2000). *Constructed Wetlands Treatment of Municipal Wastewaters*, EPA 625-R-99-010. 165 pp.

In: Progress in Aquatic Ecosystems Research
Editor: A. R. Burk, pp. 139-165

ISBN 1-59454-383-6

Chapter 7

APPLICABILITY OF HIERARCHICAL MOSAIC OF ARTIFICIAL ECOSYSTEMS (HMAE®) TO THE WASTEWATER TREATMENT

G. Ansola* and E. de Luis,
Area of Ecology, Environmental Biotechnology Research, University of León, Spain
R. Cortijo and J. M. González
Environmental Institute, University of León, León 24071, Spain

ABSTRACT

In Spain, very little interest has been shown in gaining knowledge of the purifying potential of aquatic macrophytes and testing their adequacy and viability as opposed to the conventional systems of urban sewage treatment. This has not been the case in the United States and various European countries, where this issue has received special attention since over ten years.

The treatment pilot systems are located in the Experimental Low Cost Wastewater Treatment Station of León (NW Spain).

Four helophytes have been used in different experiments: *Typha latifolia, Phragmites australis, Scirpus lacustris* and *Iris pseudacorus*. The results show a high significant differences in the elimination of COD, BOD_5 and Orthophosphate between the tanks with emergent macrophytes and the ones that function as "control" for low and medium flows.

The NPK contents in the different parts of the these plants was determinate, assessing the associated chaining with the plan phenology in natural state.

One other experiment compare the removal efficiency of four vegetation series: *Typha-Scirpus-Phragmites, Scirpus-Iris-Phragmites, Typha-Iris-Phragmites* and *Typha-Iris-Scirpus* by using 300 l/d flow rates of sewage in an experimental pilot plant. The largest removal of organic matter belongs to *Typha-Scirpus-Phragmites* and *Typha-Iris-Phragmites* show the highest removal of nitrogen. The phosphorus parameters exhibit the highest removing values for *Scirpus-Iris-Phragmites*.

* Author Corresponcence, Area of Ecology, Faculty of Biological and Environmental Sciences, University of León, 24071, León, Spain. E-mail: deggag@unileon.es; Fax.: 0034 987 29 15 68

Diurnal changes in two algae ponds for wastewater treatments has been compared too. Stratification is observed in the water stabilization pond system in summer as reflected in temperature, conductivity, dissolved oxygen and pH distribution. High rate algae pond system are completely mixed all of the time, and reflect diurnal and seasonal effects and blooms of *Monoraphidium contortum* occurred and they were higher in summer.

A constructed wetland for wastewater treatment using a Hierarchical Mosaic of Artificial Ecosystem (H.M.A.E.®) pilot system was carried out over a vegetative period in eight different flow and vegetable composition series. The system consisted of a free water pond as a first step working as primary treatment followed by a zone with *Typha* sp. and surface flow and finally a woody zone with a subsurface flow and planted with ligneous species (*Salix* sp., *Populus* sp., *Fraxinus* sp. and *Alnus* sp.). Removal efficiency reflects an optimal result and the effluent characteristics were in accordance with European Union legislation criteria for wastewater treatment systems.

An H.M.A.E.® full-scale pilot plant has been constructed in León (Spain) and their removal efficiency has been studied. The analyses conducted during the different seasons show large differences in the reduction of organic and nutrient pollution during the active growth period compared to the winter resting period.

INTRODUCTION

Wetland systems provide effective wastewater treatment and are especially efficient in the removal, and beneficial reuse, of nutrients. Constructed wetland systems, including restoration of degraded wetlands, enhance the aesthetic value of an area, and may be used by the public as a nature study area.

In the meantime, wastewater have been treated and reused successfully in agriculture, silviculture, aquaculture, and golf course and landscape irrigation. Also, more attention has been given to development of an understanding of constructed wetlands.

Furthermore, much of the wastewater treatment in a constructed wetland occurs by means of bacterial metabolism and physical sedimentation, as it does in an activated sludge. Wetland plants serve an important role by providing structure to support the algae and bacterial population that provide wastewater treatment capability and reliability in a created wetland community environment.

Constructed wetlands for wastewater treatment are classified by hydrologic regime (Hammer 1989). Free water surface systems (FWS) consist of basins or channels and barriers to seepage, and have relatively shallow water depths and low flow velocities. Subsurface flow system (SFS), sometimes called vegetated submerged bed systems (VSB), consist entirely of subsurface flow conveyed to the system via a trench or bed under drain network. As with SFS systems, they have barriers to seepage and very low flow velocities. Both FWS and SFS systems are characterized by emergent vegetation.

The European Union Directive 91/271/EEC (ECOD, 1991) establishes a series of requirements for wastewater treatment plants effluents and removal efficiency, which must be applied before the end of December 2005 for populations between 2,000 to 15,000 equivalent inhabitants and an adequate treatment for populations lower than 2,000. In León province (Spain) localities fewer than 10,000 inhabitants are about 97%, and those with less than 2,000 inhabitants represent 82.5 % of the total province population. Most of these small towns need an integral wastewater treatment.

For small urban areas and rural villages to conform with European Union norms concerning outflow quality from wastewater treatment plants, adequate technology with low investment and operation costs but with a high performance level must be found, especially as regards nitrogen and phosphorus removal. With the techniques available, nutrient outflow concentrations are far form the compulsory quality standards of the European Union for sensitive areas (10-15 mgN/l and 1-2 mgP/l).

Has been widely demonstrated in Spain that conventional (activated sludge) treatment systems applied to small rural municipalities has many operational and management problems, being at present inactive or abandoned in their great majority. As an alternative to the conventional treatment systems, low-cost and natural systems are increasingly applied in Spain, where stabilisation ponds are one of the most adequate and applied. Constructed wetlands have been, at the moment, non-considered as a treatment option by the administration but mainly due to the absence of previous experiences on pilot plants in Spain.

The use of constructed wetlands to treat non point pollution and wastewater has been investigated in numerous general studies (Hammer, 1989; Cooper and Findlater 1990, Olson and Marshall, 1992; Moshiri, 1993; Kadlec and Knight 1996; and references therein). More concrete problems have been recently researched (e.g., Pant et al., 2001; Gómez Cerezo et al., 2001); Braskerud, 2002a, 2002b; Söderqvist 2002; Pant and Reddy 2003). Although most of the key processes involved in the macrophyte-based wastewater treatment are well documented: organic mater retention (e.g. Nguyen, 2000), nutrients removal (e.g. Mitsch et al., 2000 and references therein) and pathogen reduction (e.g. Perkins and Hunter, 2000). Different technologies have been successfully used and each one is unique (Reddy and DeBusk, 1985, Mara et al, 1992, 1998, Kadlec, 1995, Cadelli et al, 1998, Younger et al, 1998).

EXPERIMENTAL DESIGN

The Experimental Low Cost Wastewater Treatment Station is in the village of Mansilla de las Mulas (León, NW Spain). It is basically an agricultural village with a population of 1637 inhabitants, located in an area with warm-cool Mediterranean weather. Annual rainfall is about 500 mm and average annual temperature is 11°C, with average low and high temperatures of -1.1°C and 28°C, respectively, (Papadakis, 1961).

Nutrient Contents in the Helophytes of Natural Ecosystem

The NPK contents in the leafs, stems and fruits of four macrophutes species (*Typha latifolia, Scirpus lacustris, Phragmites australis* and *Iris pseudacorus)* was determined, assessing the associated changing with the plant phenology.

The nutrients assimilation capacity according with the macrophyte specie it is analyzed as well as the accumulation level of them in the different parts of the plant.

There is not a clear tendency to accumulate nitrogen in these plants. *Phragmites* and *Iris* have the highest levels in their leaves. In *Scirpus* (without leaves) it accumulates in the stem. *Phragmites* shows concentrations higher than 3.5 g/100 g of dry matter (Figure 1). Other authors, (Radoux & Kemp, 1982; Neill, 1989; Breen, 1989) have achieved results similar to obtained in this test.

Phosphorus is accumulated in concentrations lower than those of nitrogen. phosphorus increases its values during the autumn, possibly due to the effect of fallen plant residues. *Typha* is the plant species with the greatest capacity for incorporating phosphorus. The fruit is the part in which the highest concentration of phosphorus appears (Figure 1). The results obtained are similar to those achieved by Black (1975) and are higher than those recorded by other authors.

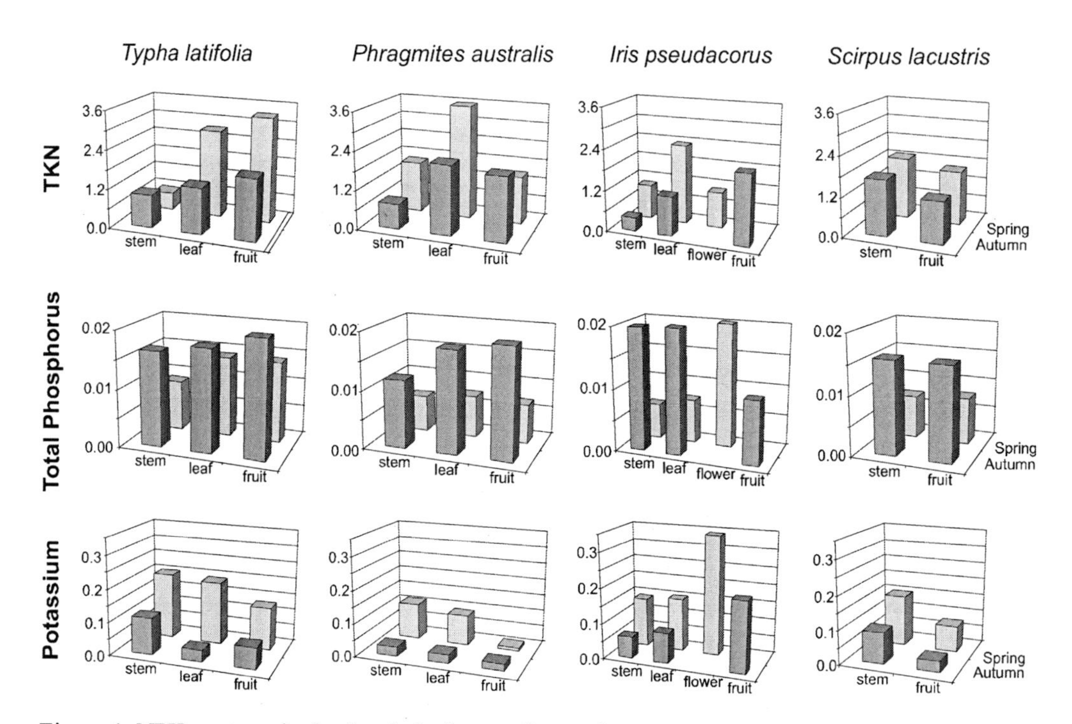

Figure 1. NPK contents in the four helophytes of natural ecosystem.

A higher concentration of potassium is found in June due to its relation with metabolic processes. The highest accumulation is found in the stems except in *Iris*, in which it appears in the fruit (Figure 1). Radoux & Kemp (1982), obtains higher values in his research, but the accumulation of potassium in different plans of the plants follows the same pattern as in this study.

Individual Helophytic Cultures Treatment

To compare the efficiency of various helophytic cultures (*Typha latifolia, Scirpus lacustris, Phragmites australis* and *Iris pseudacorus*) in removing pollutants in contrast to a "control" without plants, we have an experimental plant. All treatments were kept at the same conditions of flow rate and pollution loading rate, and at same time, to assess the differential response of each as a function of the hydraulic retention time.

The system consists of 15 small glass-fiber tanks measuring 0.85x1.3x0.55 m, each having a volume of 0.6 m^3 and an area of 1.1 m^2. The tanks were filled with a layer of fine gravel 6-8 mm in diameter and up to 25 cm in width. Water covers the gravel up to 20 cm in height. The volume is reduced to 0.29 m^3 (Figure 2).

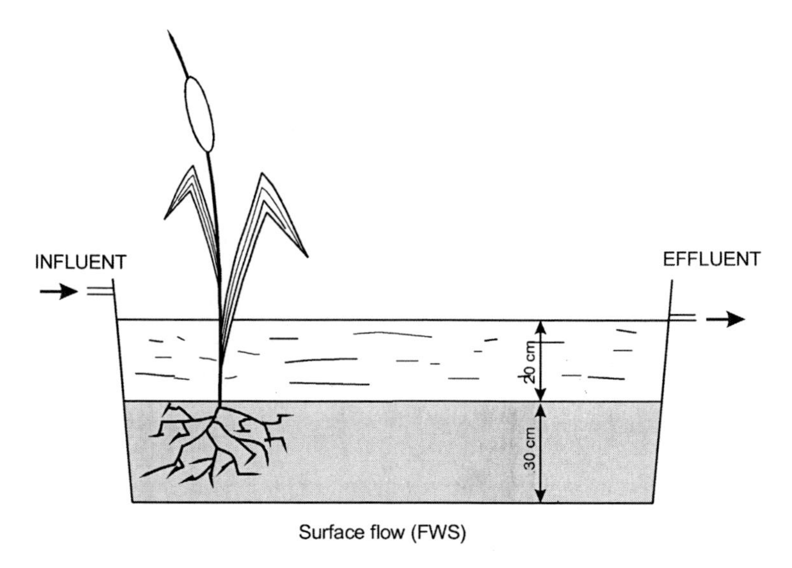

Figure 2. Surface flow system of helophytes experimental tank.

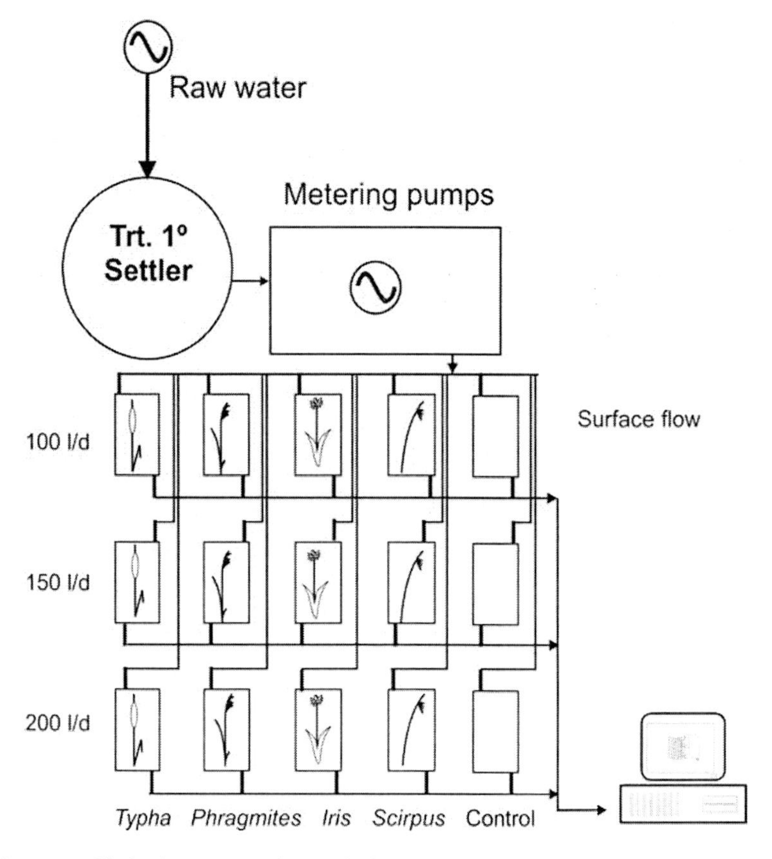

Figure 3. Diagram of helophytes experimental plant.

The tanks worked as independent systems and three of them were kept without macrophytes and considered as "control". In the other twelve, 30 feet of plant have been planted. These plants were transplanted from their natural habitat to the banks of the river Esla between October and November 1991. from November 1991 to October 1992, each species and "control" tank were tested with three different flow rates: 100,150 and 200 l/d which correspond with retention times of 3, 2 and 1,5 day, respectively (Figure 3).

Wastewater was taken from the village sewer and circulated up to a cylindrical reservoir with a total capacity of 1 m^3, in which as pre-treatment the thickest solids were removed by decanting and by having the water going through a mesh 0.25 mm de diameter. Later, the water, subsurface flowing along the surface of the substratum, always keeping at 20 cm high above it, was distributed among the tanks.

After a period of system stabilization, sampling was started in May 1992 and extended until October 1992. Fortnightly influent samples (integrated along 24 h) and effluent samples from each tank have been taken. The analyses were based on physicochemical and microbiological parameters according to the APHA-AWWA-WPRF recommendations (1989): Total suspended solids (TSS), method 2540 D. Total volatile solids (TVS); method 2540 E. Chemical oxygen demand (COD), method 5220 A. Biological oxygen demand 5 d (BOD_5), method 5210 B. Ammonium (NH_4^+), method 4500-NH_3 F. Total Kjeldahl Nitrogen (TKN), method 4500-N_{org} B. E. Faecal coliforms (FC), method 9222 D. And faecal streptococcus (FS), method 9230 C. Nitrogen in nitrite form by the colorimetric method as modified by Shinn (1941) and total phosphorus (TP) by the colorimetric methods of the ascorbic acid (Murphy & Riley, 1962).

Table 1. Mean, Minimum and Maximum Composition of the Raw Wastewater

Parameter	Minimum	Average	Maximum
pH	6.50	7.23	7.60
Conductivity (μS/cm)	390	493	636
TSS (mg/l)	8	19	92
BOD_5 (mg/l)	18.43	24.43	37.50
COD (mg/l)	66.84	130.85	254.45
Nitrite (mg/l)	0.0	0.3	0.6
Nitrate (mg/l)	0.73	50.03	275
Ammonium (mg/l)	4.76	7.17	9.95
Organic Nitrogen (mg/l)	2.06	4.36	9.68
Orthophosphate (mg/l)	4.30	8.03	21.33
Total Phosphorus (mg/l)	4.85	13.38	25.16
Total Coliforms (ufc/100ml)	400000	24062500	90000000
Faecal Coliforms (ufc/100ml)	165000	780400	2250000
Faecal Streptococci (ufc/100ml)	400	83600	460000

Table 1 shows the minimum, average and maximum values of the analyzed parameters in the influent along the test. These values on the whole slightly changeable along the sampling, are below those regarded as values frequently found in urban wastewater, so we are dealing with what can be classified as low concentration water (Metcalf & Eddy, 1998). However, the increase in the number of inhabitants in the village over the summer season brings about a rise in organic loading, nutrients and faecal pollution indicators. We also have to point out the high value obtained for nitrogen content in nitrate form which is indicative of a basically agricultural village that uses calcium ammonium nitrate as a fertilizer.

By using *Phragmites*, the removal of organic matter and nutrients is about 50% (Figure 4) and there is an increase in performance with retention time. Very similar levels were attained by Radoux & Kemp (1982) whose working conditions were similar to those used in this study: however, Vouillot & Boutin (1987) have obtained much higher performance values, 75 and 80% for COD and BOD_5 respectively.

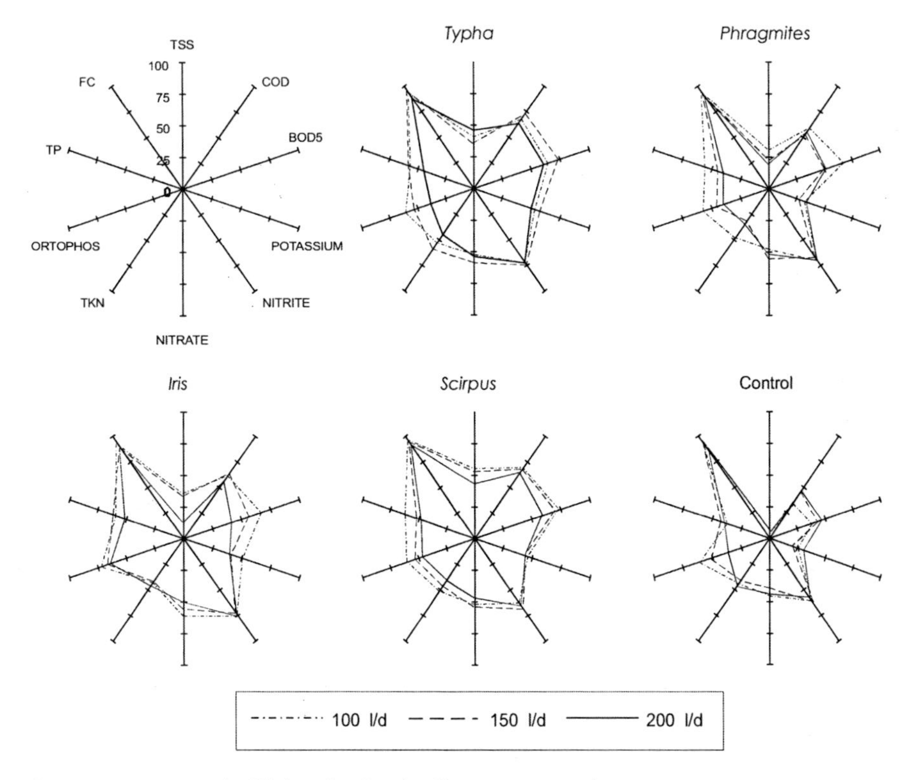

Figure 4. Average removal efficiencies for the flow rates tested.

Although without appreciable differences, the system with *Phragmites* seems to be more effective in the removal of TKN than in the removal of TP. This feature is emphasized when the flow rate increases, reaching performance values of 54% for the former as opposed to 36% for the latter (Figure 4).

The highest performance, above 90%, corresponds with the removal of the different microbiological groups assessed. This result was maintained for the other three species of helophytes and even in the "control" tanks, which seems to indicate that the presence of macrophytes did not contribute to the removal of total, faecal coliforms of faecal streptococci (Figure 4). Their decrease in the effluents could be due to a combination of several factors such as predation and physical processes. Bhamidimarri *et al.* (1991) obtained comparable performance values for faecal coliforms by using *Scirpus lacustris*.

The remainder of the helophytes tested showed greater efficiency than *Phragmites* in the removal of all physicochemical parameters analyzed.

Of the four helophytes tested, *Scirpus* and *Typha* appear to be the most suitable ones in the removal of organic matter, both sharing equivalent performance values close to 70%,(Figure 4) although clearly below the performance values attained by Bhamidimarri *et al.* (1991), who pointed out a reduction of 96% in BOD_5 using *Scirpus lacustris*. On the contrary, Radoux & Kemp (1982) noticed reductions of 59% in COD using *Typha*. In general, no connection between the performance increase and flow rate decrease was detected.

These two helophytes exhibit slightly inferior capacity for removing nutrients. In terms of TP removal, *Scirpus lacustris*, was a bit more efficient with values ranging from 47 to 61%, while treatment with *Typha* generally averaged removal efficiencies slightly lower, with a maximum of 59% for flow rate of 150 l/d (hydraulic loading of 11.8 cm/d) (Figure 4). Even smaller efficiencies for this species concerning the phosphorus have been stated by Radoux & Kemp (1982), with performance of 23% using a hydraulic loading rate of 12 cm/d.

Iris pseudacorus was the most effective species in the removal of orthophosphate with an average value of 71.2% for the 100 l/d flow. The purification efficiency for this species was 60.5% with a flow of 150 l/d; (Figure 4) Radoux (1985) id not reach more than 43.4% using the same plant.

Using *Iris pseudacorus* the removal of TP slightly increased compared with the rest of the species, whereas the TKN removal was kept on the same values as the treatment with *Typha*. However, the purification of TSS, COD and BOD_5 is reduced as opposed to *Scirpus* and *Typha*. Generally speaking, there is a correlation between the flow rate used and the performance reached.

Although the purification efficiencies do not change substantially, a greater reduction of TKN concentration took place in the treatment with *Typha* than with *Scirpus*, maximum values being 62 and 56%, respectively. Even though no significant effects of the flow rate were detected, lowest efficiencies have been noted for shorter retention times.

Ammoniacal nitrogen reached 74.7% elimination with the lowest flow experimented and using *Scirpus lacustris*. Bhamidimarri (1991) reaches 85%. In the same way, nitrate is a parameter that does not show significant differences in one of these cases, however efficiency of up to 59.3% is a reached by using *Iris pseudacorus* with a flow of 100 l/d, this one being, the inorganic form of nitrogen removed with less effectiveness.

In the "control" tanks only microbiological parameters maintain purification performance values similar to those of the tanks with helophytes. There is a general decrease in the removal of physiochemical parameters, especially to remove BOD_5, COD, TP and most particularly TSS, probably owing to the large mass of phytoplankton which developed in these tanks.

To sum up, *Scirpus, Iris* and *Typha* do not show noticeable differences in their power to remove BOD_5, COD, TKN and TP, though *Typha* appears to be the most useful at removing organic matter. Except for the microbiological parameters, the presence of helophytes has a substantial effect on the removal efficiencies of pollutants. On using *Phragmites australis* a general reduction in the performance values of all physiochemical parameters takes place.

Differences in phosphate, TP and ammonium between the low and peak flows can be observed through the variance analysis done, in which flows have been used as the first factor. When using plants as first factor in the one way ANOVA there are no significant differences between the various vegetal species, except for phosphate elimination, when comparing *Iris* to *Typha* and *Iris* to *Phragmites*. In the other parameters the significant differences appear between the tanks containing plants and the ones acting as "control" with an important significance for COD, BOD_5 and TP.

The two ways variance analysis, showed overall significance both between the plants and between the flow for orthophosphate and TP. As for COD, BOD_5 and pH significant differences were only found in plants and for ammoniac nitrogen in flows. The results of the comparisons are shown in summarized form with statistical significant differences for each parameter, designing the differences with different circle sizes (Figure 5).

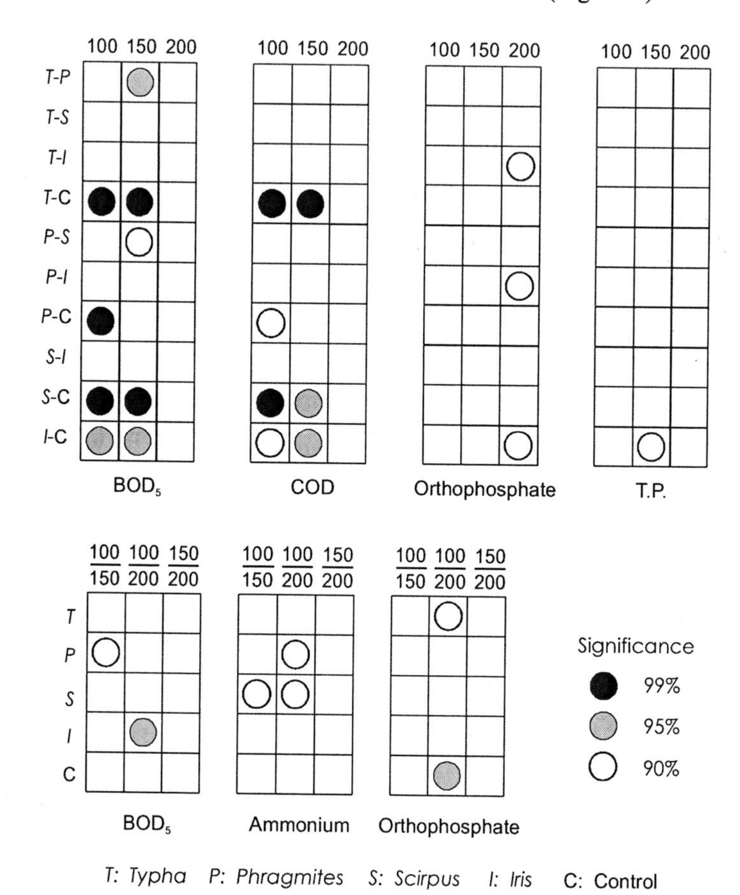

Figure 5. Diagram of statistical significant.

Helophytical Series Cultures Treatment

In the design of the test 2, we followed the model developed by the Fondation Universitaire Luxembourgeoise in Viville, Belgium, under the direction of Dr. Michel Radoux (Radoux & Kemp, 1982).

Different sequences of vegetation were established, with 4 species selected so as to find out the series of helophytes that could exhibit higher removal efficiency (Ansola *et al.*, 1995a).

The series of vegetation chosen were: *Typha-Iris-Scirpus*, *Typha-Scirpus-Phragmites*, *Scirpus-Iris-Phragmites* and *Typha-Iris-Phragmites*. One other series of 3 unplanted tanks was also installed as "control" (Figure 6). In choosing the above series we wept in mind the capacity for getting removing of organic matter at the first level of the series and disposing of nutrients at the second and third levels, characteristics which were obtained in a previous trial (Ansola *et al.*, 1995a).

The flow rate used was of 300 l/d for each of the series, which provides a loading rate of 12 g/m^{2}d in COD. With this flow rate, the hydraulic loading rate treated by each series is 2 equivalent inhabitants (Table. 2) (this approximately implies 1.5 m^2/e.i. on considering that the volume of wastewater per inhabitant per day amounts to 150 l).

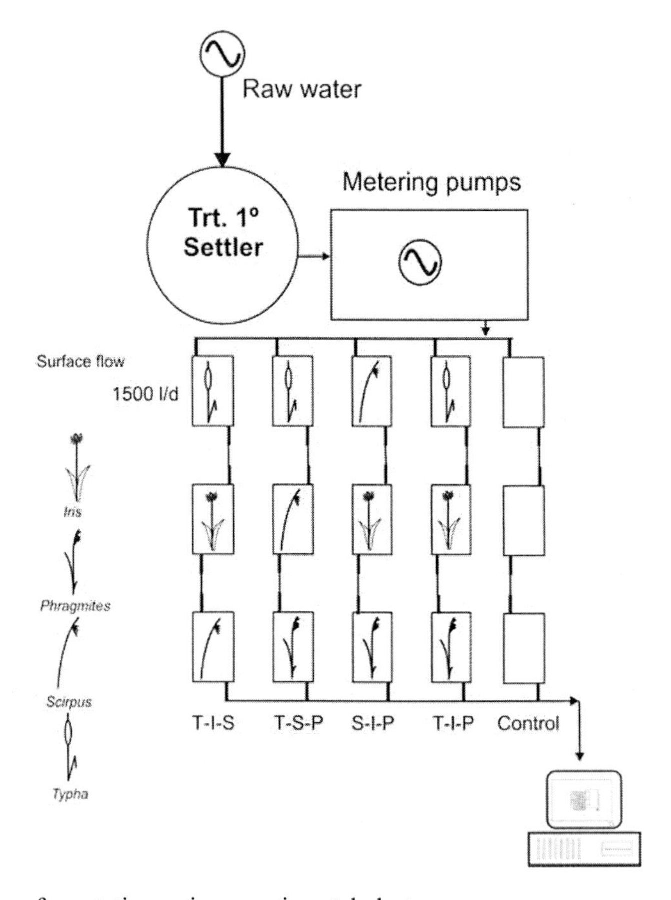

Figure 6. Scheme of vegetation series experimental plant.

Table 2. Operational Parameters

Q (l/d)	(l/m^2/d)	t.r. (d)	h.l. (cm/d)	COD (g/m^2/d)	BOD$_5$ (g/m^2/d)	NO$_3^{2-}$ (g/m^2/d)	NH4$^+$ (g/m^2/d)	TKN (g/m^2/d)	TP (g/m^2/d)
300	91	2.91	7.7	11.9	2.1	1.56	0.69	0.95	1.10

Q. - Flow rate.
t.r. - time of hydraulic retention.
h.l. - hydraulic loading.

Table 3. Mean, Minimum and Maximum Composition of the Raw Wastewater

Parameter	Minimum	Average	Maximum
BOD_5 (mg/l)	11.2	15.8	21.3
COD (mg/l)	55.3	106	197
Nitrite (mg/l)	0.03	0.2	0.56
Nitrate (mg/l)	1.25	64.5	265
Ammonium (mg/l)	3.81	18.6	73
Organic Nitrogen (mg/l)	3.35	12.1	32.5
TKN (mg/l)	7.16	30.7	93.1
Orthophosphate (mg/l)	5.60	8.01	13.7
Total Phosphorus (mg/l)	6.39	11.9	22.0
Potassium (mg/l)	1.96	14.5	38.5

Two more sampling campaigns were made in November and December, 1993, in order to assess the functioning of the system after the crop of macrophytes.

The influent was taken along 24 hours and both in influent and effluent samples the following parameters have been determined (Table 3): COD, BOD_5, potassium, nitrate, nitrite, ammonium, TKN, orthophosphate and TP.

A principal components analysis was applied using the CSS:Statistica. In order to study the possible dissimilarities in the purifying efficiency of the different vegetation series tested, we performed an analysis of the variance of a source of variation (computer program: Statview). The test of Tukey was applied to determine whether or not differences were significant, since the number of replicas has always been constant.

Typha-Scirpus-Phragmites offers the highest removal values in BOD_5 and potassium, and for all parameters a reduction of above 50% is reached. (Figure 7) The serie *Scirpus-Iris-Phragmites* attained quite similar results except for the removal of nitrite which is considerably lower. This series has reached the largest disposal values in TKN, orthophosphate an TP, in all cases exceeding 50%

With the sequence *Typha-Iris-Phragmites* there is a remarkable increase in nitrite disposal, even with values above those of the series *Typha-Scirpus-Phragmites*. Although this does not happen for the rest of the parameters, the highest values of COD and nitrate removal have been obtained with this series.

Typha-Iris-Scirpus give the lowest performance of all plant series under study, in average terms, though concerning the disposal of nitrite, it exhibits the best results of all and its reduction of orthophosphate comes close to the values reached by the series *Scirpus-Iris-Phragmites*. The "control" series could not surpass the values obtained by any plant series in any of the parameters analyzed.

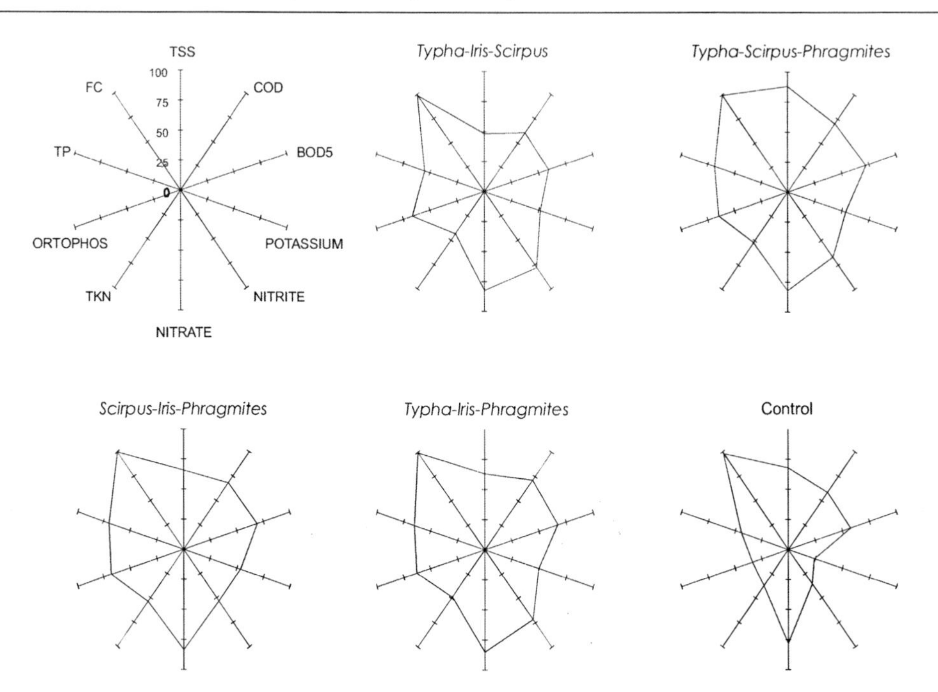

Figure 7. Average removal efficiencies for the vegetation series.

In the principal components analysis, applied to all of the data, the first two axes account for 67.12% of the total variance. Component I (47.00%) is defined in its negative part by the efficiency in the remainder of the nitrogen forms. The second component (20.12%) is defined in its negative part by the COD and BOD_5 (carbonaceous organic matter) and, to a lesser extent by TP (Figure 8).

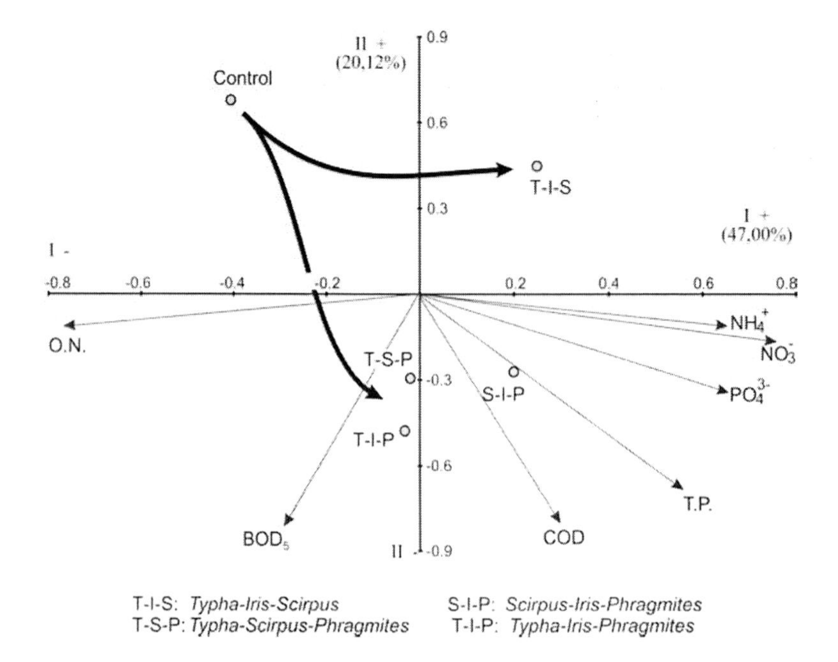

Figure 8. Principal components analysis.

In the distribution of samples within the frame defined by the first two components, two trends are clearly noticed from the most negative part component, where the "control" series stands, towards the most positive part of the same component, where the series formed by *Typha-Iris-Scirpus* lies, and towards the most negative part of component II where we find the series integrated by *Typha-Scirpus-Phragmites* and *Typha-Iris-Phragmites*, which have obtained the best performances in the removal of organic matter.

The disposition of the series *Scirpus-Iris-Phragmites* in the second quadrant, midway between both tendencies, indicates that this is the one which has obtained the highest average disposal of nearly all parameters analyzed in this trial, which is why it is placed on a halfway area, between both groups of plant series.

Regarding the first component, only the removal performance with respect to organic nitrogen is distant from the rest of the parameters, particularly ammonium, nitrate and orthophosphate, which show the highest positive correlation on this axis. This disposition enables the first component to be related to the efficiency degree of the systems in the disposal of nutrients.

Performances concerning COD and BOD_5, and on a smaller scale TP, define the negative part of the second component, thereby allowing us to correlate the sorting out of samples in relation to this component with the level of organic matter removed.

Te ANOVA globally detected significant differences with reference to orthophosphate, between the disposal performance attained by plant series formed by *Scirpus-Iris-Phragmites*, *Typha-Iris-Scirpus* and *Typha-Scirpus-Phragmites*, and that of the "control" system (test of Tukey: 2.903 with a significance of 95%, 2.791 and 2.659 with a significance of 90% respectively), there being no differences in the rest of the parameters selected for the analysis.

Ponds Treatment

Conventional – anaerobic and facultative deep ponds – are, nowadays, the main technologies associated with these processes, with variable results. Problems related to stabilization ponds include the need for a large available land area, and the presence of great amounts of algae in the final effluent, which prevents its reuse. High rate algae ponds (HRAPs), designed by W.J. Oswald of Berkeley University in the latter part of the 1960s (Oswald, 1963, 1988) form an optimal treatment system for wastewater following their initial conception as algae production reactors. The main advantage of such a system is the great reduction in land requirement, keeping a high efficiency and low operation, cost due to minimal energy requirements. HRAPs are a combination of conventional ponds (photosynthetic activity) and plug-flow algae reactors (continuously mixed). Species selection ensures the maximum nutrient removal rate and sedimentability characteristics in order to allow algae removal using simple decantation. These system are basically thin channels or basins (30-50 cm depth) through which water flow is driven by means of small propellers or blowers.

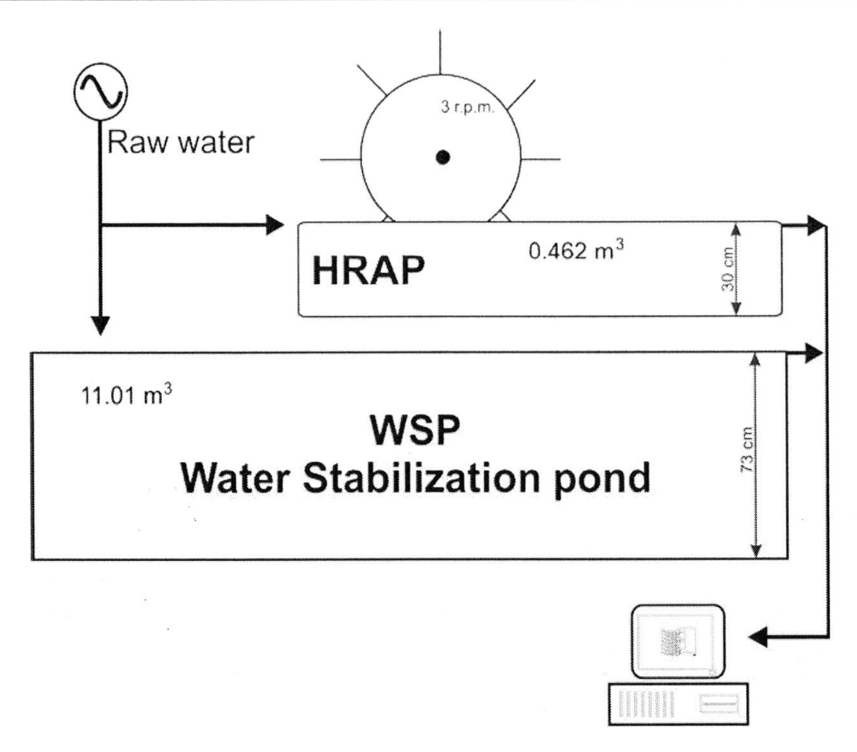

Figure 9. Ponds experimental system.

We have two pilot systems in Mansilla de las Mulas (León). The waste stabilization ponds (WSPs) have a total volume of 11.01 m^3, with an effective depth of 73 cm and a surface area of 15.09 m2, while the HRAPs have a volume of 462 l, a depth of 30 cm and surface area of 1.54 m^2. Water in the HRAPs system is impelled by small paddle wheels operating at 3 rev/min (rotational velocity) obtaining a water flow of 15 cm/s which gives optimal mixed results (Figure 9).

Raw water is taking from the municipal sewer and pumped to a setter tank and after by metering pumps feeding to each system, hydraulic detention time is for WSPs 20 days and for HRAPs 6 days into the two periods of this study- summer and winter. Final effluent is carried out to the Esla river.

Samples were taken over 1 day every 2 h. at depth intervals and were repeated in both seasons; summer and winter. Direct measurement of water temperature, pH, solar radiation, electrical conductivity and dissolved oxygen were taken at depth intervals of 10 cm. Other parameters such as alkalinity, nitrate, ammonia, total phosphorous, particulate organic phosphorous, orthophosphate, total organic carbon, chlorophyll a, faecal coliforms, faecal streptococci, clostridium, staphylococci, and total heterotrophic bacteria were analysed on laboratory from samples taken at two depths (10 and 60 cm) from the WSPs and at one depths (20 cm) from the HRAPs, due to its completely mixed characteristic. Phytoplankton were determined in both systems by counting cells using optical microscopy in samples taken from the settlement tanks.

Water influent (Table 4) was characterised by the measurement and determination of several parameters, most importantly: temperature, pH, electrical conductivity, dissolved oxygen, total suspended and volatile solids, chemical, oxygen demand, biological oxygen demand, alkalinity, nitrate, ammonia, total nitrogen, orthophosphate, total phosphorous, faecal coliforms and faecal streptococci.

Water temperature had no depth variations, except in WSPs in summer, the same is true of electrical conductivity and this is also higher in the WSPs in summer.

High pH values (up to 10) were detected in the WSPs and between 8 and 9.2 in the HRAPs as a consequence of photosynthetic activity. We found higher concentrations of chlorophyll *a* in the WSPs than un the HRAPs, and it was highest at the 60 cm depth at 12:00 h. This explains why dissolved oxygen (D.O.) follows the same pattern and also in accordance with the temperature dynamics, causes high levels of oxygen in the WSPs, (up to the HRAPs levels). Both systems had maximum concentrations of D.O. in winter. Solar radiation was at a maximum on 12:00 h. and presents a typical Gaussian distribution. Radiation was highest at the 30 cm depth in the WSPs.

Alkalinity had maximum concentration values in the WSPs, highest at the surface, and showed no seasonal variation in two systems.

Nitrate was maximum at 12:00 h. in winter at the 10 cm depth in the WSPs. Ammonia removal was best in summer, with a clear difference between the two seasons in the HRAPs.

Phosphorous was assimilated better in summer than winter in both systems and presented higher levels at the 60 cm depth in the WSPs.

Seasonal differences in total organic carbon (TOC) did not occur in the WSP, but in the HRAP it was higher in summer. There were depth differences in the WSP at 12:00 h. the TOC concentration was highest at the 60 cm depth and lowest at the 10 cm depth.

Pathogenic bacteria and faecal indicators were best removed from the WSPs where we found minimum concentration values at the 10 cm depth at 12:00 h.

Table 4. Characteristic Parameters of Raw Water Influent

Parameter	Summer	Winter
Temperature ambient (ºC)	16	8.5
Temperature water (ºC)	22.2	11.8
pH	7.9	7.9
Conductivity (μS/cm)	410	415
D.O. (mg/l)	0.6	1.4
TSS (mg/l)	17.2	20
VSS (mg/l)	17.2	20
BOD_5 (mg/l)	21	14.07
COD (mg/l)	91.28	111.6
Alkalinity (mg/l)	172.41	182.31
Nitrate (mg/l)	0.54	7.71
Ammonium (mg/l)	8.15	6.19
TKN (mg/l)	11.35	3.63
Orthophosphate (mg/l)	4.19	10.24
Total Phosphorus (mg/l)	5.76	7.75
POP (mg/l)	0.57	3.53
Faecal Coliforms (ufc/100ml)	619333	449500
Faecal Streptococci (ufc/100ml)	134100	24600

There was a greater abundance of phytoplankton in the HRAPs then in the WSPs. Monospecific blooms of *Monorraphidium contortum* occurred and they were higher in summer.

Stratification is observed in the WSP system in summer as reflected in: temperature, conductivity, dissolved oxygen and pH distribution. This system was permanently mixed in winter.

HRAP system are completely mixed all of the time, and reflect diurnal and seasonal effects.

Daily changes are reflected by the WSP dynamics.

Photosynthetic activity determines pH increases and ammonia evolution.

There is a greater abundance of phytoplankton in HRAP than WSP systems. High values of cells/ml of monospecific blooms were observed in HRAP system. These values are higher in summer than winter.

Hierarchical Mosaic of Aquatic Ecosystems

The Hierarchical Mosaic of Artificial Ecosystems (HMAE®) has been developed and it has enabled us to compare the overall wastewater treatment efficiency of using different aquatic, semi-aquatic and terrestrial ecosystems in the same climatic conditions and organic influent loads (Radoux 1988, 1989 and Cadelli et al., 1998). The HMAE® system was originally developed in Belgium (Radoux and Kemp; 1982) by applying the ecological adaptations of hydrophyte plants to flooded and polluted conditions. The system consisted of a stabilization pond (stage I) followed by a semi-aquatic ecosystem planted with helophytes (stage II) and by a terrestrial ecosystem (stage III) where ligneous species are planted.

The objective of this test was to study the applicability of this system to the wastewater and climatic conditions of the region and to know the parameters for the design of a full-scale process and to compare the two different systems: experimental design (Ansola et al., 1995a, 1995b) and pilot design on a full scale.

The HMAE® methodological and technological process developed by the Viville Experimental Plant (FUL, Belgium, Radoux and Kemp, 1982) was used to design the Experimental Low Cost Wastewater Treatment Champ is in the village of Mansilla de las Mulas (León, NW Spain).

A free water ecosystem (water stabilization pond, WSP, volume = 11 m^3 and hydraulic residence time = 1.1 d) is used as a first stage treatment and supplies two series with different flow. Each series is formed by three glass fiber tanks (each with a total volume of 0.6 m^3, a surface of 1.1 m^2 and a depth of 0.55 m). The first tanks are filled up to a 25 cm height with 6-8 mm diameter gravel and the water flows over the substrate (semi-aquatic ecosystem with surface regime, free water system FWS) and rise to a height of 20 cm. The effective water volume is reduced to 0.29 m^3 (Ansola *et al.*, 1995a).

Eight series of three glass fibre basins of 0.85 x 1.3 x 0.55 m, were installed with a capacity of 0.6 m^3 and a surface of 1.1 m^2 each basin. They were preceded by a pond like primary treatment (WSP) with a detection time of 24 hours. First basins on series were filled to 30 cm high with 6-8 mm diameter gravel bed. Water is over 20 cm of gravel (Figure 2). These basins had a superficial flow and they were planted with 30 units of *Typha latifolia* collected from their natural system (Ansola *et al.,*1995b).

Two following steps on series were totally filled with inert gravel, They had a subsurface flow and were planted with 15 units/m^2 of *Salix atrocinerea*, *Alnus glutinosa*, *Populus alba* or *Fraxinus excelsior*, each being one year old (Figure 10). The different series, composed of a surface of 4.4 m^2 each one, were as follow (Table 5).

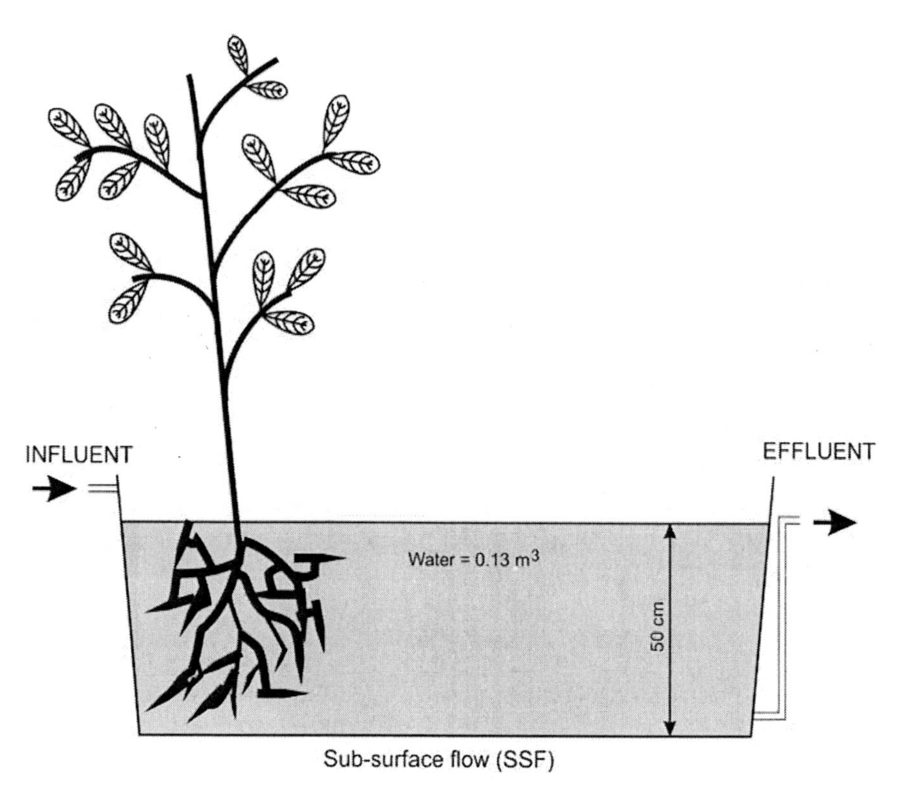

Figure 10. Subsurface flow system of experimental tank.

Table 5. Configuration of Experimental Series and their Designed Water-flow

Series 1	Pond-*Typha-Salix-Salix*	P-T-S-S	Water-flow = 160 l/d
Series 2	Pond-*Typha-Salix-Salix*	P-T-S-S	Water-flow = 250 l/d
Series 3	Pond-*Typha-Salix-Salix*	P-T-S-S	Water-flow = 80 l/d
Series 4	Pond-*Typha-Salix-Salix*	P-T-S-S	Water-flow = 200 l/d
Series 5	Pond-*Typha-Alnus-Populus*	P-T-A-P	Water-flow = 160 l/d
Series 6	Pond-*Typha-Alnus-Populus*	P-T-A-P	Water-flow = 250 l/d
Series 7	Pond-*Typha-Fraxinus-Fraxinus*	P-T-F-F	Water-flow = 160 l/d
Series 8	Pond-*Typha-Fraxinus-Fraxinus*	P-T-F-F	Water-flow = 250 l/d

Raw wastewater was taken and pumped directly from a municipal sewer to supply the system and distributed to the series basins by peristaltic pumps in continuous work. The hydraulic load was the difference between series. For the series 1, 5 and 7 the operative water-flow was 160 l/d for a surface of 4,4 m^2. The series 2, 6 and 8 had a water-flow of 250 l/d with the same surface (Figure 11). There were two special series, 3 and 4 that represented the experimental operations designed flows as a reply to the minimum and maximum flow of the full-scale pilot plant constructed in Bustillo de Cea. (León, Spain) (Ansola *et al*, 1998).

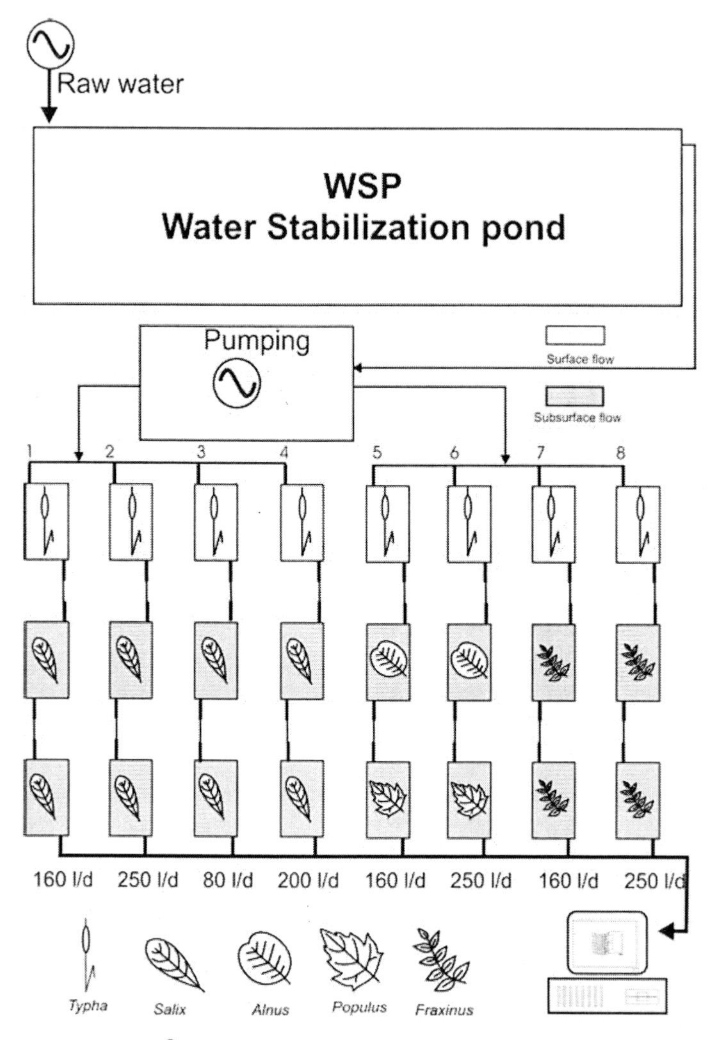

Figure 11. Diagram of HMAE® of experimental pilot system.

The average electrical conductivity detected was 436 µs/cm, which corresponds to a typically domestic wastewater (Metcalf & Eddy, 1998) and low mineral composition (Henze *et al*, 1997), with a maximum of 498 µs/cm and a minimum of 412 µs/cm. Concentration of total suspended solids (TSS) varied between 287 mg/l and 124 mg/l, the average concentration was 164 mg/l, all these values less them of 350 mg/l, which allowed for the quantification of this wastewater as weak concentration (Metcalf & Eddy, 1998) or diluted wastewater type (Henze *et al*, 1997). The organic load also expressed as the total Chemical Oxygen Demand (COD) varied between 271 and 112 mg O_2/l, with average concentration of 156 mg O_2/l, which also confirms its classification as weak concentration wastewater since the average concentration for this parameter in the typical domestic wastewater (Henze *et al*, 1997) is located between 250 and 500 mgO_2/l for very diluted types. Biochemical Oxygen Demand (BOD_5) varied between a maximum of 93 mgO_2/l and a minimum of 72 mgO_2/l, with a average concentration for the study time of 83 mgO_2/l, values below of 110 mg O_2/l, very diluted wastewater type (Henze *et al*, 1997).

Nutrients concentrations showed the same dilution effect of the wastewater used as an influent (Table 6). Nitrate presented a range of variation, with a maximum of 3.9 mg/l and a minimum of 2.4 mg/l, below the typical domestic wastewater (15 - 29 mg N/l; Metcalf & Eddy, 1998). The total nitrogen concentration in the influent, measured as Total Kjeldahl nitrogen (TKN), had a maximum of 19.6 mg/l and a minimum of 7.8 mg/l, bellow of 20 mg/l described (Henze *et al*, 1997) for very diluted wastewater type. With regard to phosphorus concentrations, (TP) was presented with a minimum of 4.2 mg/l and a maximum of 5.0 mg/l, close to 4 mg/l for very diluted types (Henze *et al*, 1997).

This very diluted influent is a common pattern in the north of Spain owing to the infiltration of ground water into the sewage (González and Cabo 1998).

Table 6. Average Influent Concentration and Standard Deviation for the Measured Parameters

Parameters (units)	Average concentration	Standard deviation
Temperaure (ºC)	7.5	0.1
pH	16.6	1.8
Conductivity (µs/cm)	463	22.9
DO (mg O_2/l)	2.1	0.4
TSS (mg/l)	164	40.1
TVS (mg/l)	145	19.2
COD (mg O_2/l)	156	39.1
BOD_5 (mg O_2/l)	83	7.0
Nitrate (mg NO_3^-/l)	3.4	0.4
Ammonia (mg NH_4^+/l)	5.1	0.9
TKN (mg/l)	14.6	2.6
TP (mg P/l)	4.7	0.3
FC (col./100ml)	600285	257602

The effect of pond treatment as a first step into the series, operating as a sedimentation tank with only one day of hydraulic detention time, reduced the average concentration levels of solids up to 31 mg/l for TSS and 26 mg/l for TVS. The organic load was reduced to 47 mg/l average concentration for COD, but it was not sufficient in BOD_5 with an average concentration of 38 mg/l.

The average concentration levels of nutrients after pond treatment were insufficient for requirements according to European legislation. Thus, we obtained average concentrations of 4mg/l for ammonia, 2 mg/l for nitrate, and with regard to total nutrients forms, 11 mg/l for total nitrogen and 3 mg/l for total phosphorous were obtained.

Disinfecting efficiency, owing to the short detention time applied, was insignificant depending on waste stabilisation pond possibilities, though an average concentration of 169,044 col/100ml was obtained.

Values of concentration for the most important parameters in the final effluent are given in Table 7, showing good results obtained for the final effluent quality and they agree with the requirements of the European Council legislation.

Table 7. Average Effluent Concentration for the Main Parameters in the Different Series Tested

Parameters (units)	Effluent							
	Series 1 PTSS 160 l/d	Series 2 PTSS 250 l/d	Series 3 PTSS 80 l/d	Series 4 PTSS 200 l/d	Series 5 PTPA 160 l/d	Series 6 PTPA 250 l/d	Series 7 PTFF 160 l/d	Series 8 PTFF 250 l/d
pH	7.38	7.45	7.36	7.30	7.39	7.44	7.39	7.32
DO (mg O_2/l)	3.8	3.5	5.8	3.4	3.8	3.9	3.7	3.7
Conduct. (μs/cm)	366	355	364	368	359	360	358	353
TSS (mg/l)	3.4	3.6	4.9	5.0	3.7	2.5	3.3	1.9
COD (mg O_2/l)	7.8	10.9	9.4	11.0	12.2	13.2	9.9	12.0
BOD_5 (mg O_2/l)	1.1	1.2	0.9	1.3	1.1	0.9	0.7	1.1
Nitrate (mg NO_3^-/l)	0.3	0.3	0.1	0.1	0.3	0.5	0.2	0.5
Ammon (mg NH_4^+/l)	0.4	0.7	0.1	2.0	0.2	1.0	0.1	0.2
TKN (mg N/l)	0.4	1.4	0.3	2.2	0.5	0.9	0.2	0.9
TP (mg P/l)	0.2	0.2	0.1	0.4	0.2	0.3	0.04	0.3
FC (col./100ml)	167	245	14	73	71	96	113	116

Average removal efficiency obtained in the study are shown in figure 1, these results show a high reduction level in all the series for the different analysed parameters.

The suspended solid removal was high in the first pond stage with an average removal of 81% in TSS, An average removal between 97% of series 4 (PTSS-200) and 99% of series 8 (PTFF-250) was obtained in the final effluent of the series. This last series, which was operating with the higher water-flow, is significantly different to others applying a statistical "t" test for independence variables (significant level of 95%), except with series 6 (PTPA-250), for the solids removal efficiency.

Suspended (TSS) and volatile (TVS) solids concentration in the influent and the effluent have positive correlation, therefore the degree of removal found for each one of these parameters does not vary.

Organic matter removal efficiency was represented for the obtained elimination in COD and BOD_5. In the pond stage an average removal of 69% to COD and 54% to BOD_5 was obtained. These results are not sufficient. Treatment in the following stages showing an highly removal efficiency and was more efficient with higher assayed hydraulic detection time. Thus removal levels varied between 91% in series 6 (PTPA-250) to 95% in series 1 (PTSS-160) for the COD average elimination and between 98% in series 7 (PTFF-160) to 99% in case of BOD_5.

According with the "t" test assayed for the variables independence of results obtained in COD reduction. Series 1 (PTSS-160) is significantly independent of the rest also occurs with the series 3 (PTSS-80).

With regard to the distribution of woody species in the third stage of the series for COD elimination, the use of *Salix atrocinerea* was more adequate than others species. However, with regard to the DBO_5 removal, the use of *Fraxinus excelsior* lad a greater efficiency, however, significant differences were not detected between the data of this parameter.

Nutrients (N and P) removal is very low in the pond stage, obtaining 25% for the TP and 23% for TN, with regard to the nitrogen forms studied an average removal of ammonia of 29% and 30 % in the case of nitrate was obtained in the first stage.

The nitrate assimilation in the series was produced with greater hydraulic detention times and fundamentally in the case of series that used *Salix* sp. as woody species, the maximum average removal in the series 3 (PTSS-80) with a 98% was obtained. All the results obtained for this parameter were found 85% of average removal (minimum in series 6 PTPA-250 and series 8 PTFF-250).

The removal of ammonia reached maximum average levels in larger used surfaces (99% in the case of series 3 PTSS-80) and it was minimal in greater applied water flow (81% in the case of series 6 PTPA-250). Significantly differences for this parameters were not found in the different flows tested with the series of *Fraxinus* sp (series 7 and 8) or between these with that which used the minor flow of *Salix* sp series (series 3).

In total nutrients forms (TN and TP), the use of *Fraxinus* sp in series with the lowest water flow applied (series 7 PTPP-160) obtained the best average results (98% for TN and 99% for TP), these results are significantly different to the other series studied, except for TN with series 3 (PTSS-80) and series 1 (PTSS-160) which obtained 98% in TN removal and 97% in TP removal for series 3, and 97% for TN and TP removal in series 1.

Disinfecting in the pond stage obtained average results of 71%. The use of macrophytes and woody species notably improves the results, with average disinfecting from faecal coliform bacteria in all the series studied up to 99.9% being obtained The maximum results for disinfecting were obtained with the lowest flows tested, with 99.99% maximum being observed in the series 3 PTSS-80 (99.99%), however there did not seem to exist a preference for the disinfecting in the use of one or another kind of woody species.

The operational experimental flow was calculated so that we could test the minimum and maximum flow to design the pilot plant on a full scale.

Pilot Design on a Full Scale

The location of Bustillo de Cea, in the northeast province of León (Spain), has 250 inhabitants in the winter increasing to 400 in full summer. This location has been chosen as an experimental prototype in order to draw conclusions for operations and maintenance costs to obtain a effluent in order the European Union Directive 91/271/CEE by the year 2005 (ECOD, 1991).

The climatic type of the area is Mediterranean (Papadakis, 1961) with 546.7 mm annual rainfall approximately and 10.2°C mean annual temperature.

We have suggested the setting up of a 890 m^2 full scale pilot plant for integral municipal wastewater treatment. This system is made up of four tanks of artificial aquatic ecosystems following Lallaing example (Radoux et al.,1995, 1997). The mean wastewater flow is 9.5 m^3/d in winter and twice this volume in summer. The maximum flow is over 7.7 m^3/h. The wastewater treatment plant has been sized for a maximum flow of 0.6 l/s. The rated retention time correspond to 10.5 d.The influent average water pH was 7.69; temperature was 14.43 ºC; conductivity was 824.5 S/cm; and dissolved oxygen 2.18 mg/l.

The distribution of surface has been as follows: 230 m^2 for the first basin formed by a lagoon (WSP, aquatic ecosystem) of up to 2 m in depth where the inflow enters and 1.5 m where the out-flow goes out, reconstructing an aquatic ecosystem with algae (basin 1 in figure 2); 210 m^2 for the second basin where 15 plants per square meter of *Typha latifolia* taken from their natural habitat have been planted. In this basin the water circulation is by free flow (FWS, basin 2 in figure 2); 87.5 m^2 for basin 3 which also has surface flow (FWS) and where 15 plants per square meter of *Iris pseudacorus* were planted (basin 2 and basin 3 are semi-aquatic ecosystems) and 362.5 m^2 for basin 4 which acts as a gravel filter without vegetation with sub-surface flow (SSF, terrestrial ecosystem) (Figure 12).

Table 8 shows the average characteristics of the wastewater in the location of Bustillo de Cea during the sampling period, as well as the average results obtained during the first months of the operation time and the first vegetative period of the helophytes after being transplanted from their natural habitat.

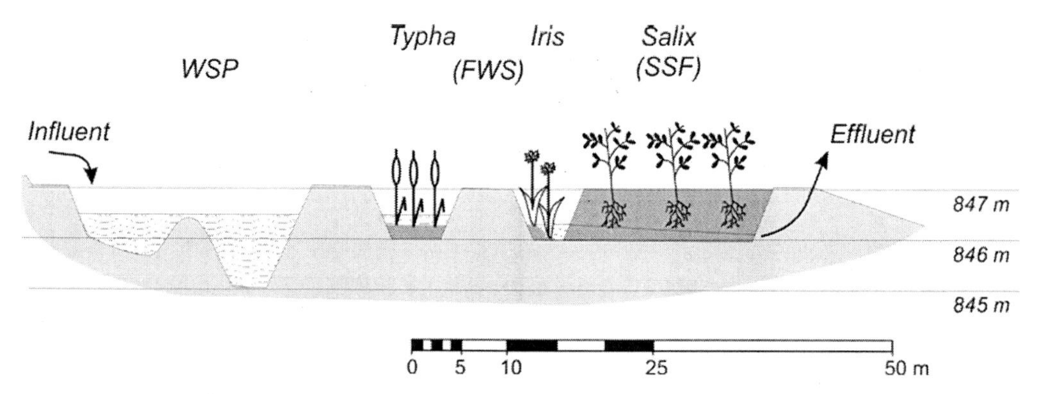

Figure 12. HMAE® system. Pilot design on a full-scale. (WSP: water stabilization pond. FWS: surface flow. SSF: subsurface flow)

Table 8. Average Values of the Influent and Effluent as Seen in the Average Performance of the Full-scale System in Bustillo de Cea (October 99 – September 2000)

	Influent	Effluent	Average removal (%)
TSS (mg/l)	188.25	34.21	60.93
BOD_5 (mgO/l)	154.80	21.70	61.44
COD (mgO/l)	294.79	81.38	59.47
NH_4^+(mgN/l)	27.80	24.15	33.04
TKN (mgN/l)	36.85	28.31	35.59
TP (mgP/l)	19.99	9.74	34.84
FC (col/100ml)	19 x10^6	367,044	96.45
FS (col/100ml)	167,978	5,385	91.73

Note: The average removal was calculated as the average of removal in each sample.

As can be seen from the above table, the average performances obtained during the whole study period have not been very high. However, there is an appreciable difference between the average values during the active growth of the wetland system (May-September) and the resting period (October-December) in which the performances fall considerably and cause the average value to go down as much as 10 times in some of the parameters as occurs in the removal of nutrients.

Table 9 shows the correlations between inlet parameters, with a significance of 95%. The greatest correlation is between TKN/NH_4^+ (0.93) as we might suppose. Relationship between conductivity and nitrogen is high positive, but not happens the same with phosphorus parameter. Have not an effect on TP, COD and BOD_5 are most correlationated with total nitrogen then with ammoniacal nitrogen and total phosphorus respectively. In order way, TP and NH_4^+ have influence on faecal contamination in that order.

Table 9. Correlations between Inlet Parameters

	TKN	NH_4^+	TP	TC	FC
pH	*.58*	*.44*	-.11	-.18	-.19
Tª	-.08	.10	*.53*	.29	.16
Cond	*.75*	*.73*	-.09	-.02	.07
D.O.	-.33	*-.48*	*-.60*	-.55	-.41
TSS	*.66*	*.62*	.15	.35	.32
COD	*.60*	*.51*	*.40*	.38	.44
BOD_5	*.44*	*.37*	*.34*	.26	.29
TKN		*.93*	.22	.25	.23
NH_4^+			.22	*.39*	*.34*
TP				*.42*	*.37*
TC					.84

Note: Significance 95%.

Seasonal differences in the nitrate removal were shown by Spieles and Mitsch (2000) and in several parameters by Radoux et al. (1995). Also they appear seasonal differences in the elimination of N and P in the work made by Tanner (2001) with *Schoenoplectus.*

The removal of organic matter is notable. In spring and summer it exceeds 70% in average removal values and it does not reach 40% in autumn. The removal of nutrients presents a similar case, with the removal of TKN, during the period of highest temperatures, rising to more than 56% while the removal of TKN and TP do not exceed 5.5% during the analyses from October to December. The difference between both periods is not so obvious in the removal of the indicating bacteria of the fecal pollution (Table 10). This is probably owing to the fact that the most intensive processes of removal are predation, competition, proteolysis or filtration (García and Bécares, 1997).

It can be seen a global reduction in all the chemical parameters. A increase, in both parameters during the pond phase, is to be seen in fecal coliforms and fecal streptococcus. This may be due to a very bad treatment of pretreatment in the first period of the analysis hence considerably increasing the average value.

On conducting a more detailed analysis and concentrating on the average removal of four parameters table 11 illustrates, that the stage represented by the tank 2 and 3 (FWS) and filled with *Typha latifolia* and *Iris pseudacorus*, reaches higher partial average performances for organic matter (49%), nutrients (30%) and faecal pollution (94%).

A statistical analysis has been conducted to observe the existence of significant differences (t-Student) in the elimination of the chosen parameters between the different phases of the treatment. The results obtained are shown in table XXX.

Table 10. Average Values of Removal Obtained during the Different Sample Periods of the Full-scale System in Bustillo de Cea

	% (October-December) average removal	% (May-September) average removal
TSS (mg/l)	35.06	78.17
BOD_5 (mgO/l)	39.13	76.32
COD (mgO/l)	37.85	73.88
$NH4^+$(mgN/l)	4.47	52.08
TKN (mgN/l)	3.57	56.93
TP (mgP/l)	5.44	54.40
FC (col/100ml)	99.08	95.14
FS (col/100ml)	89.23	92.98

Table 11. Significant Differences (t-Student) in the Different Phases of the Treatment of Full-scale System (Bustillo de Cea)

	BOD_5	NH_4^+	TP	FC
WSP-FWS	-	-	†	†
FWS-SSF	-	-	-	*
WSP-SSF	-	-	*	-

* P < 0.05.
† P < 0.01.

Between the stabilizing pond (WSP) and the tank 2 and 3 (FWS) there are significant differences (99%) in the removal of phosphorus and pathogens. A higher value of removal is produced in the FWS vegetated areas. Between the planting of *Typha* and *Iris* and the last stage (SSF), there are significant differences (95%) in the removal of fecal coliforms with a higher performance in the stage with vegetation (table 6). There are also significant differences in the removal of phosphorus (95%) between the pond (WSP) and the gravel filter (SSF), the latter obtaining higher performances of nearly 30%. In the analyses conducted no significant differences between the filters (FWS and SSF) are apparent in this parameter. It should be noted, however, that the vegetation was in a period of stabilization. This presupposes that in subsequent studies, the differences which are already apparent between both stages, will be more significant.

Table 12. Depuration Plant Cost

	$	$/m²
Land movement	14,000	15.73
Conduction and Collection box	7,730	8.68
Waterproofing and Gravel fill	37,766	42.43
Border and Enclosure	5,768	6.48
Distributor	2,911	3.27
Vegetal plantation	4,609	5.19

It has to be pointed out, however, that the data collected up to the present time, are from the first months of the operation time which presupposes that they will improve when the *Salix* are planted in the gravel bed.

Table 12 reflects the budget, by steps, of the plant construction at Bustillo de Cea. Management costs in these systems are expected to be round about 0,065 \$/m^3.

ACKNOWLEDGEMENTS

These projects have been carried out thanks to financial support provided by Diputación Provincial de León, Cea City Hall and the collaboration relationship between the University of León (Spain) and the Fondation Universitaire Luxembourgeoise (Belgium).

REFERENCES

Ansola, G.; Fernández, C. & de Luis, E. (1995a). Removal of organic matter and nutrients from urban wastewater by using an experimental emergent aquatic macrophyte system. *Ecol. Eng. 5*, 13-19.

Ansola, G.; Fernández, C. and de Luis, E. (1995b). Removal of nutrients and organic matter from urban wastewater by using an experimental constructed wetland. In: Ramadori, Cingolani,& Cameroni (ed.) *Natural and Constructed Wetlands for Wastewater Treatment and Reuse. Experiences, Goals and Limits*. pp. 39-46. Perugia.

Ansola, G.; González, J.M.; Soto, F.; García, M.; López, G.; Radoux, M.; Cadelli, D. & de Luis, E. (1998). Natural integrated systems using constructed wetlands for treating wastewater in Northwest Spain. *Procc. 6th.International Conference on Wetlands systems for water Pollution Control*. pp. 680-687. Brasil.

APHA-AWWA-WPCF. (1989). Clesceri, L.S.; Greenberg A.E. & Trussell. R.R. (ed). *Standard Methods for the examination of water and wastewater*. 17 th Edition. Baltimore. MD.

Bhamidimarri, R.; Shilton, A., Armstrong, Y.; Jacobson, P. & Scarlet, D. (1991). Constructed wetlands for wastewater treatment: the New Zealand experience. *Wat. Sci. Technol. 24*, 247-253

Black, C.A. (1975). *Relaciones Suelo-Planta*. Tomo II. Ed. Hemisferio Sur.

Braskerud, B.C. (2002a). Factors affecting nitrogen retention in small constructed wetlands treating agricultural non-point source pollution. *Ecol. Eng. 18*, 351-370.

Braskerud, B.C. (2002b). Factors affecting phosphorus retention in small constructed wetlands treating agricultural non-point source pollution. *Ecol. Eng. 19*, 41-61.

Breen, P.F. (1989): A mass balance methods for assessing the potential of artificial wetlands for wastewater treatment. *Water Research. 24(6)*, 589-697.

Cadelli D., Radoux M. & Nemcova M. (1998). Constructed wetlands in Belgium. In: Vymazal, Brix, Cooper, Green & Haberl (ed.). *Constructed wetlands for wastewater treatment in Europe*. pp. 77-93. Backhuys Publ.

Cooper, P.F. & Findlater, B.C. (eds.), (1990). *Constructed Wetlands in Water Pollution Control*. 605 pp. Oxford, England. Pergamon Press.

Corbitt, R.A. & Bowen, P.T. (1994). Constructed Wetlands for Wastewater Treatment. In: Kent, D.M. (ed). *Applied Wetlands Science and Technology*. 388 pp. 221-241. Boca Raton, Florida. Lewis Publishers, Inc.

ECOD nº L 135, (1991). *European Council Directive about urban wastewater treatment.* DOCE (91/271/CEE). pp.40-52.

García M. and Bécares E. (1997). Bacterial removal in three pilot scale wastewater treatment systems for rural areas. *Wat. Sci. Technol. 35(11-12),* 197-200.

Gómez Cerezo, R.; Suárez M.L. & Vidal-Abarca M.R. (2001). The performance of a multi-stage system of constructed wetlands for urban wastewater treatment in a semiarid region of SE Spain. *Ecol. Eng. 16,* 501-517.

González, J.M. and Cabo, A. (1998). *Libro Blanco sobre calidad de agua en el Bierzo (León, Spain).* 222 pp. Universidad de León. León.

Hammer, D.A. (ed.). (1989). *Constructed Wetlands for Wastewater Treatment: Municipal, Industrial and Agricultural.* 831 pp. Chelsea, Michigan. Lewis Publishers, Inc.

Henze, M.; Harremoës, P; La Cuor Jansen, J. and Arvin, E. (1997). *Wastewater treatment.* Springer, Berlin

Kadlec R.H. (1995). Overview: Surface flow constructed wetlands. *Wat. Sci. Technol. 32(3),* 1-12

Kadlec, R.H. and Knight, R.L. (1996). *Treatment wetlands.* 893 pp. Boca Raton, Florida. Lewis Publishers, Inc.

Mara, D.D., Mills S.W., Pearson H.W. & Alabaster G.P. (1992). Waste stabilization ponds: A viable alternative for small community treatment systems. *J.CIWEM. 6(2),* 72-78.

Mara, D.D., Cogman, C.A., Simkins, P. & Schembri, C.A. (1998). Performance of the Burwarton estate waste stabilization ponds. *J. CIWEM. 12(8),* 260-264.

Metcalf & Eddy Inc. (1998). *Ingeniería de aguas residuales. Tratamiento, vertido y reutilización.* 1485 pp. Madrid. McGraw-Hill.

Mitsch, W.J., Horne, A. & Nairn, R.W. (eds), (2000). Nitrogen and Phosphorus Retention in Wetlands. *Special Issue of Ecol. Eng. 14*, 1-206.

Moshiri, G.A. (ed.) (1993). *Constructed wetlands for Water Quality Improvement.* 632 pp. Boca Raton, Florida. Lewis Publishers, Inc.

Murphy, J. and Riley, J.P. (1962). Methods of Seawater Analysis. In: Grasshoff, Ehrhardt & Kremling (ed). *Analytical Chemistry 27*:31-36.

Neill, C. (1989). Effects of nutrients and water levels on emergent macrophyte biomass in a prairie marsh. *Ca. J. Bot. 68,* 1007-1014.

Nguyen, L.M. (2000). Organic matter composition, microbial biomass and microbial activity in gravel-bed constructed wetlands treating farm dairy wastewaters. *Ecol. Eng. 16,* 199-221.

Olson, R.K. & Marshall, K., (1992). The role of created and natural wetlands in controlling non point source pollution. *Special Issue of Ecol. Eng. 1*, 1-170.

Oswald, W.J. (1963). High rate pond in waste disposal. *Dev. Ind. Biotechnol, 4,* 112-119.

Oswald, W.J. (1988). Micro-algae and wastewater treatment. In: Borowitza & Borowitza (ed). *Microalgal Biotechnology*. 305-328. Cambrige University Press.

Pant, H.K., Reddy, K.R. & Lemon, E. (2001). Phosphorus retention capacity of root bed media of subsurface flow constructed wetlands. *Ecol. Eng. 17,* 345-355.

Pant, H.K. & Reddy, K.H. (2003). Potential internal loading of phosphorus in a wetland constructed in agricultural land. *Water Research, 37*, 965-972.

Papadakis, P. (1961). *Climatic tables for the world.* Buenos Aires.

Perkins, J.; & Hunter, C. (2000). Removal of enteric bacteria in a surface flow constructed wetland in Yorkshire, England. *Water Research, 34(6),* 1941-1947.

Radoux, M. (1985). *Epuration de rejets domestiques par marais reconstitues.* Fondation Universitaire Luxembourgeoise (ed). 113 pp. Arlon.

Radoux, M. (1988). Epuration comparée des eaux usées domestiques par trois plantations hélophytiques et par lagunage à microphytes sous un même climat temperé. *Ecological Applic. 9(1),* 25-38.

Radoux, M. (1989). Epuration des eaux usées par hydrosère reconstituèe. *Journèe D'Etude Cebedeau-IAWPRC* pp. 62-68.

Radoux, M.; and Kemp, D. (1982). Aproche écologique et experimentale des potentialités épuratrices de quelques hélophytes: *Phragmites australis* (cav.) Trin. Ex Steud. *Typha latifolia* L. et *Carex acuta* L. *Trib. Cebedeau. 465-466(35),* 325-340.

Radoux, M.; Cadelli, D. and Nemcova, M. (1995). A mosaic of artificial ecosystems as a wastewater treatment plant: Evaluation of the pilot plant of Lallaing (France). In: Ramadori, Cingolani & Cameroni (ed.) *Natural and Constructed Wetlands for Wastewater Treatment and Reuse. Experiences, Goals and Limits.* pp. 275-284. Perugia.

Radoux, M., Cadelli, D. and Nemcova, M. (1997). A comparison of purification efficiencies of various constructed ecosystems (aquatic, semiaquatic and terrestrial) receiving urban wastewaters. In: Vymazal, J. (ed.), *Nutrient Cycling and Retention in Wetlands. Wetlands Ecology and Management 4,* 201-217. Kluwer, Dordrecht.

Reddy K.R. & DeBusk, W.F. (1985). Nutrient removal potential of selected aquatic macrophytes. *J. of Environmental Quality. 14(4),* 459-462.

Shinn, M.B. (1941). Methods of Seawater Analysis. In: Grasshoff, Ehrhardt & Kremling (ed). *Analytical Chemistry* 13-33.

Söderqvist, T. (2002). Constructed wetlands as nitrogen sinks in southern Sweden: An empirical analysis of cost determinants. *Ecol. Eng. 19,* 161-173.

Spieles, D.J. and Mitsch W.J. (2000). The effects of season and hydrologic and chemical loading on nitrate retention in constructed wetlands: a comparison of low-and high-nutrient riverine systems. *Ecol. Eng. 14,* 77-91.

Tanner, C.C. (2001). Growth and nutrient dynamics of soft-stem bulrush in constructed wetlands treating nutrient-rich wastewaters. *Wetlands Ecology and Management 9,* 49-73.

Vouillot, M. & Boutin, C. (1987). Les systémes rustiques d'epuration: aspects de l'experience Française; possibilités d'application aux pays en voie de développement. *Trib. Cebedeau. 518,* 21-31.

Younger P.L., Large A.R.G. & Jarvis A.P. (1998). *Hydrology in a changing environment* Volume I pp. 495-502.

In: Progress in Aquatic Ecosystems Research
Editor: A. R. Burk, pp. 167-217
ISBN 1-59454-383-6

Chapter 8

Fuzzy Modelling of Dryland Salinity Hazard in Western Australia

***Graciela Metternicht*[*] *and David Malins*[**]**
Department of Spatial Sciences, Curtin University of Technology
Perth WA, Western Australia

Abstract

The two principal forms of logic that can be used as a basis for spatial modelling are Boolean and fuzzy logic. This paper documents research investigating the use of fuzzy logic modelling to predict the distribution of dryland secondary salinity. In considering the modelling of salinity, classical set theory would demand that all areas be classified as being saline or non-saline. Fuzzy logic allows for this unrealistic division to be removed, modelling the uncertainty of the boundary between these regions and the measurements on which they are based by allowing degrees of membership to classes, that better reflect the natural variation of soil properties. To this end, an existing approach for predicting the distribution of salinity, Fuzzy Landscape Analysis GIS (FLAG), developed by Roberts *et al.* (1997) is implemented. An attempt is made to optimize the predictive power of FLAG through the inclusion of geological and remote sensing data. FLAG models salinity distribution within the framework of fuzzy logic. Results from this investigation are compared with the outputs of a predictive model of salinity based on probability theory and Boolean logic.

The attempt to optimize FLAG was not as successful as expected. Of the FLAG based predictions, CC (created from topographic indices) produced the most accurate result. Of the optimized FLAG predictors FLAG_VEG (that incorporates vegetation data) produced the best result. The prediction of areas at risk of salinity derived from the probabilistic model was more accurate when validated against the same ground truth data used to validate the results of this investigation. However, the comparison of CC and FLAG_VEG with the probability-based prediction indicated that the differences between them were not significant at a 95 percent confidence level. It is thought that the quality of

---

[*] g.metternicht@curtin.edu.au
[**] david.malins@student.curtin.edu.au

the data available to create the Digital Elevation Model hampered the performance of the FLAG based predictions.

INTRODUCTION

Dryland secondary salinity, defined as the build up of salt in the soil usually as a result of a rising watertable caused by human-induced changes in land use, is one of the greatest environmental threats facing Australia (Cooperative Research Centre for Plant Based Management of Dryland Salinity, 2002). Within the South-west agricultural region of Western Australia, 1.8 million ha of land have already been affected, mainly due to the replacement of deep-rooted native vegetation with shallow-rooted agricultural crops (Nulsen and McConnell, 2002). The National Land and Water Resources Audit (2000) found approximately 5.7 million ha of Australia are at risk or affected by dryland salinity. The Audit found that in 50 years this area might increase to over 17 million ha.

As remedial action requires reliable information to help set priorities and choose the most appropriate course of action (Metternicht and Zinck, 2003), this paper summarizes an investigation that compares a model for predicting dryland secondary salinity that is based on fuzzy logic and one that is based on Boolean logic and probability theory. This comparison was conducted to assist in drawing conclusions on the applicability of using fuzzy logic to model salinity and highlight differences between probability-based and fuzzy logic-based modelling.

As the research builds on the work of Roberts *et al.* (1997) who developed Fuzzy Landscape Analysis GIS (FLAG), the first aim was to perform fuzzy landscape analysis using the fuzzy modelling methodology adopted in FLAG. An extension of FLAG recommended by Roberts *et al.* (1997) involving the use of vegetation layer was implemented along with the use of geological data as recommended by Laffan (1996a).

The second aim of the research is to statistically compare the fuzzy modelling of salinity as performed by FLAG with an existing 'crisp' (probability based) method for modelling salinity. The probability-based study used for the comparison is the result of the Land Monitor (2003) project. The methodology is described in Evans and Caccetta (2000a,b). Having performed the comparison, areas of agreement and disagreement between the results were analysed to determine possible causes of disagreement. The aim of the last step was to gain insight into possible means of extending and improving fuzzy landscape analysis of salinity. The hypothesis is that the use of fuzzy landscape analysis to model salinity will produce both accurate and realistic results, with minimum data requirements (e.g. a Digital Elevation Model).

BACKGROUND

Factors Affecting Dryland Salinity

Jolly et al. (2002) state that under natural conditions, catchment discharge approximates recharge entering the groundwater system in a dynamic hydrological equilibrium. Due to land use changes, vegetation discharge is no longer in operation in large areas of Australia, resulting in increased recharge. This means that discharge must also gradually increase until the catchment reaches a new dynamic equilibrium, with discharge occurring where the aquifer cannot transmit the additional recharge, i.e. the aquifer capacity is exceeded. The spatial distribution of discharge can be extremely complex and is triggered by a range of factors, including: (i) the aquifer becoming more constricted due to thinning or narrowing (e.g. basement highs); (ii) a groundwater gradient decrease due to changes in topography (e.g. break of slope); and (iii) the permeability of the aquifer decreasing (e.g. dykes, faults, thinning of sediments) (Jolly *et al.*, 2002).

Barrett-Lennard and Nulsen (1989) concluded that topography alone is not sufficient to predict the location of all salinised areas, a finding supported by Clarke *et al.* (1998a), who also found that regolith depth and the location of cleared areas do not fully explain the location of salinised areas. Where agreement does exists is on the importance of geology in influencing the location of salinised areas.

Geological barriers such as dykes, faults bedrock highs and perched aquifers intersected by topographic change have been shown to be important features when considering groundwater flow and show a spatial association with salt affected land (Engel *et al.*, 1987; Lewis, 1991; Clarke *et al.*, 1998a, Salama *et al.* 1993). Likewise, clay saprolite formed above dolerite dykes is generally less permeable than the surrounding regolith (the weathered product of granite and granitic gneiss) (Engel *et al.*, 1987). This clay acts as linear hydraulic barrier to lateral groundwater flow resulting in the discharge of saline water into the surface soils (Engel *et al.*, 1987).

Faults and fractures modify groundwater flow as they can act as carriers or barriers (or both) of groundwater (Clarke *et al.*, 1998b). Clarke *et al.* (1998a, p.89) showed that the relationship between faults and salinised land is causal, not just spatial, with higher hydraulic conductivity inside the fault zone as the mechanism underlying this relationship. According to Clarke *et al.* (1998a), faults in the Western Australian Wheatbelt have a zone of influence between 2 km and 4 km either side of the centre line of the fault, with the risk of salinity increasing upslope from these structures (George, 1998).

Fuzzy Sets and Fuzzy Modelling

Openshaw and Openshaw (1997) describe the crisp representation of continuous phenomena as both 'artificial and misleading.' They also point out that most classes in the real world are in fact continuous. Robinson (2002) states that when identifying locations that meet some suitability criteria 'Boolean methods of representation and query processing wrongly reject a larger number of locations than does the fuzzy model.' These methods assume that attributes can be measured and described exactly, though many attributes can not be recorded exactly because of measurement errors and spatial variation (Heuvelink and Burrough, 1993). Boolean logic is also poor at mimicking human reasoning because as Zadeh (1987 cited in Openshaw and Openshaw, 1997, p.269) states '*The human brain... uses fuzzy logic.*' Whether it is data from the real world or expert opinion being incorporated into a Boolean logic model, there is a requirement to move from fuzzy logic to Boolean logic and in interpreting or applying the results move back to the fuzzy paradigm. This represents another drawback to the use of Boolean logic.

Quelch and Cameron (1994) state that significant uncertainties are present in any analysis of risk. Uncertainty relates both to vagueness (the difficulty in making sharp distinctions in the world) and ambiguity (where choice between alternatives is unspecified) and is caused by lack of information, abundance of information, conflicting evidence, ambiguity, measurement, and subjective belief (Klir and Folger, 1988; Kangas and Kangas, 2002). Fuzzy logic provides a framework for dealing with both vagueness and ambiguity (Klir and Folger, 1988).

There is agreement among several authors (Pal and Bezdek, 2000; Mendel, 2000; Klir and Folger, 1988; Klir, 2000; Kangas and Kangas, 2002; Metternicht, 2001) on the value of fuzzy logic for dealing with vagueness due to linguistic imprecision, often referred to as fuzziness. Furthermore, in fuzzy sets analysis, it is quite acceptable to use limited data together with the analyst's subjective opinion to quantify an uncertain input parameter (Quelch and Cameron, 1994). There is also agreement on the applicability of using fuzzy logic to represent imprecise information and on its ability to provide a better representation of our true knowledge of the parameters than a single value or a probability density function (Quelch and Cameron 1994; Zadeh, 1992; Yager, 2002; Gonzalez et al. 1999).

Fuzzy modelling has already proved itself a powerful tool for modelling in environmental applications. These applications include the prediction of soil strength (Changying and Junzheng, 2000), soil salinity mapping (Metternicht, 1998), delineation of agroecozones (Liu and Samal, 2002), landscape regionalization (Hall and Arnberg, 2002), soil-landscape modelling (Bruin and Stein, 1998), landscape vegetation modelling (Roberts, 1996), the prediction of coral reef development (Meesters et al., 1998) and the identification of degraded forest soils (Riedler and Jandl, 2002). Kampichler et al. (2000), Kollias and Kalivas (1998) and Liu and Samal (2002) all positively comment on the accuracy and realism of fuzzy modelling. Further, Robinson (2002) mentions that a comparison of fuzzy versus non-fuzzy approaches shows that fuzzy logic-based approaches typically provide more detailed, useful information.

Fuzzy Membership Functions

Membership to a fuzzy set is defined by any real number from 0 to 1, indicating a continuous scale from full non-membership to full membership (Fisher, 2000; Bonham-Carter, 1994). Thus, a fuzzy set does not have sharply defined boundaries (as shown in Figure 1) and an element may have partial membership in several classes (Jensen, 1996).

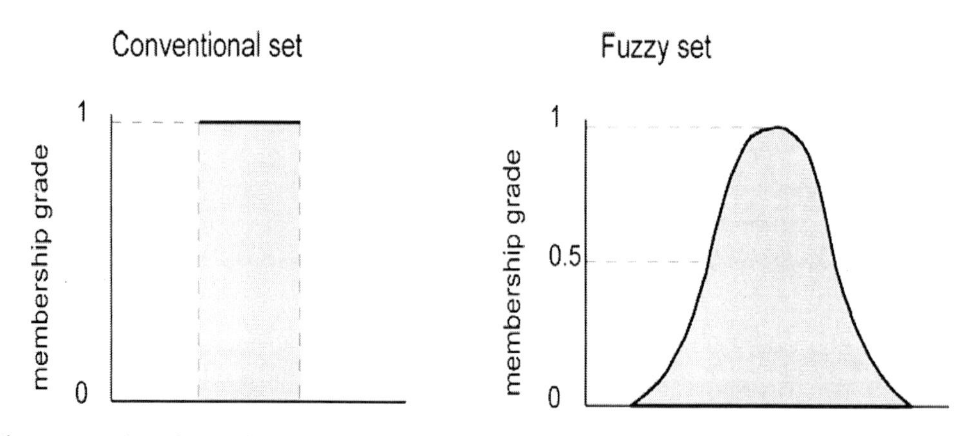

Figure 1: A plot of a Boolean set and a fuzzy set in terms of fuzzy membership.

The two major approaches of defining fuzzy membership are the semantic import model and the similarity relation model (Fisher, 2000). The similarity relation model simulates a pattern recognition approach where the data is searched for membership values by a fuzzy *c* means algorithm or fuzzy neural network. On the other hand, the semantic import model uses either, expert knowledge to define critical points that mark the limits of transition from membership 0 to 1, or membership functions of different shape (e.g. linear, sigmoidal, triangular, trapezoidal, Gaussian) to obtain the grade of membership to a particular set as discussed in Klir and Folger (1988), Dombi (1990), and Burrough and McDonnell (1998). Robinson (2003) categorizes these standard membership functions as of open form or closed form.

Fuzzy Operators for Data Integration

Among the fuzzy operators, the following are the most frequently applied to integrate the memberships of various data sets (An *et al.*, 1991). For modelling to be performed using fuzzy sets, rules of the form "*if...then*" are used to convert inputs to outputs – one fuzzy set to another (Kosko and Isaka, 1993). Just as with Boolean sets the interaction and combining of fuzzy sets is performed with operators including union, intersect, difference and inverse (Fisher, 2000). An overview of these operators is given by Burrough (1989) and Bonham-Carter (1994), and more detail is provided by Kandel (1986) or Klir and Folger (1988).

Fuzzy AND

Fuzzy AND is a logical intersection which combines the fuzzy memberships of two or more information layers using the minimum operator. It can be defined as

$$\mu_{combination} = MIN\left(\mu_{1,}\ \mu_{2,}\ \mu_{3,}\ .\ .\ .\ .\ \right) \quad (1)$$

where μ_1, μ_2 ... are the fuzzified information layers containing fuzzy membership values at a particular location. This indicates that the operator generates a map displaying minimum membership value for each location. Thus, the resultant image is a conservative estimation of a set of memberships that tend to produce small values.

Fuzzy OR

Fuzzy OR combines the fuzzy memberships of two or more information layers using the maximum operator. The output membership values are controlled by the maximum membership values of the input bands for any particular location. This can be defined as

$$\mu_{combination} = MAX\left(\mu_{1,}\ \mu_{2,}\ \mu_{3,}\ .\ .\ .\ .\ \right) \quad (2)$$

Fuzzy Algebraic Product

The fuzzy algebraic product combines information layers by multiplying the fuzzy memberships at a particular location, as defined by the following equation.

$$\mu_{combination} = \prod_{i=1}^{n} \mu_i \quad (3)$$

where μ_i, is the fuzzy membership for the i-th information layer, and i = 1,2,3, ...*n* information layers are to be combined. The combined fuzzy membership values tend to be very small with this operator due to the effect of multiplying several numbers less than 1. The output is always smaller than, or equal to, the smallest contributing membership value, and is therefore “decreasive”. Although fuzzy algebraic product gives an output that is decreasive in nature, it uses every membership value to produce the result (Bonham-Carter, 1994).

Fuzzy Algebraic Sum

Fuzzy algebraic sum is complementary to the algebraic product which can be defined as

$$\mu_{combination} = 1 - \prod_{i=1}^{n}(1 - \mu_i) \tag{4}$$

The result is always larger (or equal to) the largest contributing membership value, and thus it has an ‘increasive’ effect (Bonham-Carter, 1994).

Fuzzy Gamma Operation

The gamma (γ) operator is defined in terms of the fuzzy algebraic product and the fuzzy algebraic sum by the following equation.

$$\mu_{combination} = (Fuzzy\ algebraic\ sum)^{\gamma} * (Fuzzy\ algebraic\ product)^{1-\gamma} \tag{5}$$

where γ is a parameter chosen in the range (0, 1) (Zimmerman, 1985).

The basic steps of fuzzy logic based modelling within a GIS are shown in Figure 2. The de-fuzzification step involves converting fuzzy sets to classical sets by the use of the “*α-cut*” operator (Roberts et al., 1997). Given the fuzzy set ***A*** defined on **X** and any number $\alpha \in [0,1]$, the α-cut, ${}^{\alpha}A$, is the crisp set ${}^{\alpha}A=\{x|A(x) \geq \alpha\}$(Robinson, 2003). In words of Klir and Yuan (1995), the α-cut of **A** is the crisp set ${}^{\alpha}A$ that contains all the elements of **X** whose membership grades in **A** are greater than or eqaul to the specified value of α. When modelling using fuzzy logic, fuzzy membership values must reflect the relative importance of each map within the model (Bonham-Carter, 1994).

In considering the modelling of salinity, classical set theory would demand that all areas be classified as being saline or non-saline (Metternicht, 1998). Fuzzy logic allows for this unrealistic division to be removed. Instead, the uncertainty of the boundary between these regions and the measurements on which they are based can be modeled. Degrees of membership in classes are allowed, reflecting the natural variation of soil properties. Fuzzy logic allows this reality to be modeled taking into account the fuzzy nature of the real world. Fuzzy classification provides more information because it distinguishes smaller differences between cells, whereas Boolean classification can only classify cells as saline or non-saline (Heuvelink and Burrough, 1993).

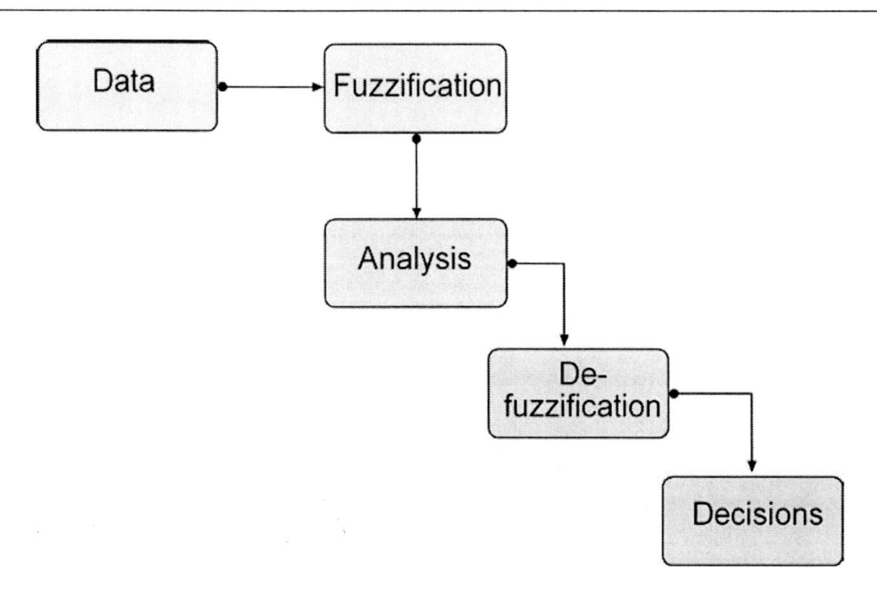

Figure 2: Basic method for performing fuzzy modelling within a GIS (After: Steiner, 2001).

Fuzzy Landscape Analysis GIS

The salinity hazard modelling strategy used in this investigation is based on fuzzy logic. The name of this model is FLAG (Fuzzy Logic Analysis GIS). FLAG was developed by the Commonwealth Scientific and Industrial Research Organization (CSIRO), requiring only topographic information, in the form of a hydrologically sound Digital Elevation Model (DEM), as input. It rapidly predicts dryland salinity hazard using a number of fuzzy topographic indices (0–1 scale) that include locally low areas (LOWNESS), plan curvature (CONCAVITY) and contributing area defined as relative height in the study area (UPNESS) (Dowling *et al.*, 2003).

The model requires making significant assumptions about the environment being modelled, as follows (Roberts *et al.,* 1997):

1. Topography is the surface expression of all factors that have acted on a site over time including geology, soils, climate, vegetation and land use. Thus the site history is a complex set of influences, which have shaped and continue to act on the topographic surface, often in an interrelated way;
2. The water table conforms to the topographic surface except it is smoother, and exhibits less total variation in relief than the surface. Thus the water table is closer to the surface at low points in the landscape, and further from the surface at high points in the landscape, while following in general the topographic surface;
3. The contributing area for and point is given by the set of points connected by a continuous, monotonic uphill path.

These assumptions need to be justified for the model to be applicable. For instance, it is known that vegetation, both living and dead, slows wind speed, reduces rainfall impact, binds soil to slow erosion and provides obstacles to the overland flow of water. In these ways vegetation acts to influence the shape of the land. Therefore when landcover changes from native vegetation to crops and pasture, as it has occurred over extensive areas of Australia, the pattern and extent of erosion also changes. Annual crops that dominate the Wheatbelt of Western Australia provide protective cover for only a portion of the year. Perennial crops provide protection for a large portion of the year and can better reduce erosion. These facts provide evidence to support the first assumption made by FLAG.

Regarding the second assumption, Fetter (2001) states that the position of the water table often follows the general shape of the topography, although the water table relief is not as great as the topographic relief. This view is supported by Freeze and Cherry (1979) who mention that the water table is coincident with the ground surface in the valleys and forms a subdued replica of the topography on the hills. Fetter (2001) also states that ground-water discharge zones are in topographical low spots. Studies conducted by Salama *et al.* (1994), Williamson (1998) and Evans *et al.* (1990) all report discharge areas being strongly influenced by topography and discharge occurring at low points in the landscape.

The third assumption is that for saturated sub-surface flow all points that are connected by a continuous, monotonic uphill path will exert some hydraulic head on the location below, and thus contribute to potential discharge at that point. These points make up the contributing area for a location. The importance of contributing area in hydrologic modelling is supported by works of Dawes and Hatton (1993), Bevan and Kirkby (1979) and Moore *et al.* (1991).

Other Salinity Mapping Techniques

Many salinity mapping techniques have already been implemented. Please et al. (2002) provide an overview of 'past and present projects undertaking mapping, modelling and prediction of locations of dryland salinity, including salinity hazard mapping'. Metternicht and Zinck (2003) provide a review of remote sensing approaches for identifying salinity as well as means of modelling temporal and spatial changes in salinity. A brief review of salinity mapping and modelling including techniques outlined in the aforementioned articles and others is provided along with a summary of their weaknesses.

Please et al. (2002) mention that the most common indirect method used to map salinity relies on the measurement of the electrical conductivity of the ground via the use of Electromagnetic Induction techniques (EM). The assumption is that a high EM reading is caused by the presence of salt in the soil being measured (Please et al., 2002). Further information on the use of electromagnetic techniques can be found in Nicoll (1993).

Extensive research on the application of satellite imagery covering the visible to infrared regions of the spectrum for the mapping of saline areas has been conducted (Metternicht and Zinck, 2003). The technique achieves generally good results when discriminating only two surface types, namely, saline and non-saline (Evans and Caccetta, 2000a). Metternicht and Zinck (2003) provide a summary of previous applications of this data and highlight the constraints and problems surrounding the use of remote sensing data. Reflectance in the visible and middle infrared bands only reveals information about the first millimetres of the top horizon of bare soil. The characteristics of which are often different from the layer below (Mougenot et al., 1993). There is significant variation in the reflectance spectrum of salt affected soil. This is caused by variations in structure and the type and amount of inclusions, complicating remote detection. Also, features quite different in nature and dynamics, to salt affected land, can generate similarly high levels of reflectance (Metternicht and Zinck, 2003).

Metternicht and Zinck (2003) state that in general, moderately to highly saline areas are easily detected, while low salinity levels and initial stages of salinisation frequently cause identification failures.

One study that uses satellite imagery as an indirect indicator for predicting areas at risk of salinization is the Land Monitor project. This is a multi-agency project following on from collaborative research work between CSIRO Mathematical and Information Sciences, Department of Land Administration and Agriculture Western Australia, under the National Dryland Salinity Program (Allen and Beetson, 1999). Its aims include mapping the current extent of salinity and the prediction of areas at risk from future salinisation. Landsat Thematic Mapper and terrain attributes derived from a very high quality DEM are the main inputs. The variables derived from these inputs are: average upslope height, average upslope slope, average downhill slope, height above nearest salt within a watershed, height above nearest stream within a watershed, water accumulation, flow path length, percentage upslope cleared area and total upslope cleared area (Evans and Caccetta, 2000a). Salinity is predicted by creating a decision tree that relates input data variables to salinity risk for areas in which ground truth data are available (Evans et al., 1996). The decision tree is then used to extrapolate a prediction of at 'risk areas' (Evans and Caccetta, 2000a).

Land Monitor does not consider the effect of geology and is limited in its modelling of the influence of vegetation. It does not consider plant species, whether the plants are annual or perennial, the depth of plant roots or whether that plants are tolerant to salt. Most importantly Leaf Area Index (LAI) (the area of leaf surface per unit of soil surface) is not considered. Many authors including Stirzaker (1996), Raper (1998), Greenwood et al. (1982, 1985 cited in Raper, 1998), Schofield et al. (1989 cited in Raper, 1998), Bell et al. (1990 cited in Raper, 1998) and Vertessy et al. (2000) all agree that leaf area is the dominant factor in determining transpiration.

Apart from the already discussed FLAG model, there are several other salinity modelling techniques that have been developed. CATSALT is a suit of models developed by the New South Wales Depatment of Land and Water Conservation. The models are integrated into a comprehensive modelling framework, which can be used to assess the cumulative impact of landuse changes made across catchments (Beale, 2001 cited in Watson, 2002). CATSALT can simulate water flow and stream salinities for medium sized catchments from 500 to 2000 km^2 under current conditions (Cresswell et al., 2003). FLOWTUBE is a simple numerical one-dimensional groundwater flow model that is based on Darcy's Law, which determines flux from factors of conductivity, thickness, width and slope (Dawes, 2001 cited in Watson, 2002). Although FLOWTUBE performs well in homogenous uniform isotropic media, such as sand and gravel aquifers, and massive clay deposits, if preferred pathways or barriers are present its performance is limited (Cresswell et al., 2003).

MODFLOW is a three-dimensional, finite difference, groundwater model described by MacDonald and Harbaugh (1988 cited in Clarke et al., 1998b). The model estimates discharge from the watertable. It has a number of disadvantages including the necessity to calculate a number of parameters externally, its non-physically based estimation of vegetation and its computational and data intensive nature (Clarke et al., 1998b).

MAGIC is a model developed by the Western Australian Water and Rivers Commission that can model groundwater hydrology and dryland salinity. It is a two layer physically based mathematical model. The input data is satellite imagery from the Landsat Thematic Mapper, topography, rainfall and evaporation data. The model simulates water use by vegetation. The effect of geology is not taken into account by the original MAGIC model though Clarke et al. (1998c) describe an extension that considers geology.

The majority of salinity mapping techniques (including the strategy adopted in the Land Monitor project) use 'crisp' Boolean logic to delineate boundaries between saline and non-saline areas. The importance of geology on the spatial distribution of salinity has already been discussed in this chapter. However, many of the commonly used models are single layer and do not model the effect of geology on the development of dryland salinity (Clarke et al., 1998c). An understanding of geological variation is necessary and computer programs based only on elevation do not predict the observed salinity (Clarke et al. 1998a). Also mentioned was the important role vegetation plays in the development of salinity. An ideal model would therefore need to simulate vegetation in a realistic, physically based way (Clarke et al., 1998c).

Study Area and Data Set

Study Area

The study area chosen is the Lake Toolibin Catchment in the Wheatbelt region of Western Australia 250 km south-east of the state capital Perth (Figure 3), located at 32° 55' South Latitude and 117° 36'East Longitude. The lake and its catchment are affected by dryland secondary salinity. The earliest reports were made in the 1930's at roughly the same time large scale clearing began (Hearn, 2001). In 1996, the CSIRO estimated that 12 percent of the catchment was salt-affected, although more recent estimates by Dogramaci *et al.* (2003) put that figure down to 8 percent, mainly in the valley flats (see Figure 4). Studies conducted by Ferdowsian *et al.* (1996) suggest that the potential area of salt-affected land is approximately 25-30 percent of the catchment unless the landscape water balance is altered.

General Characteristics

The catchment is approximately 48,000 ha in area (Hearn, 2001; Metternicht and Beeston, 2002), with the majority of the land used for agriculture based on mixed grazing and dryland cropping (Hearn, 2001). Small stands of remanent vegetation are scattered across the catchment ranging in size from less than one ha to between 200 to 500 ha (Metternicht and Beeston, 2002).

Most of the original vegetation across the catchment has been cleared. Originally the catchment supported extensive Eucalyptus woodland of York gum (*Eucalyptus loxophleba*), salmon gum (*E. salmonophloia*), white gum (*E. wandoo*), red morrel (*E. longicornis*), rock sheoak (*Allocasuarina heugeliana*) and jam (*Acacia acuminata*), though currently the main areas of remanent vegetation are held in stands across the catchment and along rode side verges (Hearn, 2001).

The region has a Mediterranean climate with warm very dry summers and cool wet winters (Hearn, 2001). The spatial variation of average annual rainfall varies from 379 mm at Toolibin in the centre of the catchment, to 417 mm at Dudinin in the east and 421 mm at Wickepin in the west (Hearn, 2001).

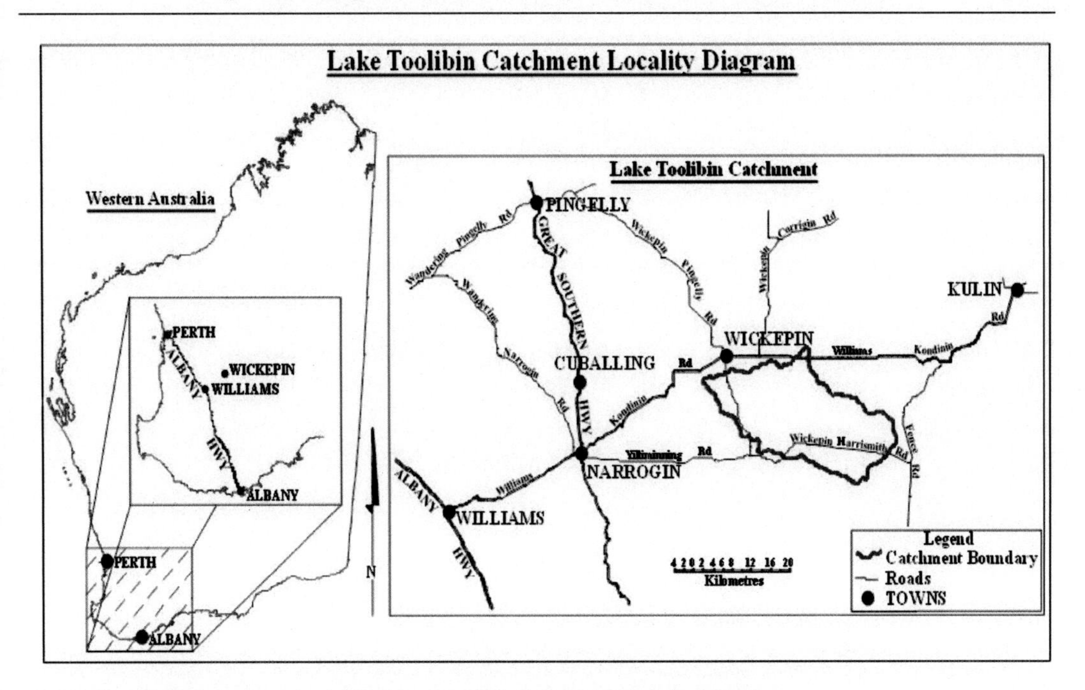

Figure 3: Catchment location diagram (modified from Drysdale, 2003).

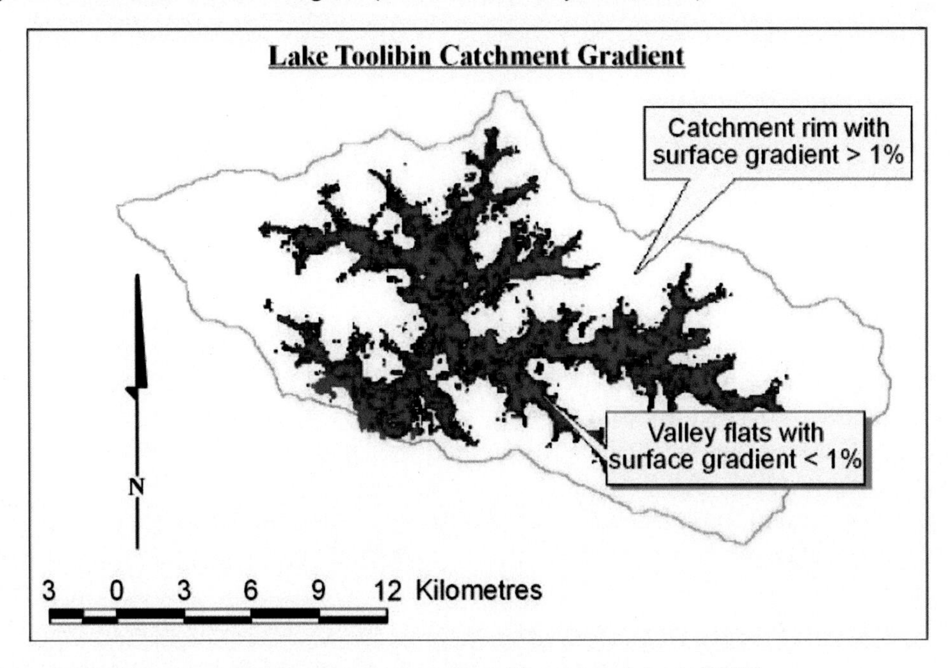

Figure 4: Catchment subdivision based on gradient (Dogramaci et al., 2003)

Topography and Groundwater

Elevation varies from 360 masl, at the north-eastern boundary of the catchment, to 298 masl at Toolibin Lake (Figure 5). Based on slope gradient, Dogramaci *et al.* (2003) divide the catchment into two areas as shown in Figure 4. The majority (approximately 65 percent) of the catchment has a gradient greater than one percent with grades as high as 6 to 8 percent and is found at the catchment rim (Dogramaci *et al.*, 2003; Hearn, 2001). The remaining (35 percent) of the catchment (known as the Toolibin Flats) found along the main drainage line in the central region of the catchment, north of Toolibin Lake, has a gradient of less than one percent with grades averaging only about 0.1 per cent to 0.2 per cent (Dogramaci *et al.*, 2003; Hearn, 2001). At areas of transition near the centre of the catchment, the flats are often bounded by very sharply defined slopes rising between 2 and 4 percent, whereas elsewhere the transition from the flats is more gradual having broad gentle slopes of 1 to 2 percent (Hearn, 2001).

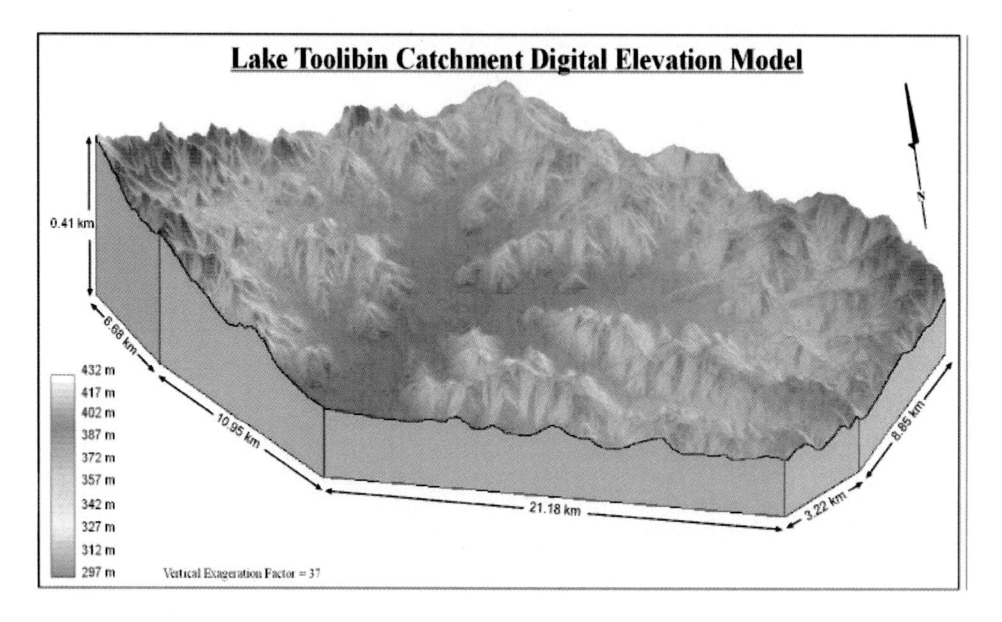

Figure 5: A Digital Elevation Model of the Toolibin catchment area

Groundwater flow is from the catchment rim toward the valley flats, eventually discharging into the Toolibin Lake (Dogramaci *et al.*, 2003). However, a more detailed study of groundwater conducted by George (1998) revealed that recharge takes place in all landform positions except in areas with strong vertical flow (discharge areas). Generally, the watertable in the low gradient (valley flats) is within 2 m of the surface and conforms to topography (Dogramaci *et al.*, 2003; George, 1998), though in the west of the catchment the relationship between elevation and water level is not consistent (George, 1998).

Dogramaci *et al.* (2003) mentions that prior to the vegetation being cleared, there was little or no recharge and hence no discharge of groundwater to the surface. After clearing, increased input from rainfall has resulted in the rise of the watertable and the discharge of brackish water into the low-lying areas of the catchment.

Geology and Hydrogeology

The Toolibin catchment is located on the Yilgarn Craton comprised of fresh or fractured Archaean granite and granite-gneiss (Clarke *et al.*, 1999; George, 1998). This basement rock forms a 'floor' to local groundwater systems functioning in the regolith above (Coram, 1999). The rocks within the catchment were subjected to deep weathering during the hotter wetter Tertiary period causing intense leaching once the climate dried, producing a deeply weathered profile within the regolith known as the lateritic profile (Clarke *et al.*, 1999), which can be described as follows (Dogramaci *et al.*, 2003: (a) the first layer above the unweathered basement rock (the Yilgarn Craton) consists of saprock or grit. This is a partially weathered layer formed at the weathering front in response to the fragmental disintegration of basement rocks rich in quartz and feldspar; (b) A completely weathered sandy-clay horizon called the saprolite or 'pallid zone' sits above the saprock, promoting the development of a semi permeable layer that impedes the upward movement of deep groundwater from the saprock horizon to the surficial sediments that cover the majority of the valley flats; (c) Above the weathered profile is a layer of surficial sediments of medium to coarse-grained quartz sands and clays, which covers the entire valley flats on the main drainage line in the middle and lower parts of the Toolibin catchment. The sediments range in thickness from 30 m in the valley (Toolibin Flats), to less than 10 m in tributary valleys (George, 1998).

An airborne magnetics survey of the catchment conducted by George (1998) revealed a large number of Proterozoic dolerite dykes, mainly trending in north-west, east-north-east and west-north-west directions (Figure 6). Also revealed were numerous discontinuous faults (Figure 7) often aligned north-south (George, 1998). These faults and dolerite dykes, locally control the position of modern natural drainage lines (Dogramaci *et al.*, 2003). The orientation of dykes and fractures, relative position of sediments and depth of regolith all help to determine the nature and rate of salinity (Dogramaci *et al.*, 2003).

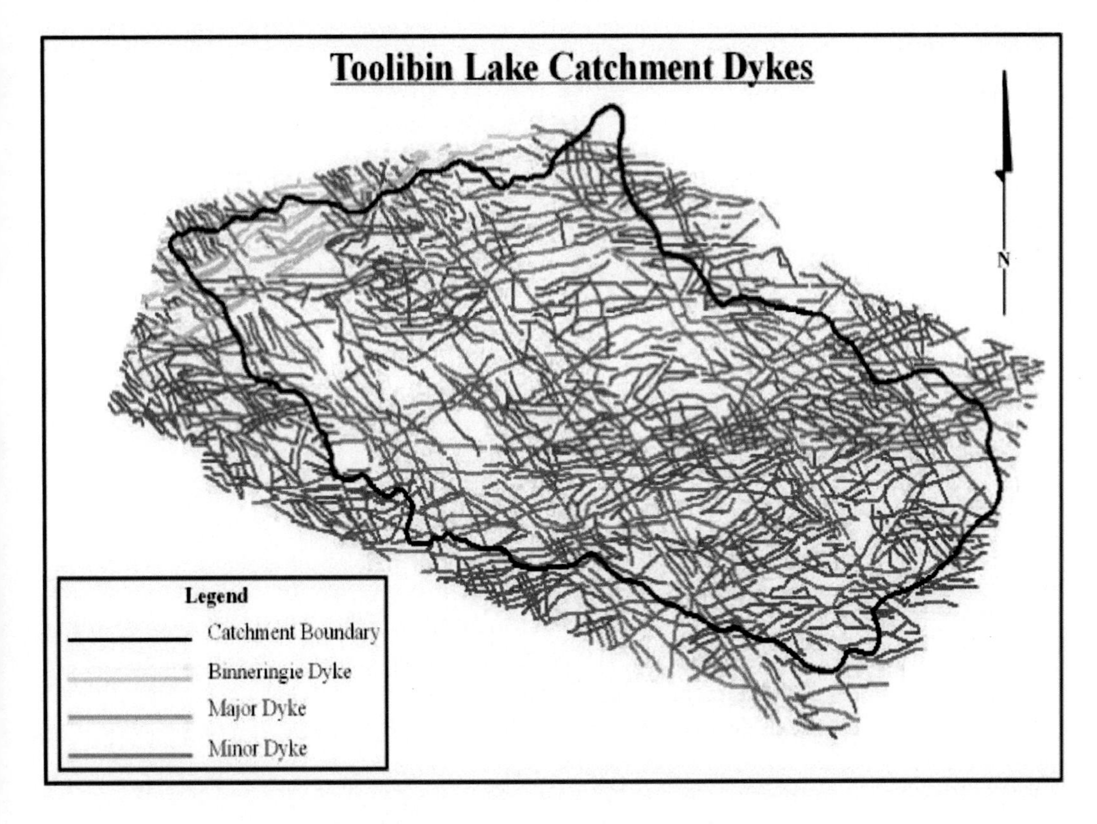

Figure 6: Dykes within the Toolibin catchment area (George, 1998)

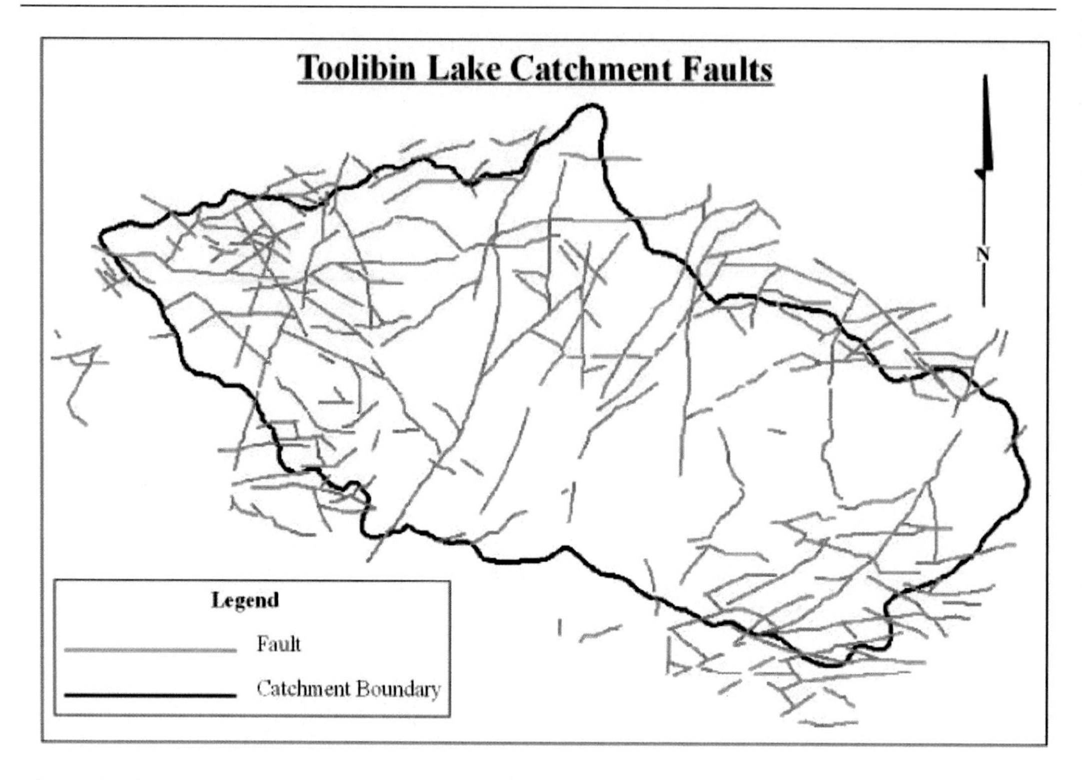

Figure 7: Faults within the Toolibin catchment area (George, 1998)

Data Sets

In its original form, FLAG requires only a digital elevation model and ground truth data. In an attempt to improve the accuracy of FLAG, two new data sets were used, namely remote sensing and geological data. A previous study of salinity at the study area, based on probabilistic theory and Boolean logic was used to compare modelling of salinity based on fuzzy logic and investigations based on Boolean logic.

Digital Elevation Model

FLAG requires a hydrologically sound DEM for the derivation of soil wetness indices. Grid elevation data were obtained from the Department of Agriculture, Western Australia with a 10 m resolution. The DEM was generated using the TOPOGRID command within ArcInfo Workstation (ESRI, 2002). The TOPOGRID command is an interpolation method specifically designed for the creation of hydrologically correct DEMs. Dowling (2000) recommends using a grid resolution of 10 m for the DEM, as this cell size was found to be similar to the expected scale of processes and spatial extent (patch size) of salinisation. Laffan (1996a) also used a cell size of 10 m as it has been found to give an acceptable compromise for spatial accuracy with topographic indices (Zhang and Montgomery, 1994). Accordingly, the DEM was generated with a resolution of 10 m.

Remote Sensing Data

The remote sensing data were used in an attempt to improve FLAG's accuracy in predicting salinity by factoring in the effect of vegetation on groundwater levels. As remnant vegetation is perennial, deep rooted and transpires groundwater throughout the year, the magnitude with which they effect groundwater depth can be significant. Crops and pastures are (generally) neither perennial nor deep rooted and have minimal impact on groundwater depth. For this reason remnant vegetation alone was used to account for the effect vegetation has on salinity. A Landsat TM scene collected in February was used to map native vegetation because of the great contrast between remnant vegetation and farmland at this time of the year (e.g. dry season). The scene was captured post-harvesting, with most of the cropping areas showing as fallow land.

Geological Data

Geophysical data were collected using SALTMAP 'a system which combines airborne electromagnetic (AEM), magnetics, radiometrics and digital elevation data (in a GIS) to attempt to define regolith conductivity and thickness, bedrock geology, soil and landform' (George, 1998, p.11). George (1998) describes SALTMAP in full. The results include the location of dykes and faults within the catchment (Figures 6 and 7), which were used in the modelling phase, as described in Methodology Section.

Ground Truth Data and Previous Studies on Salinity

Borehole data received from the Department of Agriculture, Western Australia were used as ground truth data. Included in the data were the bore cap heights, depth to bedrock from bore cap, standing or static (ground) water level and groundwater salinity level (Figure 8).

The results of this investigation were compared with those from the Land Monitor project (Figure 9), described in preceding sections. As mentioned this modelling process is based on traditional Boolean logic with all areas being considered either saline or not saline.

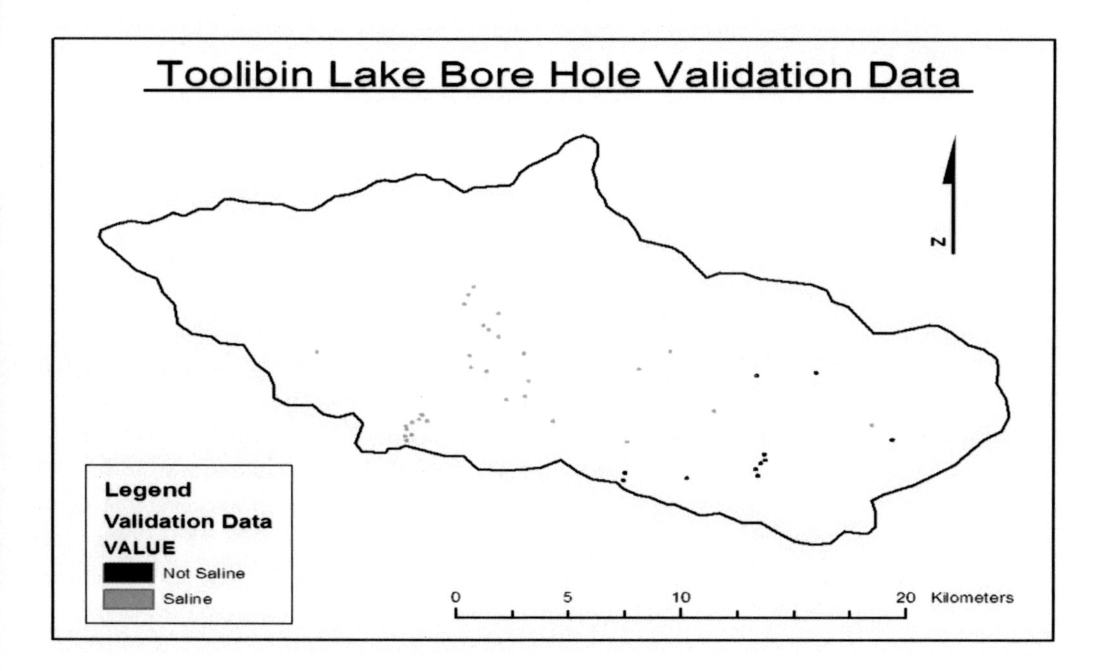

Figure 8: Location of saline and non-saline boreholes

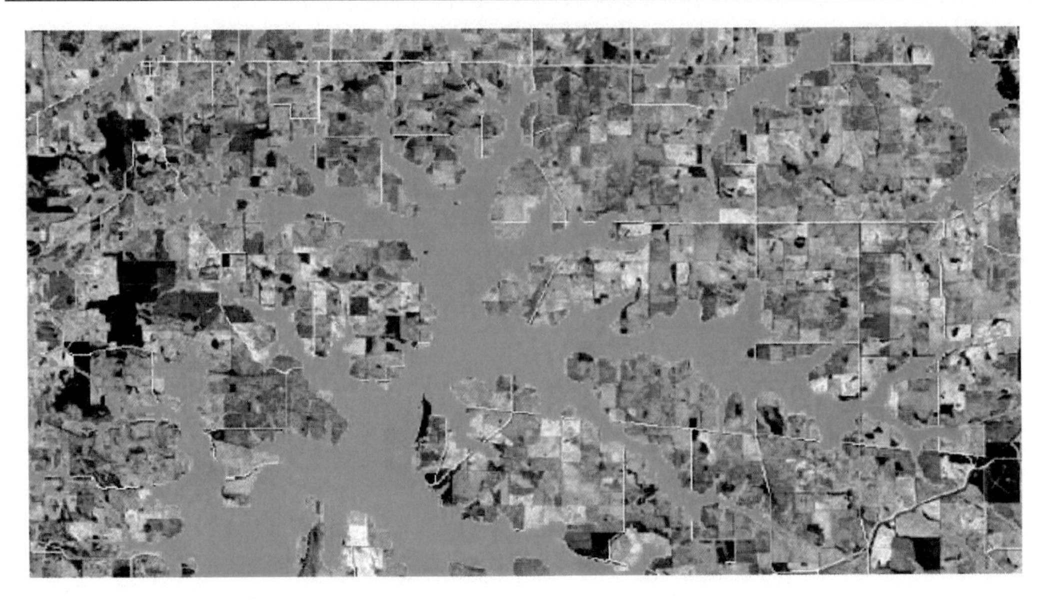

Figure 9: Toolibin salinity risk map created by the Land Monitor project overlaid on a 1993 Landsat TM Band 4 in greyscale (Evans and Cacceta, 2000b)

METHODOLOGY

Implementation of the Original FLAG Model

Figure 10 represents a simplified cartographic model of the original FLAG methodology. In this chapter, layer names given in BLOCK CAPITALS refer to layers shown in Figure 10.

The first stage is to derive a number of topographic indices from the elevation data. These indices are then combined to determine likely areas for salinity and groundwater discharge. The fuzzy maps show this threat in a continuous manner. Maps showing a continuous range of values can be difficult for end users to interpret. For this reason the final maps can be converted to binary images of discharge and non-discharge Laffan (1996a).

Derivation of Topographic Indices

The first topographic indicator derived was the set of local low points called LOWNESS. The first step in creating LOWNESS is to calculate the average local elevation using a 130 m by 130 m moving window. This creates a smoothed DEM. The difference between this and the original DEM are calculated to create a layer (LOWNESS) showing locations that are low relative to the surrounding landscape. The use of a 130 m by 130 m window was ad-hoc, to maintain some similarities with Roberts *et al.* (1997) and Laffan (1996a) and does not necessarily reflect an optimal value. As mentioned before, FLAG assumes that the groundwater table conforms to the landscape but with less variation. Therefore, low areas will be closer to the groundwater table and therefore have high values in the LOWNESS layer.

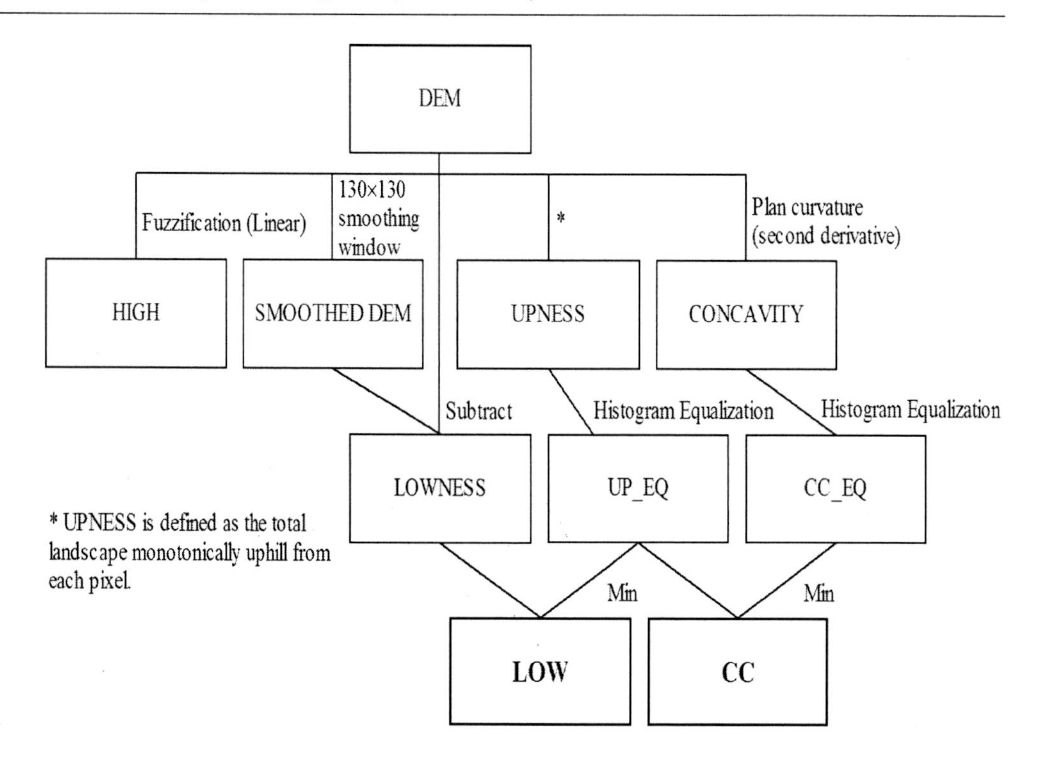

Figure 10: A simplified cartographic model of FLAG.

Next, the second derivative of the surface was calculated as an estimate of convexity or its complement *concavity*. This was calculated using a plan curvature function that fits a polynomial surface within a 3 by 3 pixel window. Based on the same assumption as LOWNESS highly concave areas are expected to be closer to the groundwater level.

The UPNESS layer represents the fraction of the total landscape monotonically uphill from each pixel. The assumption on which this indicator is based has already been discussed. UPNESS is created using two Fortran 77 programs.

Combination of Indices

The next stage was to combine the topographic indices to determine likely areas for salinity. A relativisation operator was used to enable fuzzy sets defined by a variety of means and with different ranges to be combined and compared. This procedure scales all pixels by subtracting the minimum membership value from each pixel and dividing by the range of membership values to produce fuzzy sets with membership linearly distributed between zero and one (Roberts *et al.*, 1997).

LOWNESS was relativised such that locations low in the landscape had high values (maximum of one) and locations at or above the local landscape had zero membership in the set. A similar operation was performed on the *concavity* layer. However, because of the distribution of values in the concavity layer (an extreme peak with long tails) a histogram equalisation was performed (monotonically rescaled to produce approximately equal numbers of points within each subrange of values).

UPNESS is relativised for use with relativised LOWNESS. It is also histogram equalised to produce a layer for use with the histogram equalised concavity set. Based on the model's assumptions, likely sites for salinity and discharge are assumed to be those low in the landscape and with high contributing area. Therefore relativised LOWNESS and relativised UPNESS were intersected using the fuzzy logic AND operator. This layer was relativised to produce the fuzzy discharge index LOW.

It was assumed that sites likely to have salinity were sites with high concavity and high contributing area. Therefore the histogram equalised *concavity* and UPNESS layers were intersected using the fuzzy logic AND operator. This layer was relativised to produce a fuzzy discharge index CC showing areas with high *concavity* and UPNESS. The two fuzzy sets CC and LOW are assumed to be predictors of relative soil wetness.

Converting Fuzzy Maps to Binary Maps

The fuzzy maps produced by FLAG are a prediction of relative salinity potential relative to local conditions. Training data is used to determine a threshold value whereby all fuzzy values greater than or equal to such value are deemed to have high salinity potential.

This process, known as α-cut, converts the fuzzy maps to binary maps. This allows CC and LOW to be compared with ground truth maps and the Land Monitor data, which are both binary. As the appropriate α-cut (threshold) value is not known in advance, the optimal α value is calculated using a contingency table (Figure 11) to compare actual and predicted discharge for different threshold values. Using this table four diagnostic statistics indicating the performance of different threshold values are calculated to determine the optimal α threshold value.

		Actual	
		y	n
Predicted	y	a	b
	n	c	d

Figure 11: Contingency table used to derive performance statistics.

A series of α-cuts at 0.05 intervals are performed on CC and LOW and a corresponding series of binary maps, which are subsequently compared with the ground truth maps, are produced as result. At each interval a contingency table (Figure 11) is created to compare predicted discharge and ground truth discharge. Within this table cell ***a*** contains correct positive predictions, cell ***d*** contains correct negative predictions, cell ***b*** contains errors of commission, and cell ***c*** contains errors of omission.

Using the contingency table, four diagnostic statistics are calculated to evaluate the α cut predictors. Accuracy provides a measure of overall accuracy. Efficiency tries to overcome this by focussing on errors of omission, whereas discrimination focuses on errors of commission. Power (Equation 6) ignores the correct negative predictions and attempts to combine efficiency and discrimination. Roberts *et al.* (1997) conclude that efficiency and discrimination are of little use in selecting and appropriate α cut value, but that the power statistic is effective for determining this value. By examining the plot of α against power, optimal α-cut thresholds for both LOW and CC were determined.

$$\text{power} = \left[\frac{a}{(a+b+c)}\right]^{0.5} \tag{6}$$

These statistics were calculated for both CC and LOW for each of the binary maps produced at each α-cut threshold (19 each) and plotted against α. The plots were created to assist in determining an α-cut value that best fits the ground truth data.

Borehole data were used as ground truth information to train LOW and CC. FLAG required the ground truth data delineating known areas of dryland salinity in the form of polygons. The borehole data did not explicitly indicate areas of dryland salinity and was in the form of points, thus requiring pre-processing before its use within the FLAG software.

Implementing an Optimised Version of the FLAG Model

The fact that FLAG uses only elevation data limits its performance and tests the assumptions on which it is based. Within this investigation, an optimisation of FLAG was implemented by including vegetation and geology data.

Combination of Final FLAG Indicators

To ease the integration of new data and simplify the final result CC and LOW were combined to produce one fuzzy output layer. The Section on 'Fuzzy sets and fuzzy modelling' lists the fuzzy operators useful for data integration. Given that the gamma operator (Equation 5) enables a compromise between the decreasive nature of fuzzy algebraic product and the increasive nature of fuzzy algebraic sum, and based on findings reported by Islam and Metternicht (2004), this operator was chosen for combining CC and LOW. The same methodology used to convert CC and LOW to binary maps was used to convert the layer produced by combining them to a binary map.

Modelling Remnant Vegetation

Remnant vegetation patches were extracted from the Landsat scene using the Normalized Difference Vegetation Index (NDVI) (Equation 7). This index produces values between -1 (no vegetation) and 1 (completely healthy green vegetation) (Gibson and Power, 2000). A linear membership function (see Figure 12) was used to convert the NDVI into a fuzzy layer. Looking at a histogram of the NDVI (Figure 13) a significant drop can be observed at 0.00675. Therefore, checking the index values against a Landsat False Colour Composite where patches of remnant vegetation could be easily visualized, it was assumed that pixels with values above 0.0067 were representatives of areas with remnant vegetation. As the NDVI value increases the likelihood of salinity is assumed to decrease. Consequently, from 0.00675 to 0.49 (the maximum NDVI value recorded for the area) the fuzzy membership (salinity likelihood) decreases linearly. Pixels exhibiting an NDVI value equal or lower than 0.00675 adopt a fuzzy membership of one to the fuzzy set 'salinity likelihood', whereas pixels with values equal to 0.49 are assigned a fuzzy membership degree of zero to the set.

The fuzzy linear membership function of Figure 12 assigns a fuzzy membership value between zero and one to the NDVI values comprised in the range 0.00675-0.49.

$$NDVI = \frac{near\ \text{infrared} - visible\ red}{visible\ red + near\ \text{infrared}} \tag{7}$$

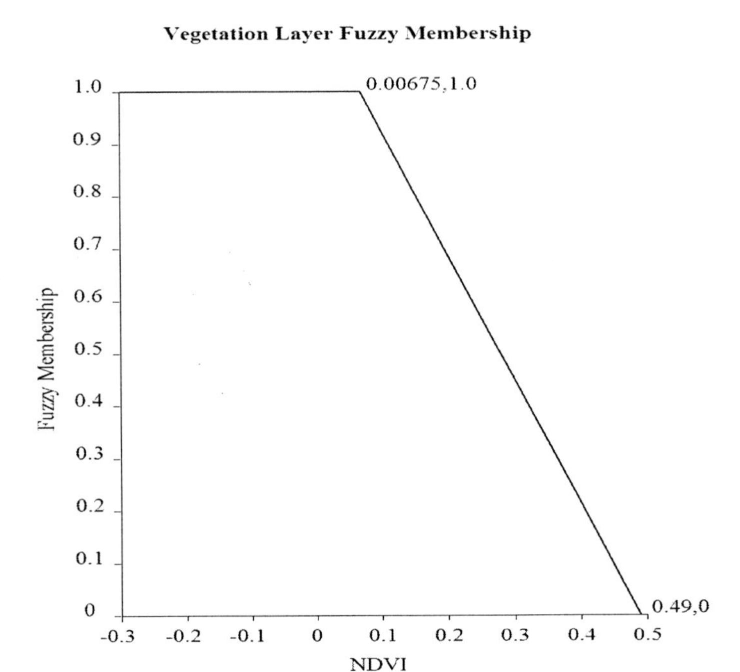

Figure 12: Fuzzy membership function of the layer 'remnant vegetation'.

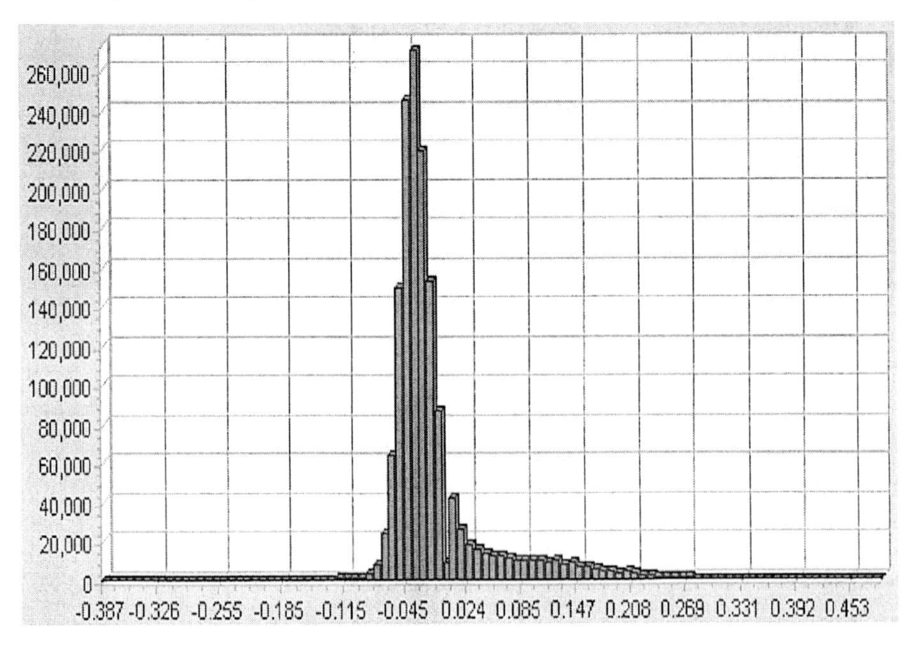

Figure 13: Histogram of the Normalized Difference Vegetation Index.

Processing of Geological Data

The first step in modelling the effect of faults on dryland salinity was to buffer the linear fault features. The buffer polygons are intended to indicate areas affected by salinity. A buffer distance of 500 m was used. Although Clarke *et al.* (1998a) suggest faults have a zone of influence on groundwater 2 to 4 km either side of a fault, Searle and Baillie (1998) indicate that the most significant influence was within 500 m of the fault line and used this buffer distance when modelling the effect of faults. Jolly *et al.* (2002) and Engel *et al.* (1987) report that salinity and waterlogging primarily occur on the upslope side of faults and dykes. Accordingly, the buffer polygons were split in half, with the half-buffer on the lower side been removed and the remaining upslope half-buffer area fuzzified.

A grid showing the distance from the faults created. This was overlayed with upslope half-buffer layer to so that these polygons contained the distance from the nearest fault. Sigmu (Laffan, 1996b) an ArcInfo AML was used to fuzzify the upslope-buffers. This program uses a sigmoidal membership function, based on the cosine function, to rescale a dataset (upslope distance) into the interval 0 to 1. A sigmoidal function (Figure 14) was chosen because a previous study by Ventriss *et al.* (1982) on the relationship between groundwater and faults and dykes mapped a traverse of groundwater level perpendicular to a dyke, showing the water level resembling a sigmoidal cosine function. The cartographic model of this procedure is shown in Figure 15.

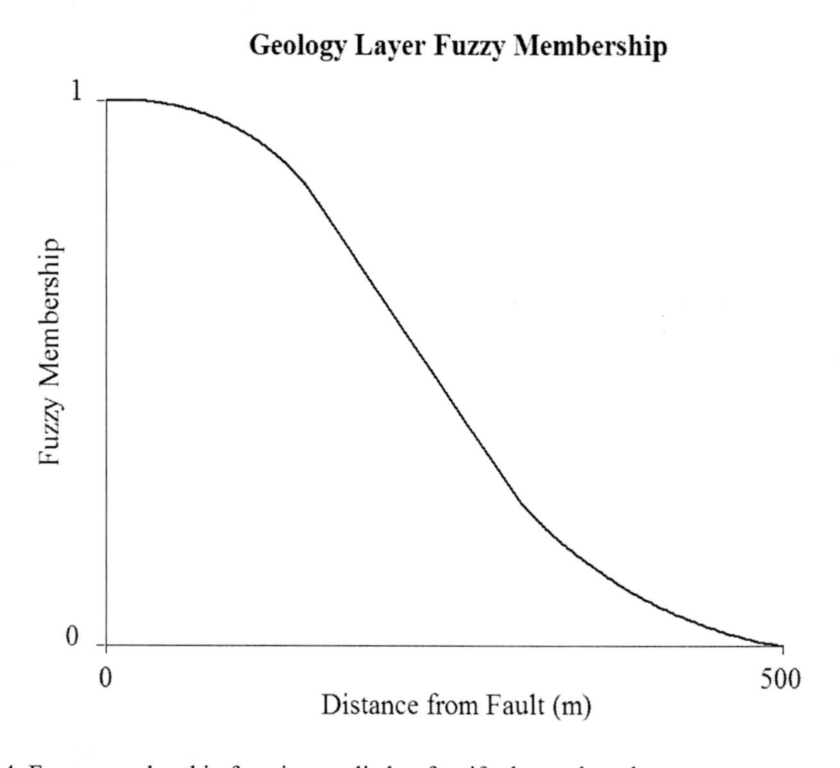

Figure 14: Fuzzy membership function applied to fuzzify the geology layer.

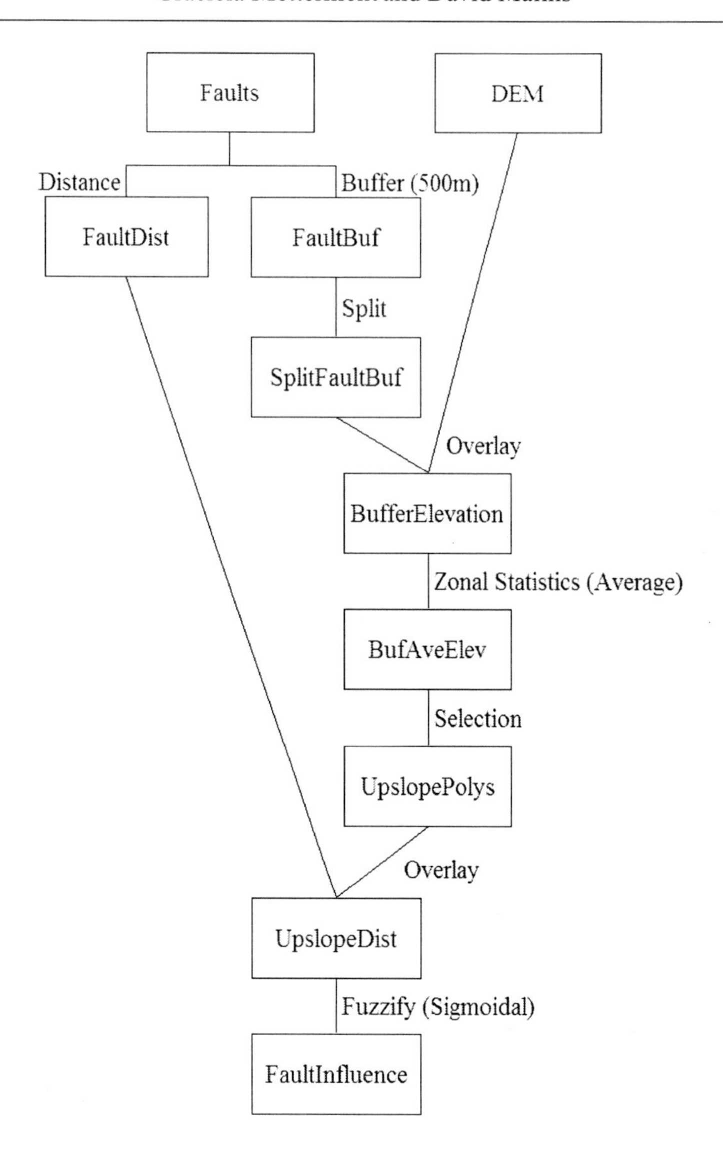

Figure 15: Cartographic Model used to create the layer of fault influence.

Integration of New Data

At this stage in the investigation there were three key fuzzy layers; geology, vegetation and CCLOW (the layer produced by the combination of CC and LOW). Three different fuzzy models were produced. The first model combined the fuzzified CCLOW and vegetation layers (Figure 16). The second model integrated the CCLOW and fuzzy geologic information layer (Figure 17), and the third model (Figure 18) integrated, in a two-step process, the fuzzy vegetation and geology layers with the CCLOW layer. The results were converted to binary maps. This stage of the investigation was implemented using ArcMap (ESRI, 2002a) and Arcview (ESRI, 1999).

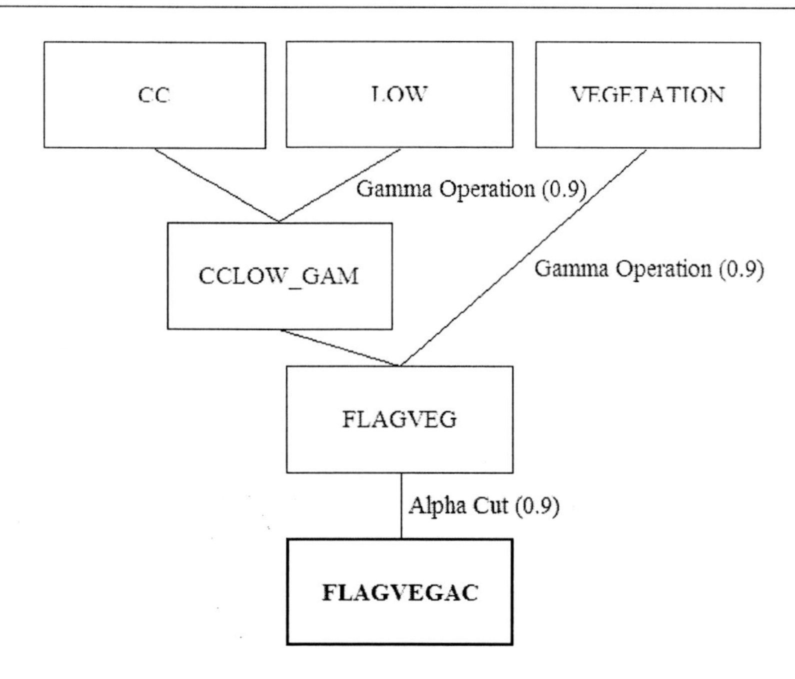

Figure 16: FLAG and vegetation cartographic model. Cartographic model that integrates the influence of vegetation into the FLAG model.

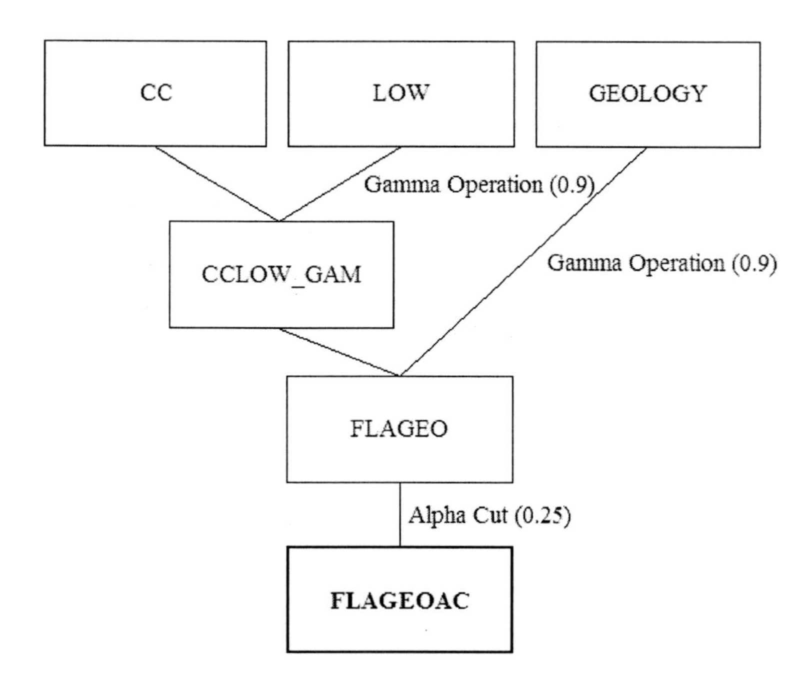

Figure 17: FLAG and geology cartographic model. Cartographic model that integrates the influence of geology into the FLAG model.

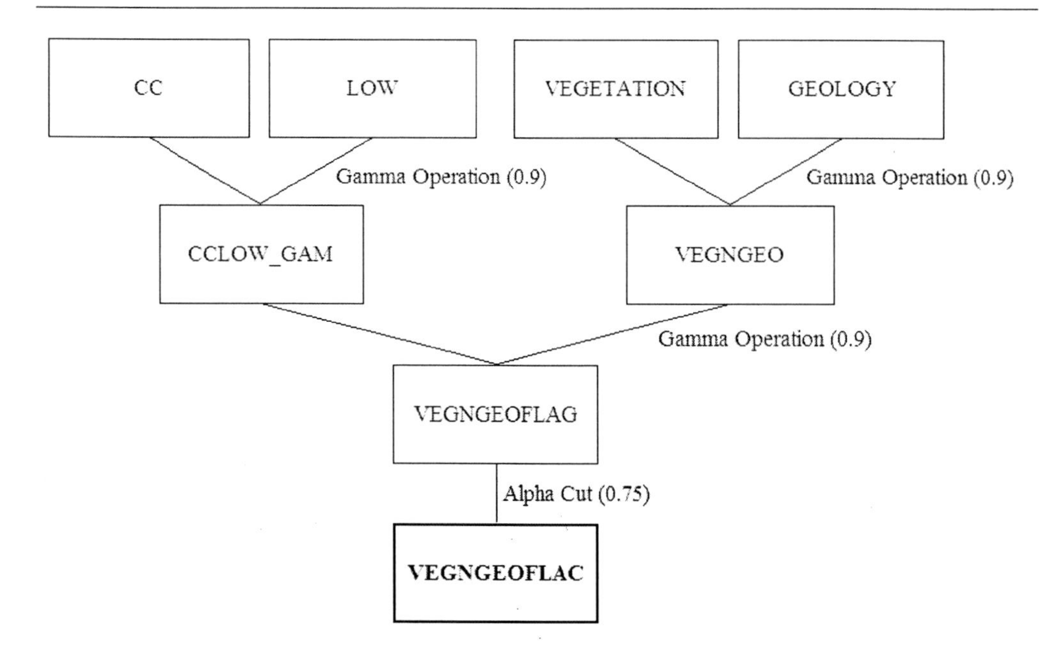

Figure 18: FLAG, geology and vegetation cartographic model. Cartographic model of the integration of FLAG, geology and vegetation.

The vegetation layer was combined with CCLOW using the gamma operator. Using this operator produced a result that effectively incorporated the evidence of both layers rather than unrealistically increasing membership regardless of low confidence, or decreasing membership regardless of high fuzzy membership. Based on advice given in Islam and Metternicht (2004) a γ value of 0.9 was used. The result was converted to a binary map using the previously described alpha cut procedure.

As with the vegetation layer, the geology layer was combined with CCLOW using the gamma operator. Again, a γ value of 0.9 was used and the result was converted to a binary map using the previously described alpha cut procedure. Having individually integrated the geology and vegetation data these layers were now combined and integrated with CCLOW as a single layer. The gamma operator was used to combine vegetation and geology to create a single layer (VEG_GEO) that represented that combined influence they have on the distribution of salinity. VEG_GEO was then combined with CCLOW using the gamma operator. This layer was then converted to a binary map.

Validation

The comparison of the ground truth classes with predicted classification constitutes what Congalton (1991) terms site-specific accuracy assessment. That is, the locational and classification accuracy are assessed together. Five layers from this investigation were compared with the ground truth data: 1) CC; 2) CCLOW; 3) the gamma combination of CCLOW and geology; 4) the gamma combination of CCLOW and vegetation; and 5) the gamma combination of CCLOW, geology and vegetation.

The starting of point for the comparison was an error matrix or contingency table. An error matrix is a square array of numbers that records the number of pixels assigned to a particular category relative to the actual category (Jensen, 1996). It is a very effective way to represent accuracy because the accuracies of each category are plainly shown along with both the errors on inclusion (commission errors) and errors of exclusion (omission errors) present in the classification (Congalton, 1991).

The ground truth data to which the predictions were compared included the borehole salinity polygons used to perform the training as well as borehole derived polygons delineating non-saline areas. Bores at which the ground water level was within 10 m of the surface and which had a groundwater salinity level exceeding 2,000mS/m were classified as locations possessing dryland salinity. The distance of 10 m is based on risk analysis performed by Short and McConnell (2001), who classify areas at which the groundwater depth is within 2 m of the surface to be at high risk of dryland salinity, and areas at which the groundwater depth is within 10 m of the surface and that have rising groundwater, to be at moderate risk. Bicknell (2003) classifies water with salinity between 2,000 mS/m and 3,000 mS/m to be very saline and water with salinity exceeding 3,000 mS/m to be extremely saline. Likewise, based on the study of Short and McConnell (2001) and Bicknell (2003), areas with a groundwater table 10 m or deeper and a salt level of 200 mS/m or less were considered at 'low to no risk' of salinity. Having isolated the point locations affected by dryland salinity, and those 'at low risk', these were buffered by 100 m to convert them to polygons.

Having created error matrices comparing each of the predictions with the ground truth data, descriptive statistics were derived. Firstly the four diagnostic statistics that were used to evaluate the α cut predictors were calculated for each prediction. Next the accuracy of the two categories (saline and not saline) was computed. First the omission error or "producers accuracy" was calculated. This represents the probability of a reference pixel being correctly classified. It is calculated by dividing the total number of correct pixels in a category by the total number of pixels of that category as derived from the reference data (Congalton, 1991). Secondly the commission error or "user's accuracy" that indicates the probability that a pixel classified on the image actually represents that category on the ground was calculated. This is found by dividing the total number of pixels in a category by the total number of pixels that were classified in that category.

The error matrix was then used as the basis for calculating the KAPPA coefficient. KAPPA analysis produces a KHAT statistic that is another measure of agreement or accuracy (Congalton, 1991). It measures the proportion of agreement between classifications after chance agreements have been removed from consideration (Banko, 1998). KHAT increases to one as chance agreement decreases and becomes negative as less than chance agreement occurs (Banko, 1998). The KAPPA statistic has the advantage of being statistically testable against the standard normal distribution. It tests the null hypothesis that there is no agreement between the two classifiers. The KAPPA statistic is recommended by Rosenfield and Fitzpatrick-Lins (1986), Congalton (1991), Fitzgerald and Lees (1994) and Banko (1998). The KHAT statistic is calculated using Equation 8.

$$\hat{K} = \frac{p_o - p_e}{1 - p_e} \tag{8}$$

where $p_o = \sum_{i=1}^{r} x_{ii} / N$ represents the overall proportion of agreement and $p_e = \sum_{i=1}^{r} (x_{i+} \times x_{+i}) / N^2$ represents the overall proportion of chance expected agreement. Using the error matrix the variance of the KAPPA statistic can be calculated. Finally, the KAPPA test statistic is converted to the standard normal Z score using Equation 9, and tested against the normal distribution.

$$Z = \frac{\hat{K}_1}{\sqrt{\hat{\mathrm{var}}\left(\hat{K}_1\right)}} \tag{9}$$

A simple test was then conducted to determine if the accuracy of the models was dependent on the landform, with the aim to determine: a) the percentage of the predictions falling in high/low areas of the catchment that are correct, and b) the percentage of the predictions in steep/flat areas of the catchment that are correct. Areas with a greater than the average catchment elevation were grouped into one class and areas with less than average catchment elevation into another. Overlaying this map with a map showing the correct and incorrect predictions enabled the first question to be answered. Furthermore, areas with a slope less than or equal to one percent were grouped into one class and areas with a slope greater than one percent were grouped into another. Overlaying this map with a map showing the correct and incorrect predictions enabled the second question to be answered.

Comparison of Modelling Techniques

The aim of this phase of the investigation was to compare the fuzzy logic based prediction of salinity developed in this investigation with the prediction of salinity based on Boolean logic and probability theory made in the Land Monitor project. The two best predictions (CC and FLAGVEG) made in this investigation (as determined by KAPPA analysis) were compared with the Land Monitor prediction. To this end, each of the two FLAG based predictions was overlayed with the Land Monitor prediction to fill in contingency tables comparing the pairs of layers.

As when comparing the predictions with the ground truth data, a contingency table was used to calculate statistics (overall accuracy, efficiency, discrimination, power, produces accuracy and consumers accuracy) to compare the two data sets.

A test can also be conducted to determine if two independent KHAT values, error matrices, and therefore sets of data are significantly different. Equation 10 is used to calculate the test statistic for testing if two independent error matrices are significantly different. The null hypothesis is that there is no disagreement between the KHAT values.

$$Z = \frac{\left|\hat{K}_1 - \hat{K}_2\right|}{\sqrt{\hat{var}\left(\hat{K}_1\right) + \hat{var}\left(\hat{K}_2\right)}} \tag{10}$$

The landform maps were used to assist in the understanding of the location of areas of agreement and disagreement. Both of these maps were overlayed with each of the maps of agreement and disagreement between the three predictions being compared. From these maps percentages could be calculated to reveal the landform types where agreements or disagreements occur.

Discussion of Results

Basic FLAG Model Results

CC and LOW are the two final discharge indices from the implementation of the original FLAG model. Maps of these two indicators are shown in Figures 19 and 20 respectively. As it was expected, low lying areas have high salinity potential and elevated locations have low salinity potential. CC tended to show high salinity potential at locations that were low relative to the rest of the catchment and low relative to the surrounding landscape. LOW tends to show high salinity potential only at locations that are low relative to the surrounding landscape. The plots of α against power constructed for CC and LOW as described in the previous section, were examined to determine the optimal α-cut value needed to generate binary maps of these layers (Figures 21 and 22). A value of 0.35 was used on CC and value of 0.05 on LOW. The α-cut (binary) versions of these maps are shown in Figure 23 and Figure 24, respectively.

Extended FLAG Results

This investigation extended the original FLAG model through the inclusion of remote sensing and geology data. To ease the integration of this new data and simplify the final result CC and LOW were combined together using the gamma operator. The result (CCLOW_GAM) is shown in Figure 25. The combination of CC and LOW via the gamma operator captures the intent of FLAG by predicting salinity at local and catchment scale low points as well as highly concave locations. This was then converted to a binary map using an α-cut value of 0.35 derived from plots of the same type as shown in Figures 21 and 22. The result is shown as Figure 26.

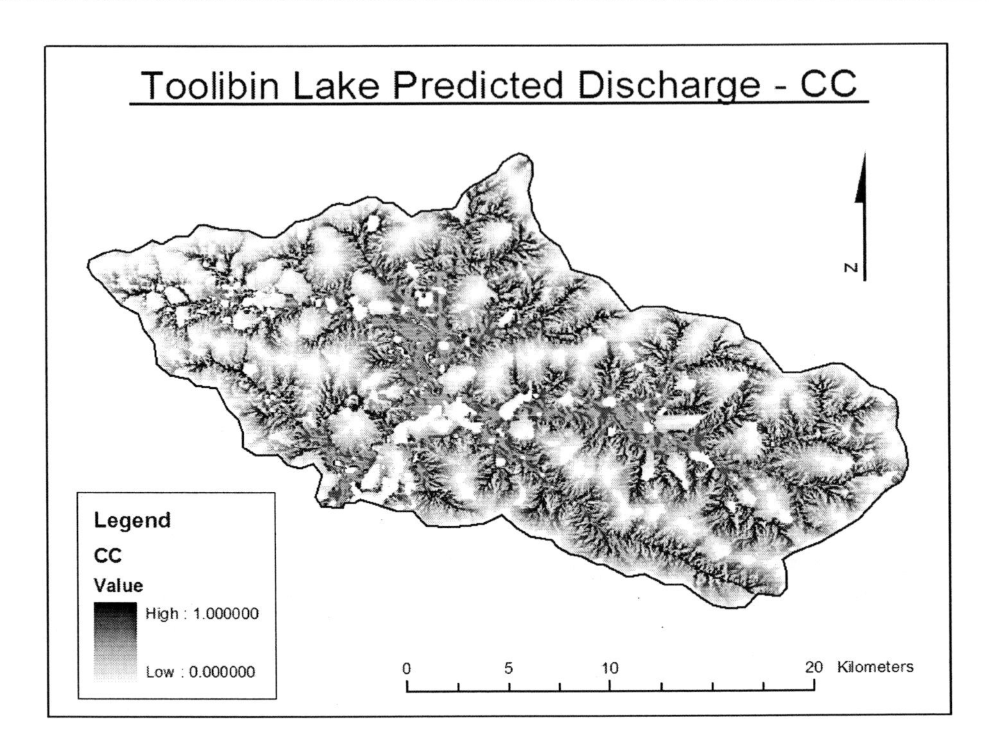

Figure 19: Toolibin catchment fuzzy discharge as predicted by the CC FLAG filter

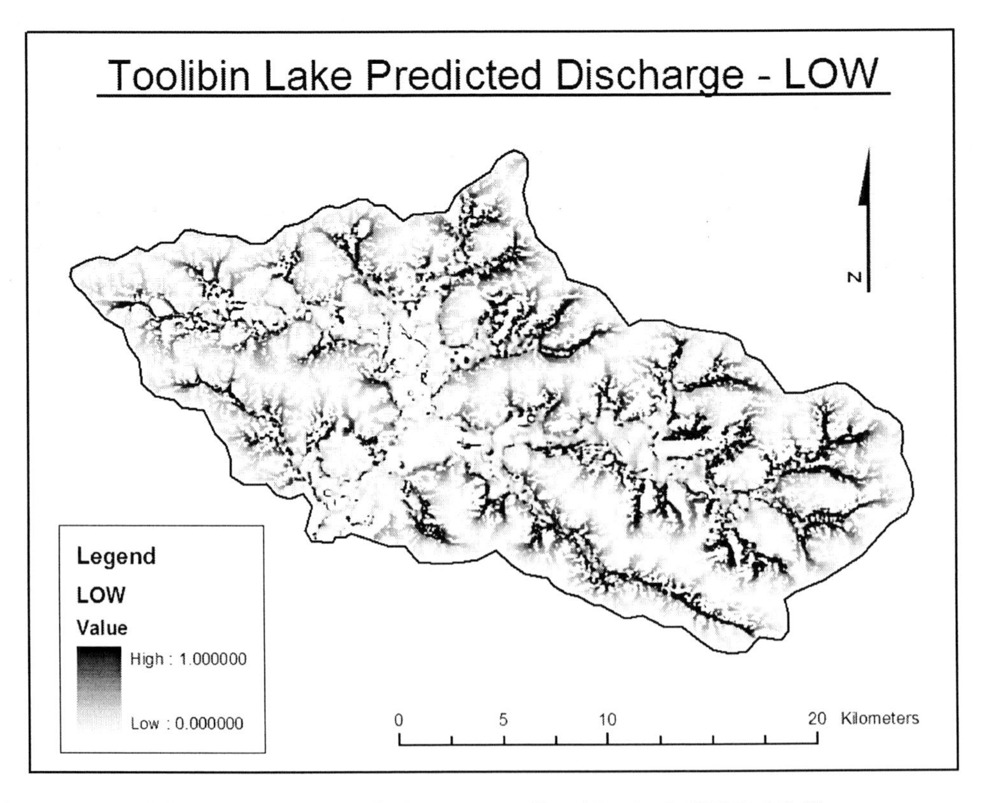

Figure 20: Toolibin catchment fuzzy discharge as predicted by the LOW FLAG filter.

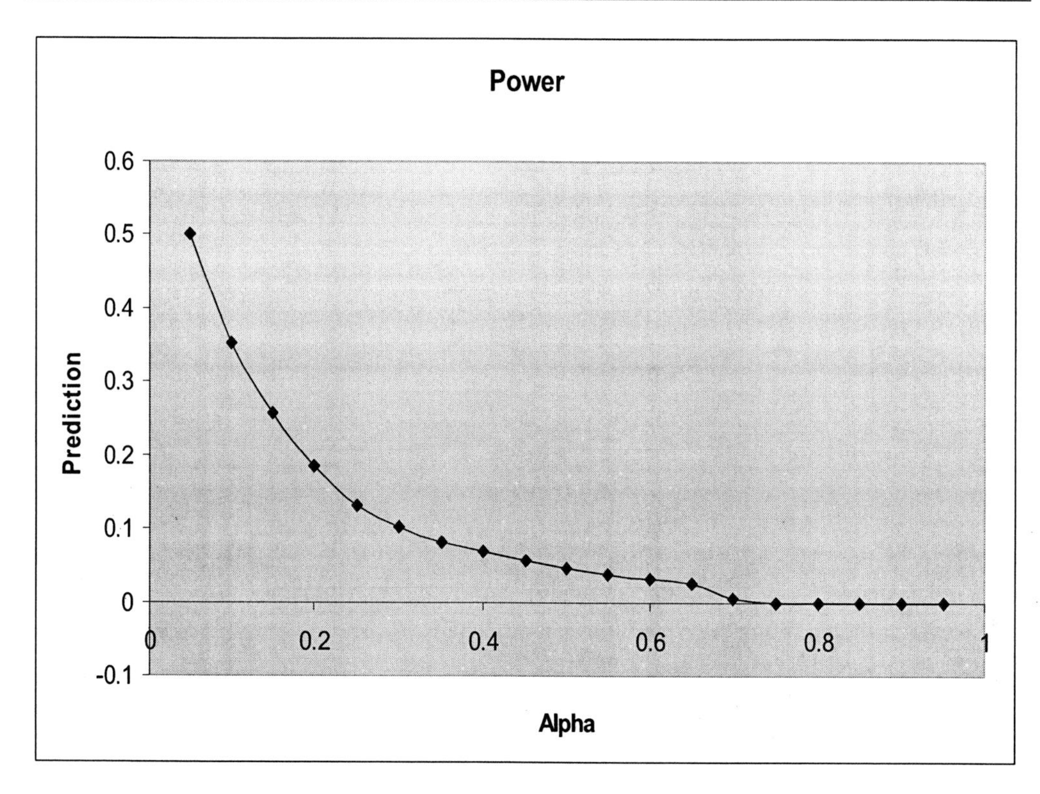

Figure 21: Graph showing a plot of predictive power against α for LOW

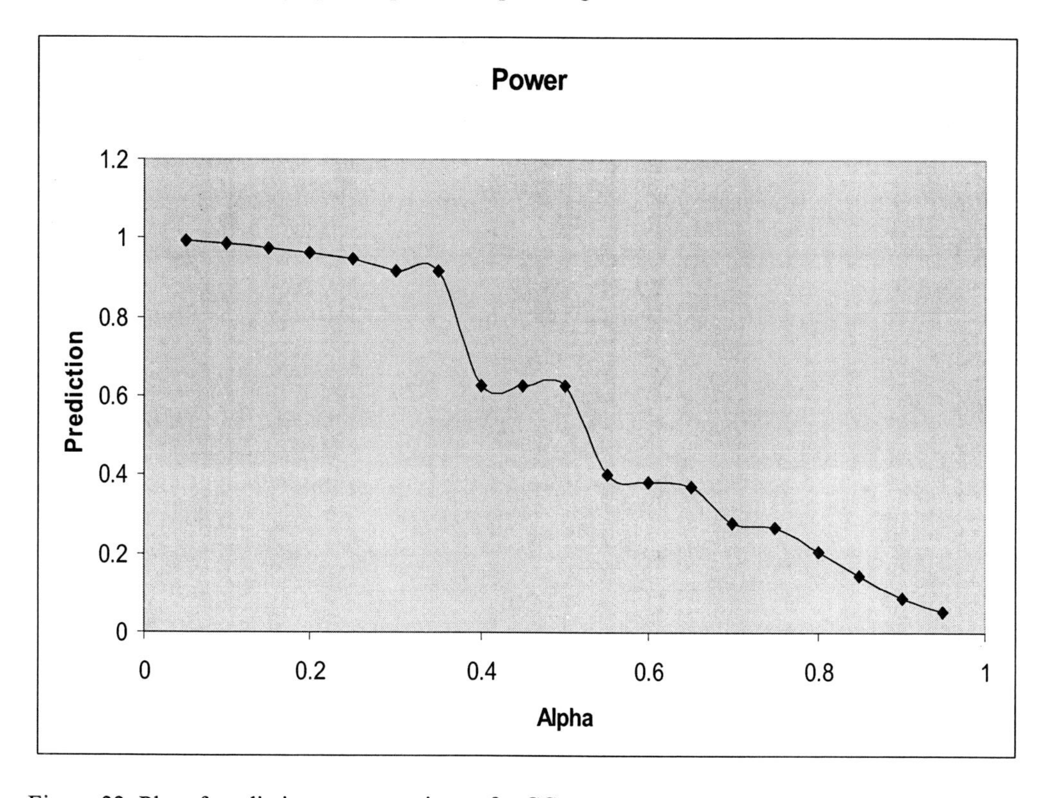

Figure 22: Plot of predictive power against α for CC.

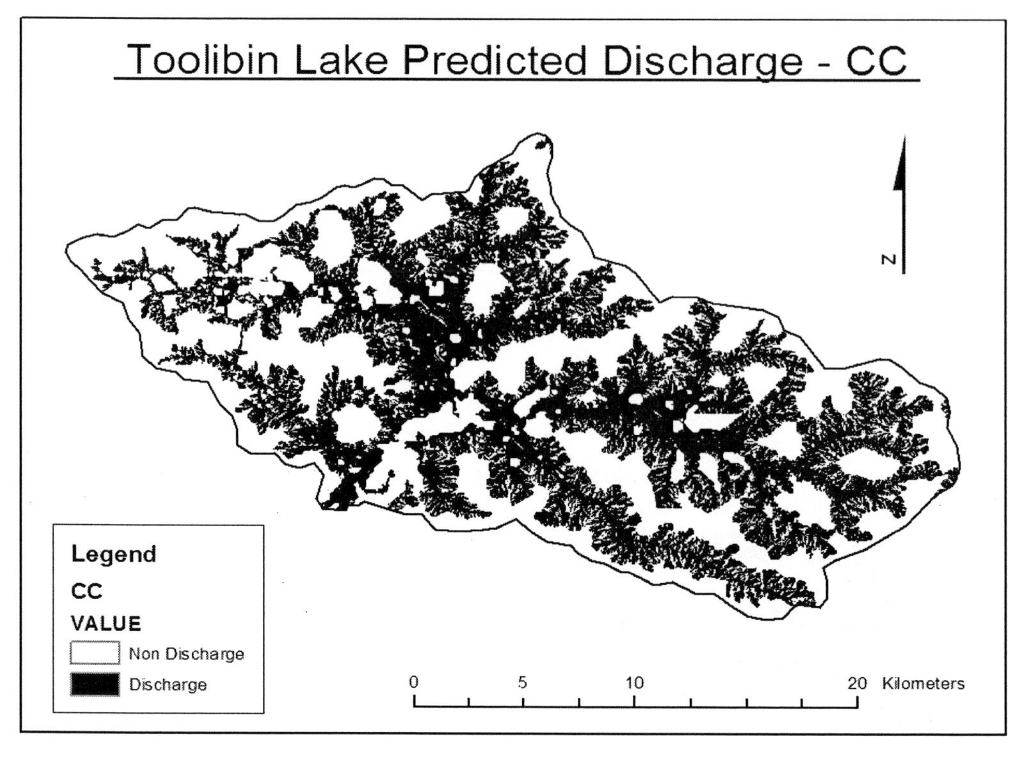

Figure 23: α-cut CC prediction of discharge.

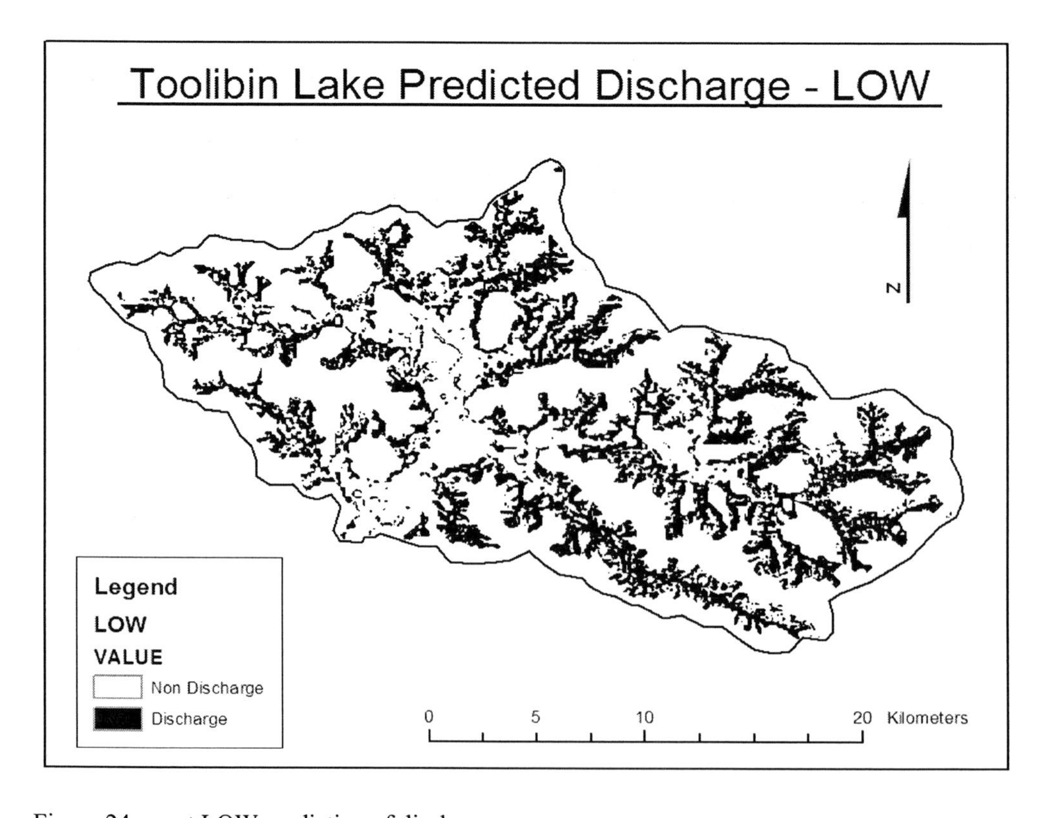

Figure 24: α-cut LOW prediction of discharge

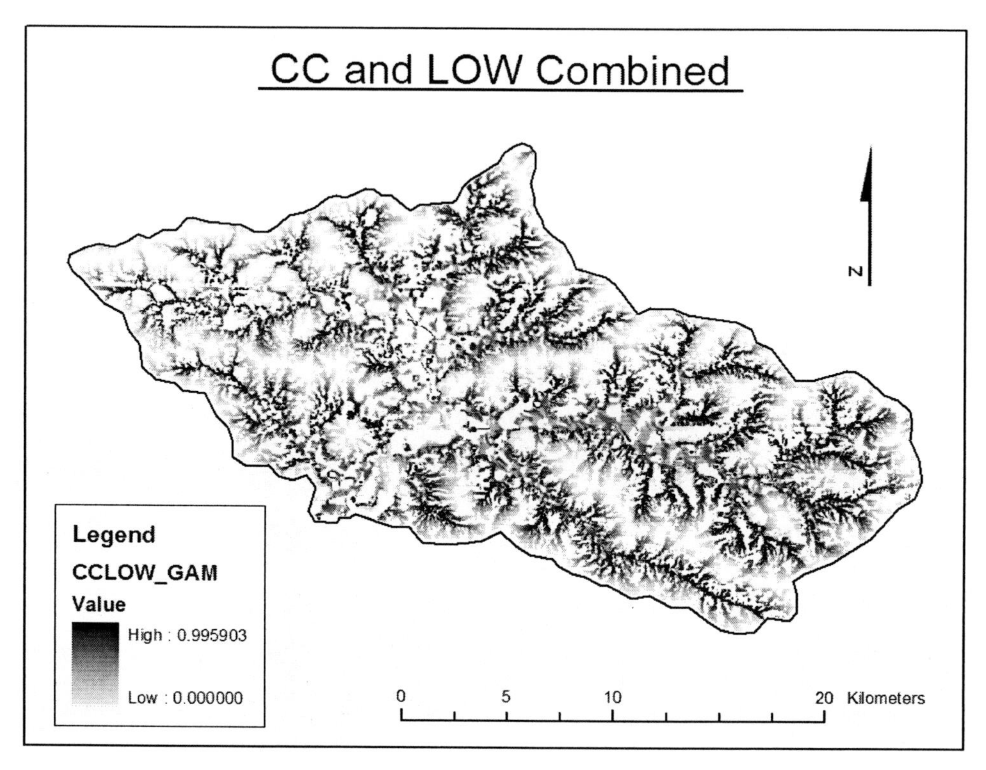

Figure 25: CC and LOW combined using the gamma operator

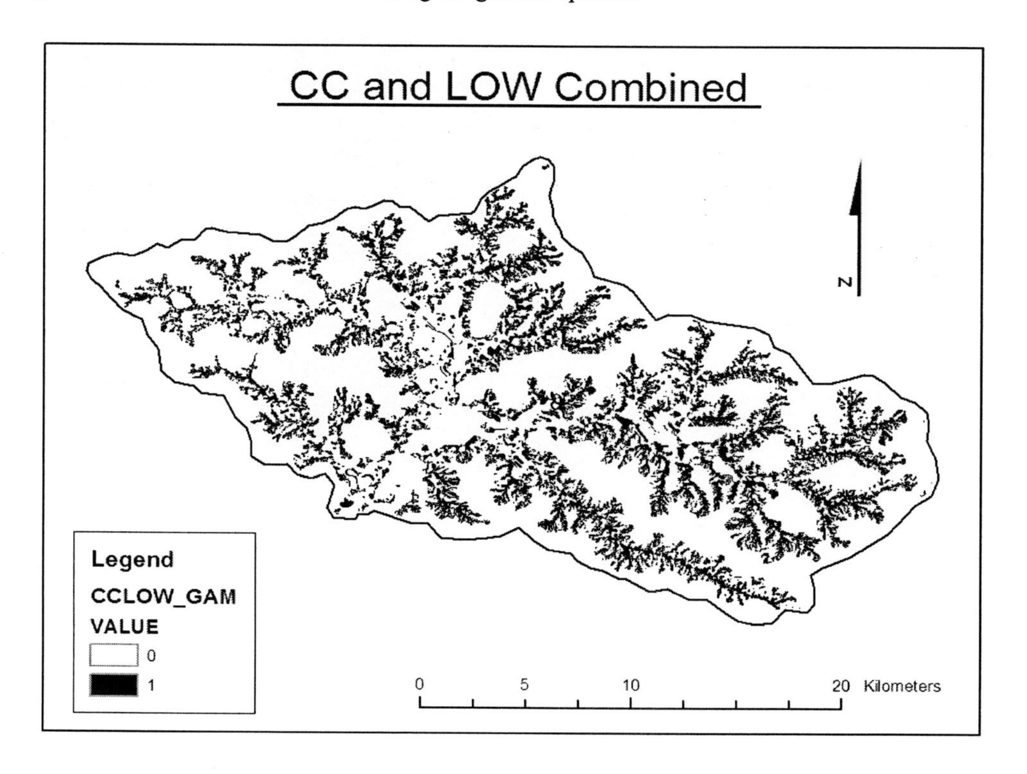

Figure 26: α-cut CCLOW prediction of discharge

The attempted optimisation of FLAG by inclusion of vegetation information in the modelling phase was undertaken using a Landsat TM scene as the source data. This image was used to extract the location of native remnant vegetation using the NDVI, and the output of this image transformation was then fuzzified for integration with the FLAG data. The fuzzified map of vegetation influence is shown in Figure 27. This was then incorporated with the FLAG data (Figure 28) and converted to a binary prediction of salinity (Figure 29). An α-cut value of 0.9 was used to perform the conversion. The graph of alpha against predictive power is shown in Figure 30. Unlike the other predictive layers the α value with the greatest power was not used. As it can be seen in the graph, the power level remains relatively constant until an α level of 0.9 at which it falls dramatically. Using an α-cut value of 0.15, that had the equal highest power value, produced a result that, upon visual inspection (see Figure 31), seemed to overestimate discharge potential.

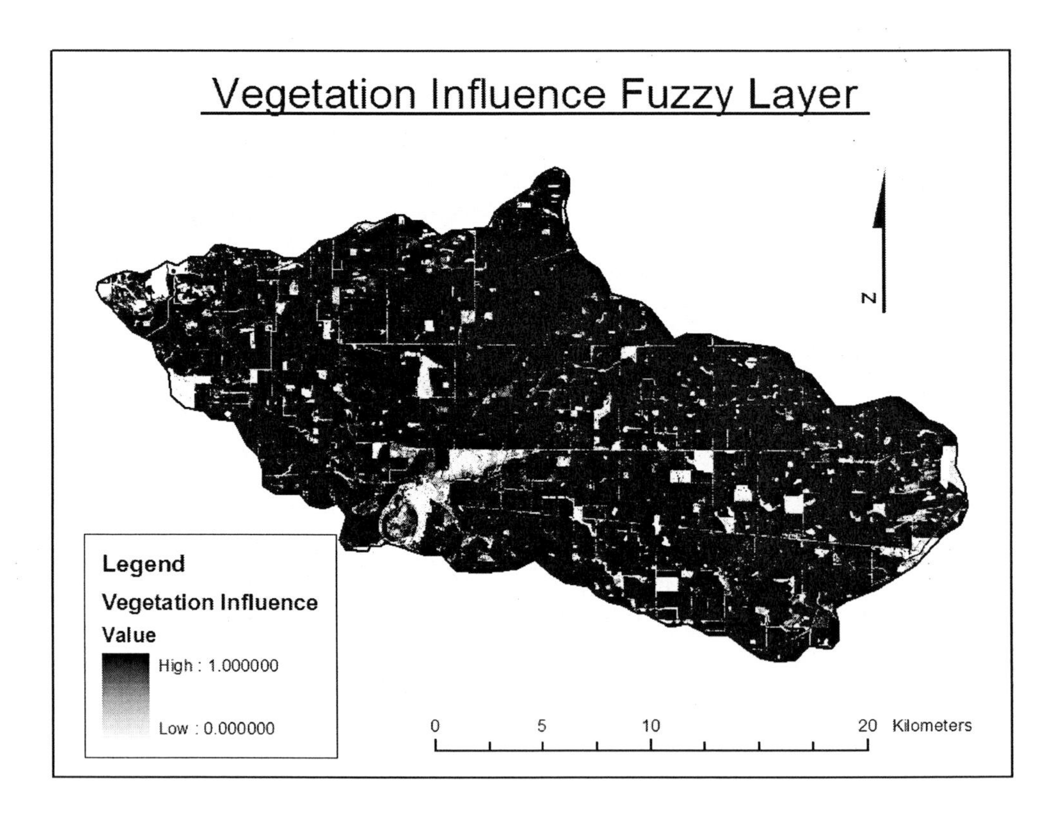

Figure 27: Fuzzy layer representing the influence of vegetation on discharge potential

The fuzzy layer representing the influence of faults on discharge is shown in Figure 32. Discharge potential can be seen to decrease away from the faults. A fuzzy map showing the gamma combination of the fuzzy geology layer and CCLOW is shown in Figure 33. This fuzzy prediction was defuzzified using an α-cut at 0.25 and the result is shown in Figure 34.

As already discussed, vegetation and geology were combined together into a single layer (Figure 35). This layer was combined with FLAG to produce a fuzzy prediction of discharge based on vegetation, geology and FLAG (Figure 36). This was converted to a binary prediction of discharge (Figure 37) using the optimal α-cut value interpreted from a plot of α against predictive power.

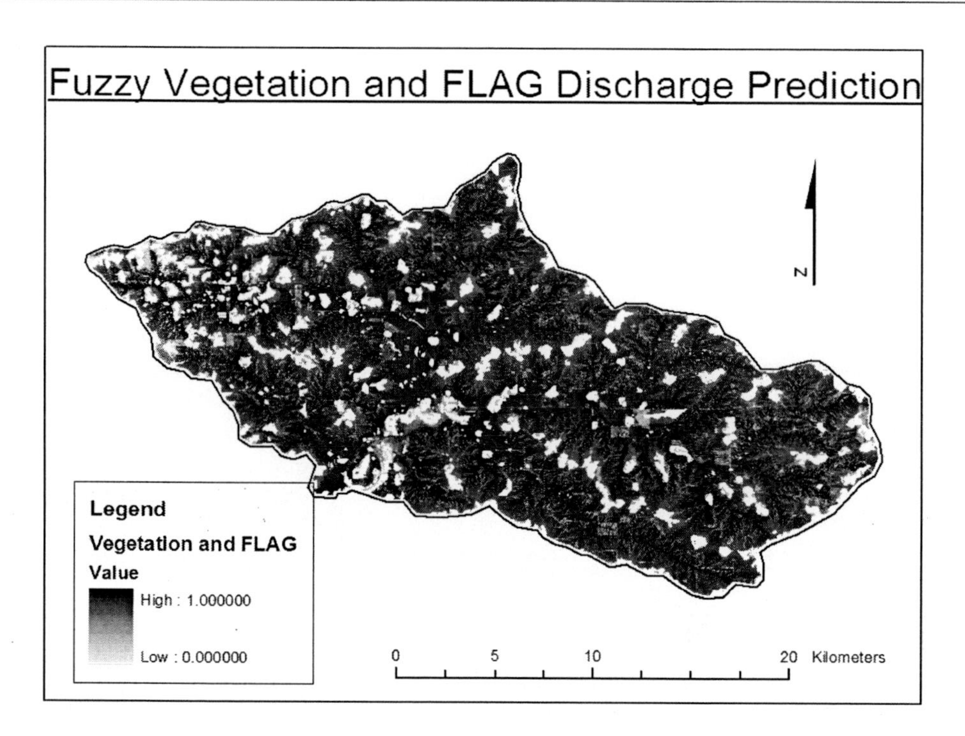

Figure 28: Combined FLAG and vegetation fuzzy prediction of discharge potential

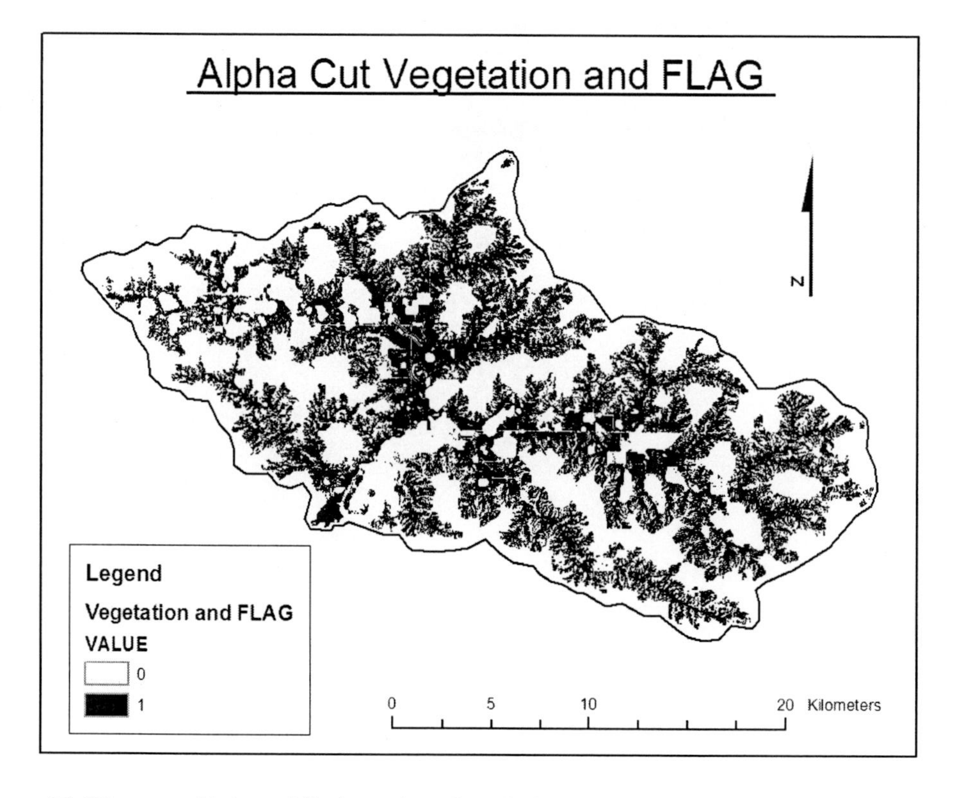

Figure 29: Binary prediction of discharge based on the integration of vegetation and FLAG. α-cut value of 0.9

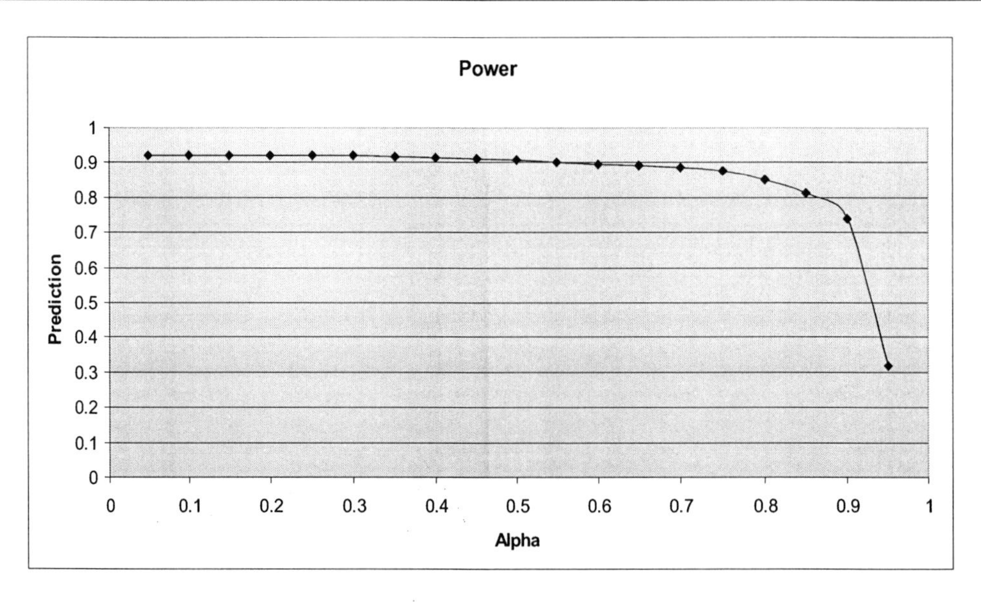

Figure 30: Plot of predictive power against alpha

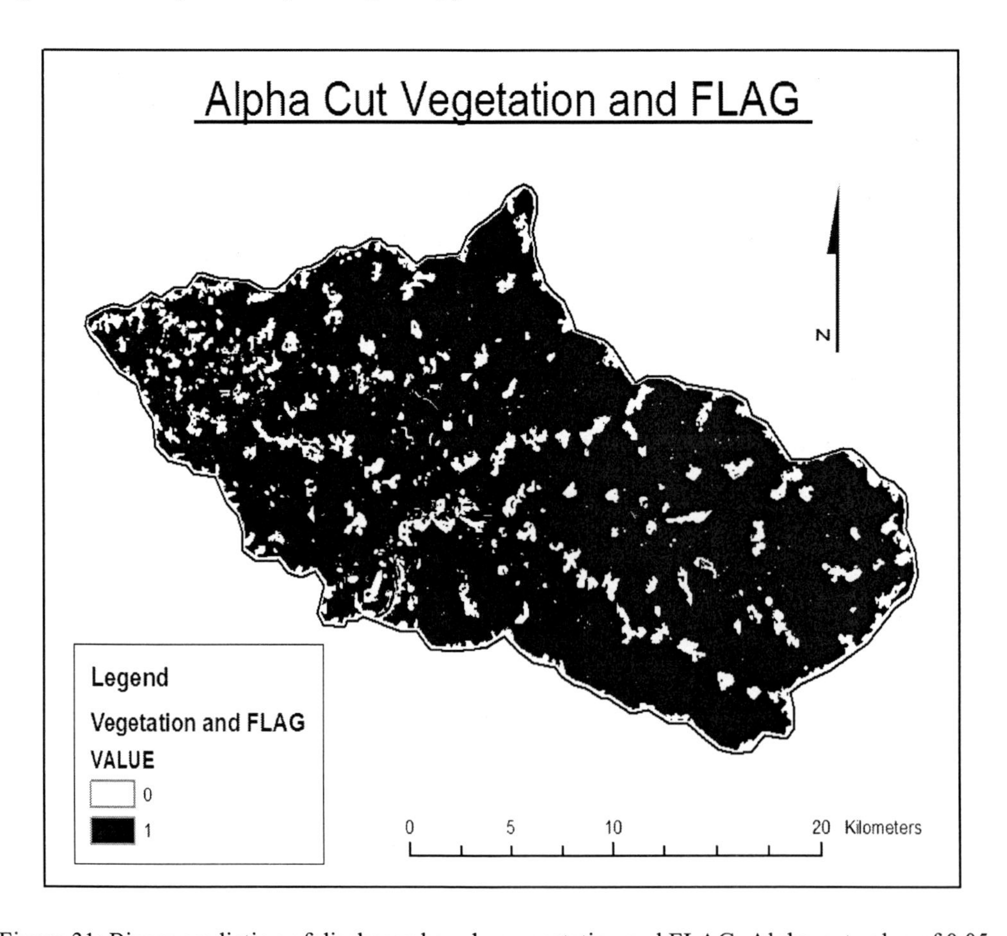

Figure 31: Binary prediction of discharge based on vegetation and FLAG. Alpha-cut value of 0.05.

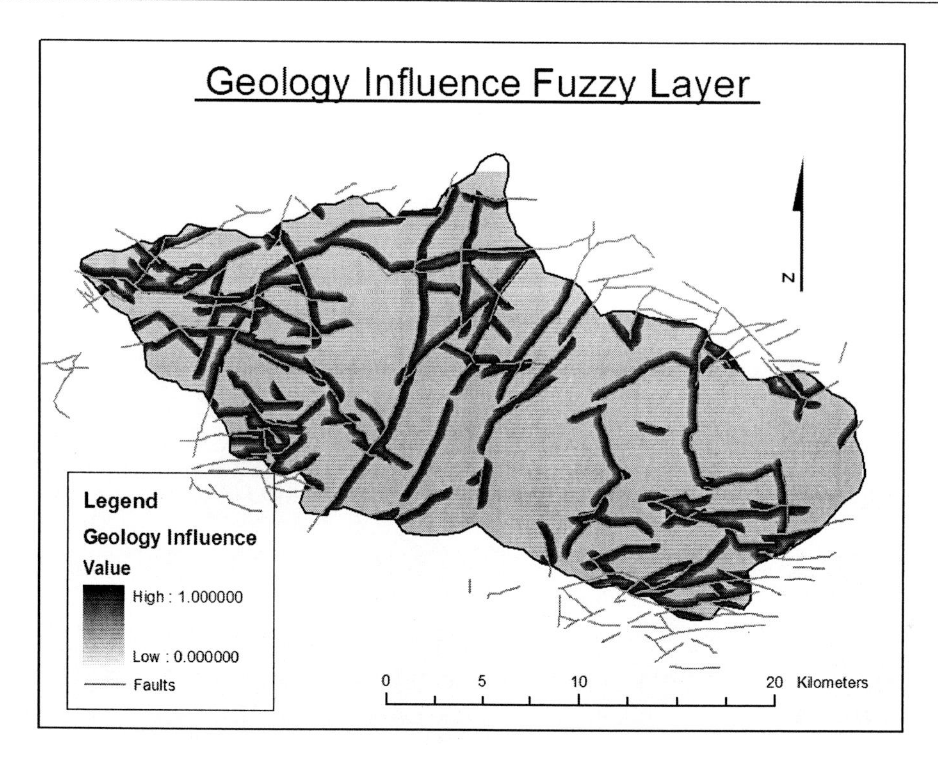

Figure 32: Fuzzy layer representing the influence of faults on discharge potential

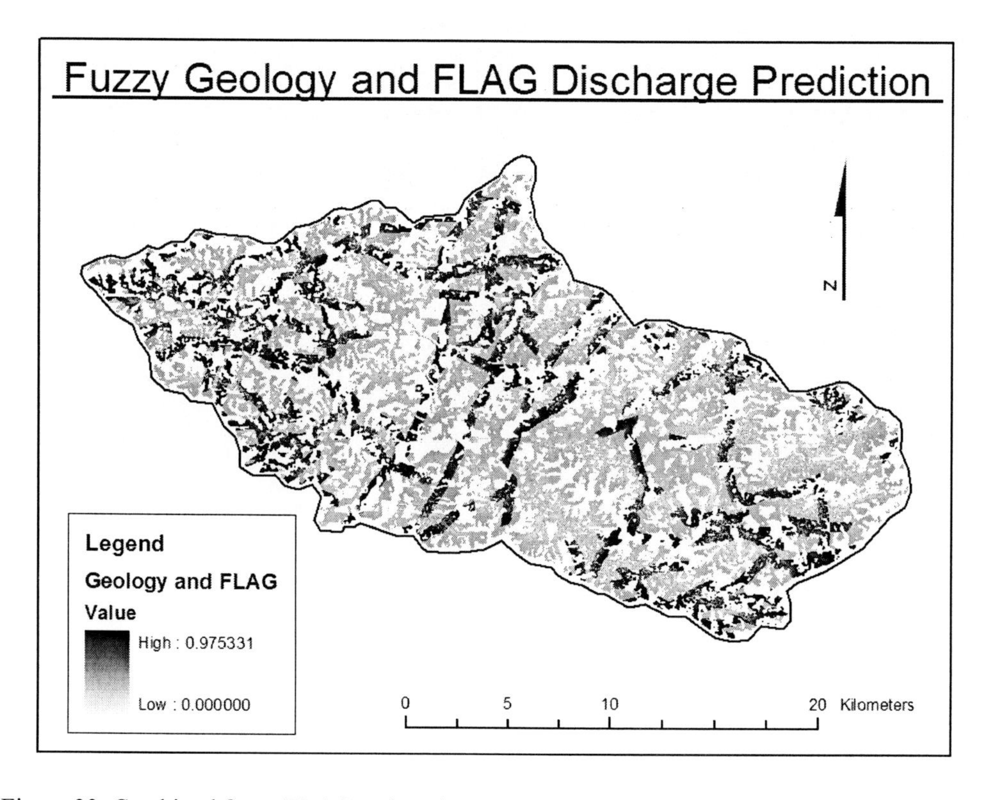

Figure 33: Combined fuzzy FLAG and geology prediction of discharge potential

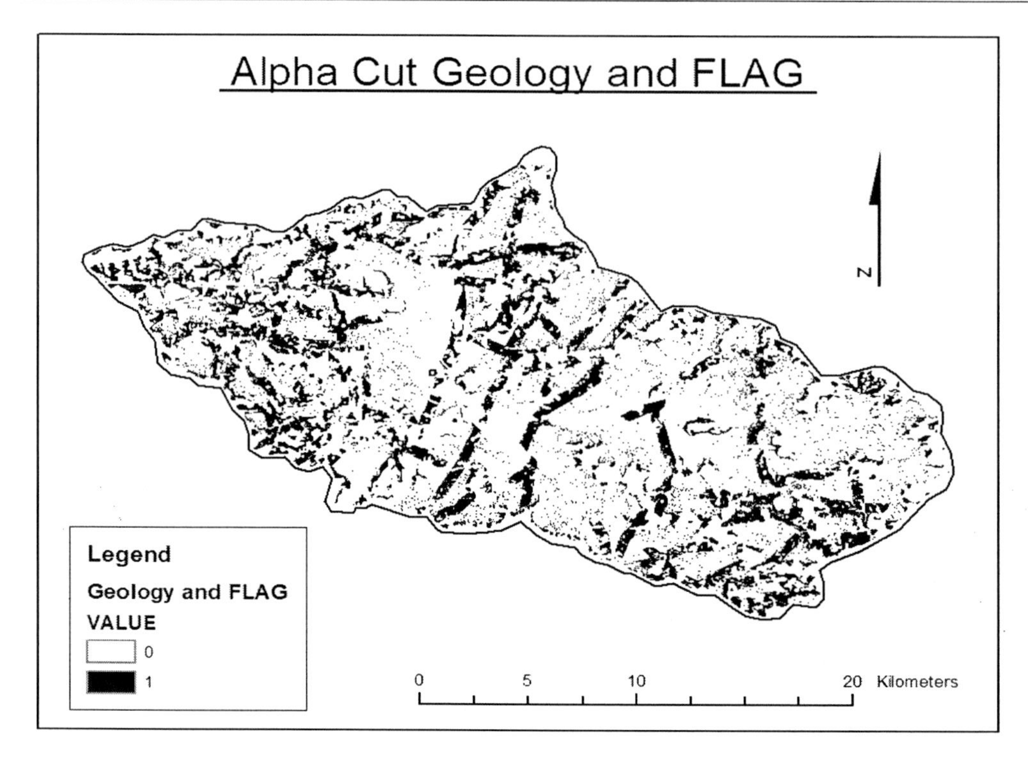

Figure 34: Combined binary FLAG and geology prediction of discharge potential

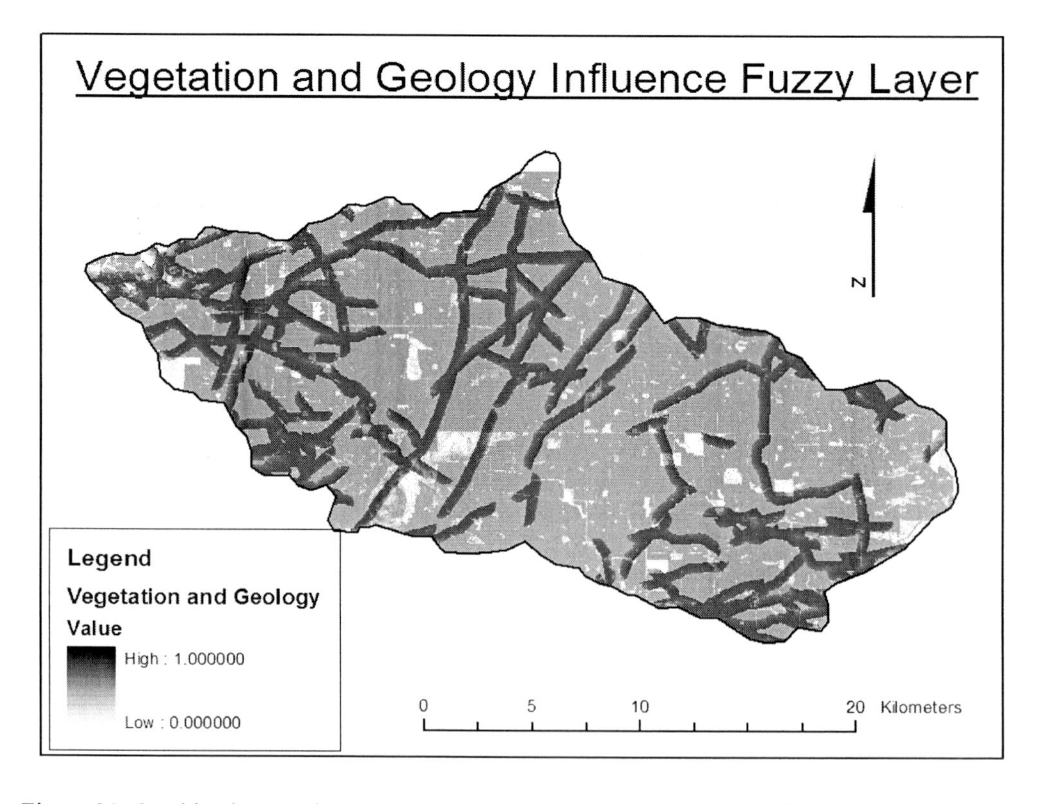

Figure 35: Combined vegetation and geology prediction of discharge potential

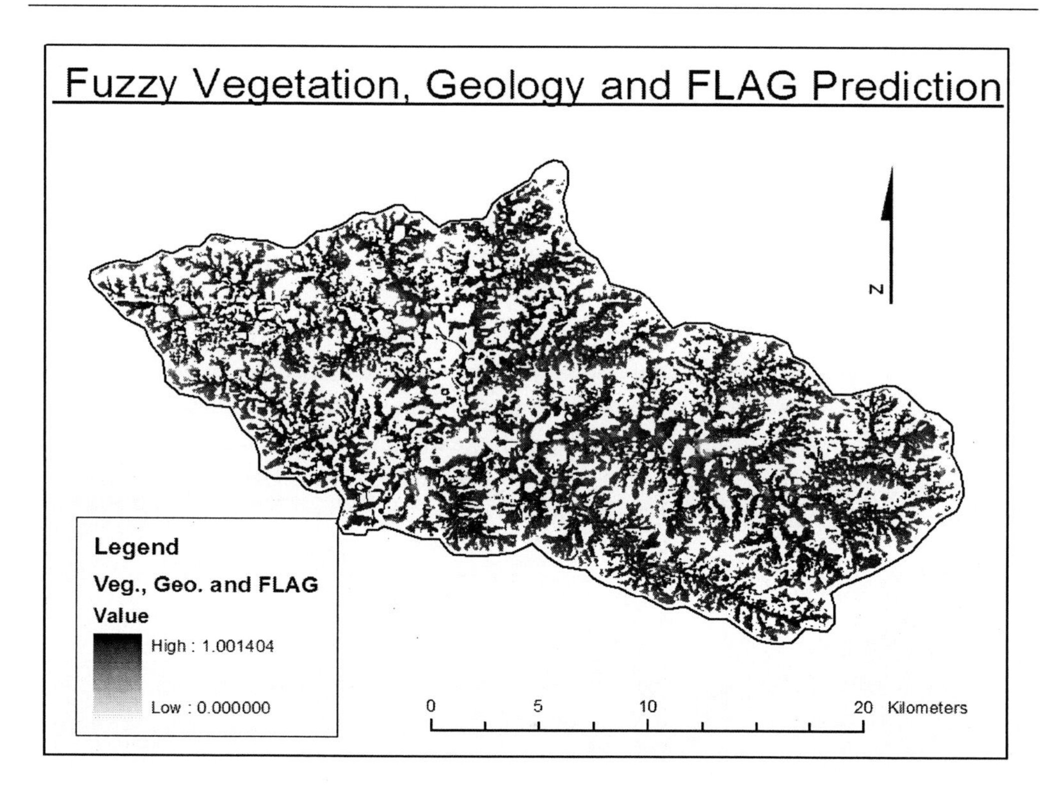

Figure 36: Combined vegetation, geology and FLAG fuzzy prediction of discharge potential

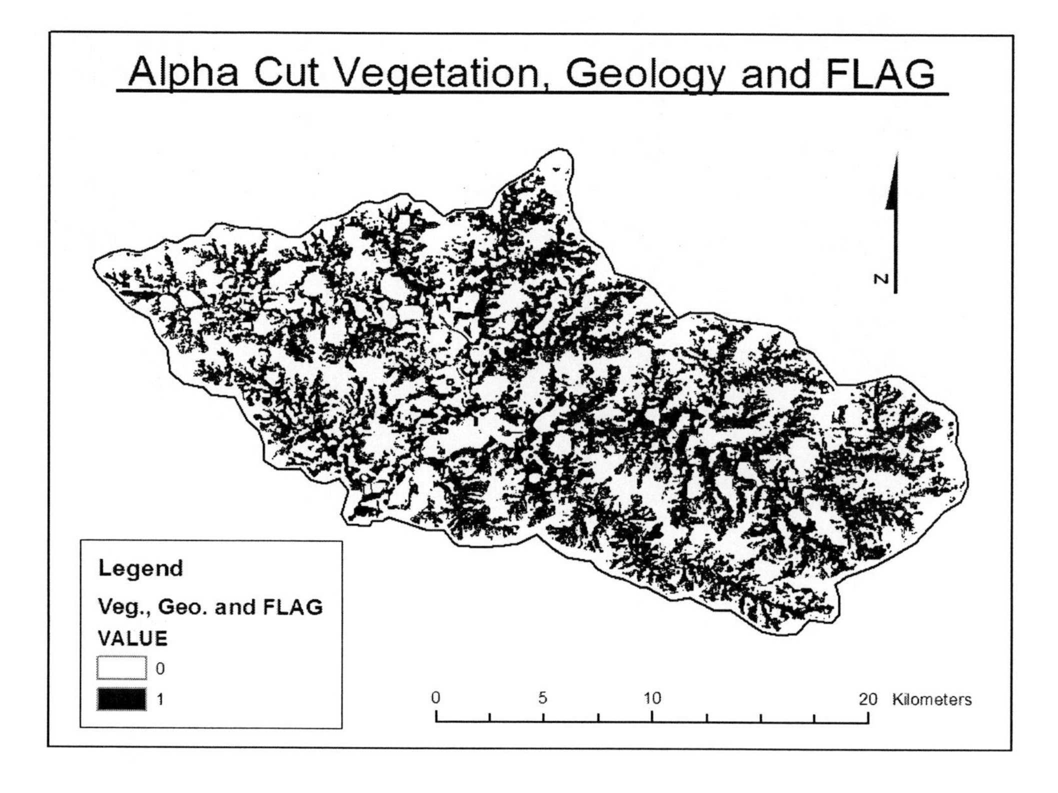

Figure 37: Combined vegetation, geology and FLAG binary prediction of discharge potential

Validation

The predictions of discharge potential made in this investigation were statistically compared with ground truth data to measure their accuracy. Five layers resulting from this investigation were compared with the ground truth data: 1) CC; 2) CCLOW; 3) the gamma combination of CCLOW and geology (e.g. FLAG_GEO); 4) the gamma combination of CCLOW and vegetation (e.g. FLAG_VEG); and 4) the gamma combination of CCLOW, geology and vegetation (e.g FLAG_GEO_VEG).

The contingency table and diagnostic statistics comparing CC and ground truth data are shown in Table 1. CC has high overall accuracy, efficiency discrimination and power, as well as high producers and user's accuracy for saline areas. The later means that most saline areas have been correctly identified as saline (73 percent) and most areas classified as saline are saline (60 percent). However, only 57 percent of areas classified as non-saline are actually non-saline. Landis and Koch (1977) suggest ranges of agreement for the KAPPA test statistic. These are poor ($K < 0.4$), good ($0.4 \leq K \leq 0.75$) and excellent ($K > 0.75$). These ranges were adopted in this investigation. Table 2 shows that CC has good accuracy (KHAT 0.45), whereas CCLOW was found to be a poor predictor of salinity when validated against the ground truth data, as exemplified by the KHAT value of 0.089, and therefore not considered for further analysis. Table 2 also presents the variance of the KHAT statistic and Z statistic used for determining if the classification is significantly better than a random result. At the 95 percent confidence level, the critical value is 1.96. The Z statistic for CC is 47.96 and the classification is therefore significantly better than random. Likewise, table 3 reveals that CC was accurate across most landform types particularly high locations in the landscape, with the least accuracy reported at low landscape locations.

Table 1. Contingency Table and Diagnostic Statistics for the Fuzzy CC Model

		Actual Saline	Non-Saline	Users Accuracy
CC	Saline	5,323	843	0.863
	Non-Saline	1,909	2,594	0.576
Producers Accuracy		0.736	0.755	
Accuracy (overall)	0.742056			
Efficiency	0.736034			
Discrimination	0.863283			
Power	0.811908			

Table 2. Khat Statistics, Variance and Z Values for the Fuzzy Models CC, CCLOW, and the Integration of FLAG with Vegetation and Geology Information; and the Crisp Model of Land Monitor

CC Low	0.089	0.007	11.980
FLAG and geology	0.029	0.006	4.787
FLAG and vegetation	0.422	0.023	18.008
FLAG, geology and vegetation	0.184	0.022	8.232
Land Monitor	0.663	0.009	72.698

Table 3. Landform Accuracy Statistics for the Fuzzy and Crisp Models Outputs, Compared against Ground Truth Data

	Landform type			
	Rim	**Flat**	**High**	**Low**
CC	77.82%	72.79%	88.67%	69.81%
FLAG and vegetation	76.35%	70.48%	88.13%	67.28%
Land Monitor	62.24%	91.46%	100.00%	78.16%

The contingency table and basic descriptive statistics comparing FLAG_GEO and the ground truth data show low overall accuracy, efficiency and power (Table 4). Table 4 also reveals that 82 percent of non-saline areas and 21 percent of salines areas have been correctly identified. Table 2 shows that the FLAG_GEO model has poor accuracy (KHAT 0.02). The Z statistic for FLAG_GEO is 11.98 and the classification is therefore better than random at the 95 percent confidence level. Although geology was included to try and improve the performance of FLAG, the results show that this was not achieved. Poletaev (1992) states that faults tend to occur in topographic depressions. Discharge was assumed and modelled to be greater at faults; however it appears that the increased likelihood of discharge at depressions as already accounted for in FLAG. This may mean that the inclusion of fault data added no new information to the model. The underperformance of FLAG_GEO may also have been caused by the choice of fuzzy membership function or the choice of fuzzy operator used to integrate the layers. For example, it may be more appropriate to combine FLAG and GEO using the fuzzy AND operator, and fuzzify the geology influence layer using a linear membership function.

Table 4. Contingency Table and Diagnostic Statistics for the Fuzzy FLAG Model Optimised with the Inclusion of Geologic (e.g. Faults) Data

		Actual Saline	**Non-Saline**	**Users Accuracy**
FLAG_GEO	Saline	1,557	603	0.721
	Non-Saline	5,675	2,834	0.333
Producers Accuracy		0.215	0.825	
Accuracy (overall)	0.412			
Efficiency	0.215			
Discrimination	0.721			
Power	0.446			

The contingency table and basic descriptive statistics comparing FLAG_VEG and the ground truth data are shown in Table 5. FLAG_VEG has quite high overall accuracy, efficiency discrimination and power, as well as high producer's and user's accuracy for saline areas. This means that most saline areas have been correctly identified, with low commission errors (e.g. high producer's accuracy). However, only 55 percent of areas classified as non-saline are actually non-saline. Table 2 shows that FLAG_VEG has good accuracy (KHAT 0.42). It also presents the variance of the KHAT statistic and Z statistic used for determining if the classification is significantly better than a random result. The Z statistic for FLAG_VEG is 18.008 and the classification is therefore significantly better than random at the 95 percent confidence level. Table 3 reveals that FLAG_VEG performs well across all landform types, with a slightly lower performance on flat and low areas. As with geology, the underperformance of FLAG_VEG may have been caused by the choice of fuzzy membership function or the choice of fuzzy operator used to integrate the layers.

Table 5. Contingency Table and Diagnostic Statistics for the Fuzzy FLAG Model Optimised with the Inclusion of Fuzzy Vegetation Data

		Actual Saline	Non-Saline	Users Accuracy
FLAG VEG	Saline	806	130	0.861
	Non-Saline	343	419	0.55
Producers Accuracy		0.701	0.7632	
Accuracy (overall)	0.721			
Efficiency	0.701			
Discrimination	0.861			
Power	0.794			

Table 6 presents the contingency table and basic descriptive statistics of FLAG_GEO_VEG, showing showing this model has moderate overall accuracy, efficiency and power. Table 6 reveals that 67 percent of non-saline areas were correctly identified. This predictor is better at identifying saline areas, as 54 percent of saline areas were correctly identified as being saline, and 77 percent of the areas labelled as saline are actually saline. However, table 2 shows a low KHAT for FLAG_GEO_VEG (0.183). The table also presents the variance of the KHAT statistic and Z statistic used for determining if the classification is better than a random result. The Z statistic for FLAG_GEO_VEG is 8.23 and the classification is therefore better than random at the 95 percent confidence level.

Table 6. Contingency Table and Diagnostic Statistics for the Fuzzy FLAG Model Optimised with the Inclusion of Fuzzy Vegetation and Geology Data (Integration in One Step)

		Actual Saline	Non-Saline	Users Accuracy
FLAG GEO VEG	Saline	626	182	0.77475
	Non-Saline	523	367	0.41236
Producers Accuracy		0.54482	0.668488	
Accuracy (overall)	0.584806			
Efficiency	0.544822			
Discrimination	0.774752			
Power	0.685801			

Of the five FLAG based predictions of discharge potential, CC produced the most accurate result (see Table 2). The attempt to optimize FLAG through the inclusion of remote sensing and geology data did not improve its accuracy. Of the optimised FLAG predictors FLAG_VEG produced the best result. FLAG was most accurate on high and steep areas located near the catchment rim (Table 3). The reduced performance of FLAG on flat level terrain may be due to the assumptions of the model being 'stretched out' at these locations. FLAG has historically performed better in hilly terrain Laffan (1996a) and Roberts *et al.* (1997). All models report good accuracy at identifying saline areas, though they tend to underestimate their extent (e.g. high commission errors are reported for non saline areas). FLAG_GEO produced the highest underestimation of saline areas.

Comparison of FLAG Performance against Crisp Modelling

As a first stage in the comparison, the prediction of areas at risk of salinity derived from the Land Monitor model was validated against the same ground truth data used in the last stage of the investigation. Only the two most accurate models from this investigation, CC and FLAG_VEG were compared with the Land Monitor prediction.

Table 7 shows that Land Monitor has high overall accuracy, efficiency discrimination and power, with high produces and user's accuracy for saline areas. This means that most saline areas have been correctly identified as saline and most areas classified as saline are saline. The performance in non-saline areas is lower, with 67 percent of commission errors for non saline areas (e.g. underestimation of saline areas). Table 2 shows that Land Monitor has good accuracy (KHAT 0.66) based on the KHAT characterisation introduced by Landis and Koch (1977). It also presents the variance of the KHAT statistic and Z statistic used for determining if the classification is significantly better than a random result. The Z statistic for FLAG_VEG is 72.7 and the classification is therefore significantly better than random at the 95 percent confidence level. Table 3 reveals that Land Monitor performs well across all landform types, however performance is lower at the steep catchment rim. The KHAT score and overall accuracy for Land Monitor is higher than any of the FLAG based predictors.

Table 7. Contingency table and diagnostic statistics for probabilistic Land Monitor model

		Actual Saline	Non-Saline	Users Accuracy
Land Monitor Prediction	Saline	5,445	0	1
	Non-Saline	1,787	3,437	0.657
Producers Accuracy		0.752	1	
Accuracy (overall)	0.832			
Efficiency	0.752			
Discrimination	1			
Power	0.867			

The KAPPA analysis result for the pairwise comparison between the CC and Land Monitor predictions (Table 8) shows a result of 1.55, which is less than 1.96 (the critical value at the 95 percent confidence level) revealing that the two matrices are not significantly different. Furthermore, table 9 reveals a strong agreement in spatial prediction of salinity in the high and steep areas of the catchment (70 and 77 percent respectively). Likewise, table 10 summarises the KAPPA analysis result for the pairwise comparison of the FLAG_VEG and Land Monitor error matrices. The result (1.34) is less than 1.96 (the critical value at the 95 percent confidence level) suggesting that the two matrices are not significantly different.

Table 9 reveals that just as with CC and Land Monitor there is strong agreement between the predictions in the high and steep areas of the catchments (70 and 76 percent respectively), with a progressive disagreement towards the low lying areas of the catchment (55 percent).

Table 8. KAPPA Analysis Result for the Pairwise Comparison of the CC and Land Monitor error Matrices

Pairwise Comparison	Z statistic
Land Monitor Prediction vs CC	1.55

Table 9. Comparison of Surface Area Agreement between the Two Best Fuzzy Models and the Probabilistic Land Monitor Model, across Different Landform Types

	Landform type			
	Rim	Flat	High	Low
CC and Land Monitor	70.78%	60.62%	77.69%	57.70%
FLAG VEG and Land Monitor	69.82%	56.46%	76.45%	55.29%

Table 10. KAPPA Analysis Result for the Pairwise Comparison of the FLAG_VEG and Land Monitor Error Matrices

Pairwise Comparison	Z Statistic
Land Monitor vs FLAG_VEG	1.34

The Land Monitor prediction had better overall accuracy and a higher KAPPA statistic value when validated against the same ground truth data used to validate the results of this investigation. It also provided the highest accuracy in the two classes, saline and non-saline. Comparing the accuracy of CC, FLAG_VEG and Land Monitor over the different landforms (Table 3), Land Monitor performs better across nearly all landform types, except for the rim area of the catchment, where the CC and FLAG_VEG models outperform Land Monitor. The catchment rim has steeper slopes than the rest of the catchment. CC is designed to detect concave areas that are common in the catchment rim and LOW and UPNESS are designed to detect low areas at a local and catchment scale. These characteristics of FLAG make it more effective at the catchment rim. While the Land Monitor model appeared to produce a better result when compared against the same ground truth data as FLAG, a comparison of the two best FLAG predictions with the Land Monitor prediction indicated that the differences between them were not significant at a 95 percent confidence level.

Factors Influencing the Accuracy and Quality of the Solution

The only data used in FLAG is a DEM. While the optimized versions of FLAG created in this investigation use other data, they were only included to fine tune FLAG, the DEM is still the key data. Therefore the quality of results from FLAG and optimised versions of FLAG are highly dependent on the accuracy and quality of the DEM, which in turns greatly depends on the data source and the interpolation technique used (Hutchinson and Gallant, 1999). A visual inspection of the outputs from the CC and LOW models reveal the existence of errors in the

DEM, with the appearance of artefacts (Figure 38). These errors can be seen in the central, bottom left and bottom right portions of Figure 38. These areas are within the flat areas of the catchment. Both CC and FLAG_VEG were less accurate in this region. Unfortunately these errors could not be removed through the smoothing of the DEM or the derived indicators.

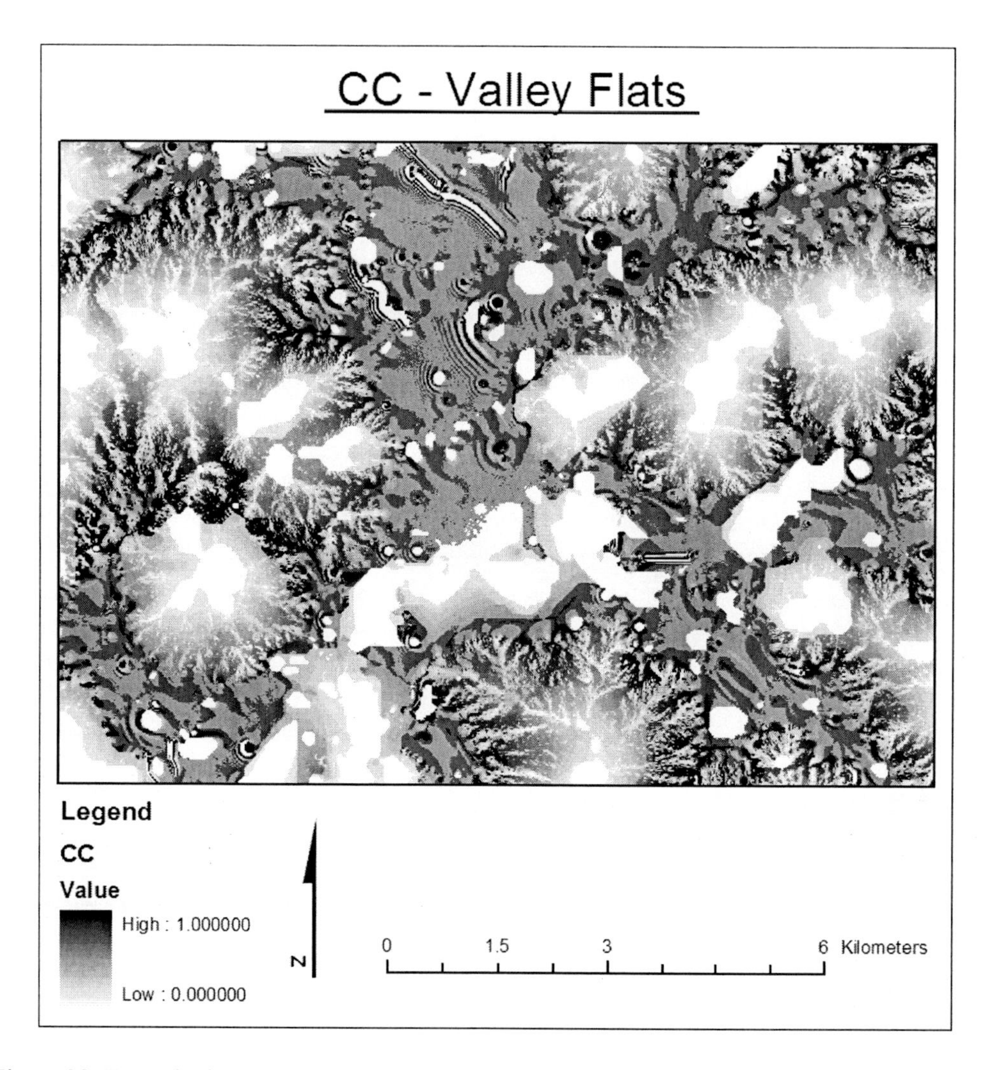

Figure 38: Errors in the output of the CC model.

Possible causes of errors within a DEM model are discussed by Carrol and Morse (1996) and Wilson et al. (2000). The causes include the fact that grid heights represent the average elevation within the grid, that the selection of spot height collection points is often based heavily on convenience and the existence of random and rounding errors in DEM data. Florinsky (1998) found that high errors in gradient, aspect, horizontal landsurface curvature and vertical landsurface curvature are typical for flat areas. This could justify the lower performance of FLAG and its optimised versions on flat areas.

CONCLUSIONS

The investigation presented in this chapter implemented and extended a model for predicting the location of dryland salinity using fuzzy modelling. The model (FLAG) was created by Roberts et al. (1997) and was extended here through the inclusion of geological and vegetation data. The result from this fuzzy logic based model was compared to a previous prediction of dryland salinity created by a model based on Boolean logic and probability theory. Based on the results of this investigation the following conclusions can be made:

- CC (based on concavity and the fraction of the total landscape monotonically uphill from each pixel) was the most accurate fuzzy model for salinity prediction. It was particularly accurate on modelling the distribution of saline areas in high and steep areas of the catchment (e.g. 89 and 78 percent of agreement as compared to ground truth data);
- Vegetation and linear geological features have a significant impact on the distribution of dryland salinity. It was reasoned that the original FLAG model could be optimized by incorporating fuzzy layers representing the impact of vegetation and geology on dryland salinity. To this end, a fuzzy layer was created to explore the influence of geological faults on the distribution of dryland salinity. Using a fuzzy gamma operator, this layer was combined with the fuzzy discharge indices CC and LOW. This layer was shown to have poor prediction accuracy (e.g. a KHAT value of 0.029). Likewise, a fuzzy layer was also created to represent the influence remnant natural vegetation has on the distribution of dryland salinity. A fuzzy gamma operator was used to integrate this layer with CCLOW. This improved model (e.g. FLAG_VEG) was found a good predictor of salinity (e.g. KHAT value of 0.42). A fuzzy output was also created combining FLAG (e.g. CCLOW), geology and vegetation at once. This was a poor predictor of salinity, indicating that the performance of the fuzzy models is influenced by the number and sequence on which the fuzzy layers are integrated.
- The optimized versions of FLAG did not improve the accuracy of FLAG. The underperformance of the optimized versions of FLAG may have been caused by the choice of fuzzy membership function or the choice of fuzzy operator used to integrate the layers. Further research needs to be conducted to evaluate whether the use of different membership functions and/or fuzzy operators affect the performance of the model. Overall, the performance of all the FLAG based models was reduced by errors arising in the DEM.
- The salinity prediction output resulting from the Land Monitor project (Evans and Caccetta, 2000a) was the crisp output against which the fuzzy models were compared. A higher level of accuracy was achieved by the Land Monitor model, as compared to the FLAG based methods. However, when the two most accurate FLAG predictions were compared with the Land Monitor prediction using KAPPA statistic analysis, none of them was found significantly different to the Land Monitor prediction, at the 95 percent confidence level. Comparing the accuracy of CC, FLAG_VEG and Land Monitor over different landforms, Land Monitor performs

better across nearly all landform types. The only exception is that CC and FLAG_VEG both outperform Land Monitor in the steeper catchment rim area.

Using fuzzy logic allows dealing with the vagueness and ambiguity within the data and knowledge. The continuous nature of salinity discharge in the real world can be represented by using fuzzy modelling. From our study it is clear that the use of fuzzy logic based modelling to predict the location of dryland salinity has many potential advantages, especially in areas where the occurrence of salinity is not limited to flat, low lying areas, and where data sets are limited.

The accuracy of FLAG may be improved by incorporating the influence of dykes into the model, and by reducing the errors in the DEM used as input for the modelling process. Studies by Klingseisen (2003) and Klingseisen *et al* (2004) have shown that significant errors occurring in flat areas of a DEM could be corrected using an adaptive filter as described in Caccetta (2000). Likewise, combining the fuzzy layers representing the influence of vegetation and geology with CC rather than the CCLOW may improve the performance of the optimized FLAG predictors.

ACKNOWLEDGEMENTS

The authors like to thank Trevor Dowling from the CSIRO for his advice, feedback and provision of software and Dr Shawn Laffan from the University of New South Wales for providing advice and software early in the project. Thanks are extended to the Department of Planning and Infrastructure, Western Australia, in particular Rodney Hoath, and Damian Shepherd at the Department of Agriculture, Western Australia for the timely provision of data.

REFERENCES

Allen, A. and Beetson, B. (1999) The Land Monitor Project, *Proceedings of WALIS Forum 1999*, Perth, Western Australia, March, pp. 74-77.

An, P., Moon, W.M. and Rencz, A. (1991) Application of Fuzzy Set Theory to Integrated Mineral Exploration, *Canadian Journal of Exploration Geophysics*, Vol. 27, No. 1, pp. 1-11.

Banko, G. (1998) A Review of Assessing the Accuracy of Classifications of Remotely Sensed Data and of Methods Including Remote Sensing Data in Forest Inventory, *Interim Report IR-98-081*, International Institute for Applied Systems Analysis, Laxenburg, Austria, 36 pp.

Bevan, K.J. and Kirkby, M.J. (1979) A Physically Based Variable Contributing Area Model of Basin Hydrology, *Hydrological Sciences Bulletin*, Vol. 24, No. 1, pp. 43-69.

Bicknell, D. (2003) *Salinity Measures, Units and Classes*, Department of Agriculture, Western Australia, http://agspsrv34.agric.wa.gov.au/environment/salinity/measurement/salinity_levels_for_revegetation.htm, Last Accessed: April 3, 2004.

Barrett-Lennard, E. G. and Nulsen, R. A. (1989). Dryland soil salinity-cure, containment or catastrophe? Proceedings of the 5th Australian Agronomy Conference. Western Australia: Australian Society of Agronomy pp 212-220.

Bonham-Carter, G. (1994) Geographic Information Systems for Geoscientists: Modelling with GIS, Pergamon, New York, USA, 398 pp.

Bruin, S. and Stein, A. (1998) Soil-landscape Modeling using Fuzzy c-means Clustering of Attribute Data Derived from a Digital Elevation Model (DEM), *Geoderma*, Vol. 83, pp. 17-33.

Burrough, P. and McDonnell, R. (1998) *Principles of Geographical Information Systems for Land Resource Assessment.* Oxford, Oxford University Press.

Caccetta, P. (200) Technical note – A simple approach for reducing discontinuities in digital elevation models (DEMs), CSIRO Mathematical and Information Sciences, Wembley, Perth, Western Australia, Report No. 2000/231.

Carrol, D. and Morse, M.P. (1996) A National Digital Elevation Model for Resource and Environmental Management, *Contemporary Cartography*, Vol. 25, No. 2, pp. 43-49.

Changying, J. and Junzheng, P. (2000) Fuzzy Prediction of Soil Strength Based on Water Content and Composition, *Journal of Terramechanics*, Vol. 37, April, pp. 57-63.

Clarke, C.J., Bell, R.W., Hobbs, R.J. and George, R.J. (1999) Incorporating Geological Effects in Modelling of Revegetation Strategies for Salt-Affected Landscapes, *Environmental Management*, Vol. 24, pp. 99-109.

Clarke, C.J., George, R.J., Bell, R.W. and Hobbs, R.J. (1998a) Major Faults and the Development of Dryland Salinity in the Western Wheatbelt of Western Australia, *Hydrology and Earth Systems Sciences*, Vol. 2, No. 1, pp. 77-91.

Clarke, C.J., Mauger, G.W., Bell, R.W. and Hobbs, R.J. (1998b) Computer Modelling of the effect of Revegetation Strategies on Salinity in the Western Wheatbelt of Western Australia 2, *Australian Journal of Soil Research*, Vol. 36, pp. 131-142.

Clarke, C.J., Mauger, G.W., Bell, R.W. and Hobbs, R.J. (1998c) Computer Modelling of the effect of Revegetation Strategies on Salinity in the Western Wheatbelt of Western Australia 2, *Australian Journal of Soil Research*, Vol. 36, pp. 131-142.

Congalton, R.G. (1991) A Review of Assessing the Accuracy of Classifications of Remotely Sensed Data, *Photogrammetric Engineering and Remote Sensing*, Vol. 37, pp. 35-46.

Cooperative Research Centre for Plant Based Management of Dryland Salinity (2002) *About Salinity*, http://www1.crcsalinity.com/pages/about.asp, Last Accessed: August 17, 2003.

Coram J.E., Dyson P.R., Houlder P.A. and Evans, W.R. (1999) Australian Groundwater Flow systems contributing to Dryland Salinity, *Report to National Land and Water Resources Audit*, Bureau of Rural Sciences, Canberra, Australia, 74 pp.

Cresswell, R.G., Dawes, W.R., Summerell, G.K. Beale, G.T.H., Tuteja, N.K. and Walker, G.R. (2003) Assessment of Salinity Management Options for Kyeamba Creek, New South Wales: Data Analysis and Groundwater Modelling, *CSIRO Land and Water Technical Report 26/03 CRC for Catchment Hydrology Technical Report 03/9 MDBC Publication 12/03*, Murray-Darling Basin Commission, Canberra, Australia, 30 pp.

Dawes, W.R. and Hatton, T.J. (1993) TOPOG_IRM: Model Description, *Technical Memorandum. 93/5*, CSIRO Division of Water Resources, Canberra, Australia, 33 pp.

Dogramaci, S., George, R., Mauger, G. and Ruprecht, J. (2003) *Water Balance and Salinity Trend, Toolibin Catchment, Western Australia*, Department of Conservation and Land Management, Perth, Australia, 60 pp.

Dombi, J. (1990). Membership function as an evaluation. *Fuzzy sets and systems*. vol. 35, pp. 1-21.

Dowling, T.I. (2000) FLAG Analysis of Catchments in the Wellington Region of NSW, *Consultancy Report March 2000*, CSIRO Land and Water, Canberra, Australia, 37 pp.

Dowling, T.I., Summerell, G.K. and Walker, J. (2003) Soil Wetness as an Indicator of Stream Salinity: A Landscape Position Index Approach, *Environmental Modelling and Software*, Vol. 18, pp. 587-593.

Drysdale, G. (2003) *Personal Communication*, Curtin University of Technology, Bentley, Western Australia.

Engel, R., McFarlane, D.J. and Street, G. (1987) The Influence of Dolerite Dykes on Saline Seeps in South-western Australia, *Australian Journal of Soil Research*, Vol. 25, pp. 125-136.

ESRI (1999) ArcView 3.2, ESRI, Redlands, California, USA (computer program).

ESRI (2002) ArcInfo Workstation 8.3, ESRI, Redlands, California, USA (computer program).

Evans, R., Brown, C. and Kellet, J. (1990) Geology and Groundwater, in: *The Murray*, Mackay, N. and Eastburn, D. (eds.), Murray-Darling Basin Commission, Canberra, Australia, pp. 77-93.

Evans, F.H. and Caccetta, P.A. (2000a) Broad-scale Spatial Prediction of Areas at Risk from Dryland Salinity, *Cartography*, Vol. 29, No. 2, pp. 33–40.

Evans, F. H. and Caccetta, P. A. (2000b) Salinity Risk Prediction using Landsat TM and DEM Derived Data, *Proceedings of the 10th Australasian Remote Sensing Conference*, Adelaide, Australia, August, pp. 1-10.

Evans, F.H., Caccetta, P.C., and Ferdowsian, R. (1996) Integrating Remotely Sensed Data with Other Spatial Data Seta to Predict Areas at Risk from Salinity, *Proceedings of the 8th Australasian Remote Sensing Conference*, Canberra, Australia, March, (On CD-ROM by Auscript).

Ferdowsian, R., George, R., Lewis, R., McFarlane, D., Short, R. and Speed, R. (1996) The Extent of Dryland Salinity in Western Australia, *Proceedings 4th National Workshop on the Productive Use and Rehabilitation of Saline Lands*, Australia, March, p. 89-88.

Fetter, C.W. (2001) *Applied Hydrogeology*, 4th ed., Prentice Hall, Upper Saddle River, USA, 598 pp.

Fisher, P. F. (1996) Boolean and Fuzzy Regions, in: *Geographic Objects with Indeterminate Boundaries*, Burrough, P. A. and Frank, A. U. (eds.), Taylor & Francis, London, UK, pp. 87-94.

Fisher, P. (2000) Fuzzy modelling, in: *GeoComputation*, Openshaw, S. and Abrahart, R. (eds.), Taylor & Francis, London, UK, pp. 161-186.

Fitzgerald, R.W. and Lees, B.G. (1994) Assessing the Classification Accuracy of Multisouce Remote Sensing Data, *Remote Sensing of Environment*, Vol. 47, pp. 362-368.

Florinsky, I.V. (1998) Accuracy of Local Topographic Variables Derived from Digital Elevation Models, *International Journal of Geographical Information Science*, Vol. 12, No. 1, pp. 47-61.

Freeze, R.A. and Cherry, J.A. (1979) *Groundwater*, Prentice Hall, Englewood Cliffs, USA, 604pp.

George, R.J. (1998) Evaluation of Airborne Geophysics for Catchment Management, *The National Airborne Geophysics Project, Toolibin Western Australia*, National Dryland Salinity Program, Canberra, Australia, 95 pp.

Gibson, P.J. and Power, G.H. (2000) *Introductory Remote Sensing*, Routledge, New York, USA, 249 pp.

Gonzalez, A., Pons, O. and Vila, M.A. (1999) Dealing with Uncertainty and Imprecision by Means of Fuzzy Numbers, *International Journal of Approximate Reasoning*, Vol. 21, pp. 233-256.

Hall, O. and Arnberg, W. (2002) A Method for Landscape Regionalization based on Fuzzy Membership Signatures, *Landscape and Urban Planning*, Vol. 59, May, pp. 227-240.

Hearn, S.J. (2001) Soil Conservation and Management Strategies for the Toolibin Catchment, *Resource Management Technical Report No. 75*, Department of Agriculture Western Australia, South Perth, Australia, 29 pp.

Heuvelink, G.B.M. and Burrough, P.A. (1993) Error Propagation in Cartographic Modelling using Boolean Logic and Continuous Classification, *International Journal of Geographical Information Systems*, Vol. 7, No. 3, pp. 231-246.

Hutchinson, M. and Gallant, J. (2000) Digital Elevation Models and Representation of Terrain Shape. In: *Terrain Analysis: Principles and Applications*, edited by J.P. Wilson and J.C. Gallant. (New York: John Wiley and Sons), pp.29-50.

Islam, Z. and Metternicht, G. (2004) The Performance of Fuzzy Operators on Fuzzy Classification of Urban Land Covers, *Photogrammetric Engineering & Remote Sensing*, (in press).

Jensen, J.R. (1996) Introductory Digital Image Processing: a Remote Sensing Perspective, 2nd ed., Upper Saddle River, USA, 316 pp.

Jolly, I., McEwan, K., Cox, J., Walker, G. and Holland, K. (2002) Managing Groundwater and Surface Water for Terrestrial Vegetation Health in Saline Areas, *Technical Report 23/02*, CSIRO Land and Water, Canberra, Australia, 88 pp.

Kampichler, C., Barthel, J. and Wieland, R. (2000) Species Density of Foliage-dwelling Spiders in Field Margins: A Simple Fuzzy Rule Based Model, *Ecological Modeling*, Vol. 129, May, pp. 87-99.

Kandel, A. (1986) *Fuzzy Mathematical Techniques with Applications*, Addison-Wesley Publishing Company, Massachusetts, USA, 274 pp.

Klingseisen, B. (2004) GIS based generation of topographic attributes for landform classification. MSc thesis, School of Geoinformation, Carinthia Tech Institute - University of Applied Sciences, Villach, Carinthia, Austria, 122 pages. Online: www.cage.curtin.edu.au/~graciela/projects/cpstof/img/publications/BKthesis_2004.pdf

Klingseisen, B, Warren, G. and Metternicht, G (2004) LANDFORM: GIS based generation of topographic attributes for landform classification in Australia. In: *Applied Geoinformatics 2004, Proceedings of 16th AGIT Symposium Salzburg*, Strobl, J., Blaschke, T., Griesebner, G. (eds.), Wichmann, Heidelberg, pp- 344-353.

Klir, G.J. and Folger, T.A. (1988) *Fuzzy Sets, Uncertainty and Information*, Prentice Hall, New Jersey, USA, 355 pp.

Klir, G. and Yuan, B. (1995) Fuzzy sets and Fuzzy logic: Theory and applications. Upper Saddle River, NJ, Prentice-Hall.

Kollias, V.J. and Kalivas, D.P. (1998) The Enhancement of a Commercial Geographic System (ARC/INFO) with Fuzzy Processing Capabilities for the Evaluation of Land Resources, *Computers and Electronics in Agriculture*, Vol. 20, June, pp. 79-95.

Kosko, B. and Isaka, S. (1993) Fuzzy Logic, *Scientific American*, Vol. 269, No. 1, pp. 76-81.

Laffan, S. (1996a) *Rapid Appraisal of Groudwater Discharge Using Fuzzy Logic and Topography*, paper presented to The Third International Conference/Workshop on Integrating GIS and Environmental Modelling, Santa Fe, USA, January. Retrieved: May 28, 2003, from http://www.ncgia.ucsb.edu/conf/SANTA_FE_CD-ROM/sf_papers/laffan_shawn/s_fetxt4.html.

Laffan, S. (1996b) Sigmu, Department of Geography, School of Resource and Environmental Management, Faculty of Science, Australian National University, Canberra, Australia (computer program).

Land Monitor (2003) *Land Monitor*, http://www.landmonitor.wa.gov.au/index.html, Last Accessed: April 8, 2003.

Landis J.R., Koch G.G. (1977) A one-way components of variance model for categorical data, *Biometrics*, vol. 33, pp. 671-679.

Lewis, M.F. (1991) Lineaments and Salinity in Western Australia – Carriers or Barriers?, *Proceedings of the International Hydrology and Water Resources Symposium 1991. Challenges for Sustainable Development*, Perth, Australia, October, pp. 202-209.

Liu, M. and Samal, A. (2002) A Fuzzy Clustering Approach to Deliniate Agroecozones, *Ecological Modeling*, Vol. 149, April, pp. 215-228.

Meesters, E.H., Bak, R.P.M., Westmacott, S., Ridgley, M. and Dollar, S. (1998) A Fuzzy Logic Model to Predict Coral Reef Development under Nutrient and Sediment Stress, *Conservation Biology*, Vol. 12, October, pp. 957-965.

Mendel, J.M (2000) Uncertainty Fuzzy Logic and Signal Processing, *Signal Processing*, Vol. 80, pp. 913-933.

Metternicht, G. I. (1998) Fuzzy Classification of JERS-1 SAR Data: An Evaluation of its Performance for Soil Salinity Mapping, *Ecological Modelling*, Vol. 111, January, pp. 61-74.

Metternicht, G. I. (2001) Assessing Temporal and Spatial Changes of Salinity using Fuzzy Logic, Remote Sensing and GIS. Foundations of an Expert System, *Ecological Modelling*, Vol. 7, No. 2-3, pp. 163-179.

Metternicht, G. I. and Beeston, G. (eds.) (2002) Assessment of Changes in Remanent Native Forest at Catchment Level, *Resource Management Technical Report 248*, Government of Western Australia, Department of Agriculture, South Perth, Australia, 15 pp.

Metternicht, G. I. and Zinck, J. A. (2003) Remote Sensing of Soil Salinity: Potentials and Constraints, *Remote Sensing of Environment*, Vol. 85, Issue 1, pp 1-20.

Mougenot, B., Pouget, M., and Epema, G. (1993) Remote Sensing of Salt-Affected Soils, *Remote Sensing Reviews*, Vol. 7, pp. 241-259.

Moore, I.D., Grayson, R.B. and Ladson, A.R. (1991) Digital Terrain Modelling: A Review of Hydrological, Geomorphological and Biological Applications, *Hydrological Processes*, Vol. 5, pp. 3-30.

National Land and Water Resources Audit (2000) Australian Dryland Salinity Assessment 2000: Extent, Impacts, Processes, Monitoring and Management Options, *Australia: National Land and Water Resources Audit*, Commonwealth of Australia, Turner, Canberra, Australia, 129 pp.

Nicoll, C. (1993) *Land Assessment Using Electromagnetic Induction – A Guide to the use of Electromagnetic Induction Techniques in the Analysis of Landscapes Affected by Dryland Salinity*, CSIRO Division of Water Resources, Canberra, Australia, 18 pp.

Nulsen, B. and McConnell, C. (2002) *Salinity at a Glance*, http://www.agric.wa.gov.au/environment/salinity/intro/salinity_at_a_glance.htm, Last Accessed: September 8, 2003.

Openshaw, S. and Openshaw, C. (1997) *Artificial Intelligence in Geography*, John Wiley and Sons, West Sussex, England, 329 pp.

Pal, N.R. and Bezdek, J.C. (2000) Quantifying Different Facets of Fuzzy Uncertainty, in: *Fundamentals of Fuzzy Sets*, Dubois, D. and Prade, H. (ed.) Kluwer Academic Publishers, Massachusetts, USA, pp. 459-480.

Please, P., Evans, W.R. and Watson, W.D. (2002) Dryland Salinity Mapping in Central and South West New South Wales: Collation and Documentation of Information, *Salt Mapping 30_09_02 Salient Solutions Australia*, New South Wales National Parks and Wildlife Service, Jerrabomberra, New South Wales, 60 pp.

Poletaev, A.I. (1992) *Fault Intersections of the Earth Crust*, Geoinformmark, Moscow, Russia, 50 pp.

Quelch, J. and Cameron, I.T. (1994) Uncertainty Representation and Propagation in Quantified Risk Assessment using Fuzzy Sets, *Journal of Loss Prevention in the Process Industries*, Vol. 7, No. 6, pp. 463-473

Raper, G.P. (1998) Agroforesty Water use in Mediterranean Regions of Australia, *Water and Salinity Issues is Agroforestry No. 2 RIRDC Publication No. 98/62*, Rural Industries Research and Development Corporation, Canberra, Australia, 71 pp.

Riedler, C. and Jandl, R. (2002) Identification of Forest Soils by Means of a Fuzzy-logic based Model, *Journal of Plant Nutrition and Soil*, Vol. 165, No. 3, pp. 320-325.

Roberts, D.W. (1996) Landscape Vegetation Modeling with Vital Attributes and Fuzzy Systems Theory, *Ecological Modeling*, Vol. 90, October, pp. 175-184.

Roberts, D.W., Dowling, T.I. and Walker, J. (1997) FLAG: A Fuzzy Landscape Analysis GIS Method for Dryland Salinity Assessment, *Technical Report No 8/97*, CSIRO Land and Water, Canberra, Australia, 23 pp.

Robinson, V.B. (2002) A Perspective on Geographic Information Systems and Fuzzy Sets, *Proceedings of North American Fuzzy Information Processing Society Proceedings - IEEE*, New Orleans, USA, June, pp. 1-6.

Robinson, V.B. (2003) A perspective on the fundamentals of fuzzy sets and their use in Geographic Information Systgems. Transactions in GIS, vol. 7, pp. 3-30.

Rosenfield, G. and Fitzpatrick-Lins, K. (1986) A Coefficient of Agreement as a Measure of Thematic Classification Accuracy, *Photogrammetric Engineering and Remote Sensing*, Vol. 52, No. 2, pp. 223-227.

Salama, R.B., Farrington, P., Bartle, G.A. and Watson, G.D. (1993) The Role of Geological Structures and Relic Channels in the Development of Dryland Salinity in the Wheatbelt of Western Australia, *Australian Journal of Earth Sciences*, Vol. 40, pp. 45-56.

Searle, R. and Baillie, J. (1998) *Prediction of Landscape Salinity Hazard Using a Geographic Information System*, Department of Natural Resources, Queensland, Australia, Retrieved: January 13, 2003, from http:/www.dnr.qld.gov.au/resourcenet/land/lris/era/projects/sham/.

Short, R. and McConnell, C. (2001) Extent and Impacts of Dryland Salinity, *Resource Management Technical Report 202*, Department of Agriculture Western Australia, South Perth Australia, 25 pp.

Stirzaker, R.J. (1996) Agroforestry and Hydrology: What do we need to know?, *Report on the Agroforestry and Hydrology Workshop, Kurrajong Hotel, Canberra, 11-12 June 1996*, Rural Industries Research and Development Corporation, Barton, Australia, 16 pp.

Ventriss, H.B., Collett, D.B. and Boyd, D.W. (1982) Relationship Between Groundwater Occurrence and a Dolerite Dyke in the Northampton area of Western Australia, *Proceedings of the Australian Water Resources Council Conference on Groundwater in Fractured Rock*, Canberra, Australia, August-September, pp. 217-227.

Vertessy, R., Connell, L., Morris, J., Silberstein, R., Heuperman, A., Feikema., Mann, L., Komarzynski, M., Collopy, J. and Stackpole, D. (2000) Sustainable Hardwood Production in Shallow Watertable Areas, *Water and Salinity Issues is Agroforestry No. 6 RIRDC Publication No. 00/163*, Rural Industries Research and Development Corporation, Canberra, Australia, 105 pp.

Watson, A.J. (2002) *NSW Murray Catchment Salinity Report*, NSW Department of Land and Water Conservation Murray Region, Albury, NSW, 41 pp.

Williamson, D.R. (1998) Waterlogging and Salinisation, in*: Farming Action Catchment Reaction*, Williams, J., Hook, R.A. and Gascoigne, H.L. (eds.) CSIRO Publishing, Collingwood, Australia, pp. 162-187.

Wilson, J.P., Phillip, L.R. and Snyder, R.D. (2000) Effect of Data Source, Grid Resolution, and Flow-Routing on Computed Topographic Attributes, in: *Terrain Analysis: Principles and Applications*, Wilson, J.P. and Gallant, J.C. (eds.), John Wiley and Sons, New York, USA, pp. 133-161.

Yager, R.R. (2002) Uncertainty Measures using Fuzzy Measures, *IEEE Transactions on Systems, Management and Cybernetics*, Vol. 32, No. 1, pp. 13-20.

Zadeh, L.A. (1992) Knowledge Representation in Fuzzy Logic, in*: An Introduction to Fuzzy Logic Applications in Intelligent Systems*, Yager, R.R. and Zadeh, L.A. (ed.) Kluwer Academic Publishers, Massachusetts, USA, pp. 1-26.

Zhang, W. and Montgomery, D.R. (1994) Digital Elevation Model Grid Size, Landscape Representation, and Hydrologic Simulations, *Water Resources Research*, Vol. 30, No. 4, pp. 1019-1028.

Zimmermann, H. J., 1985. *Fuzzy Set Theory and its Application*, Kluwer Publication, Boston

INDEX